THE TOPOLOGY OF FIBRE BUNDLES

PRINCETON LANDMARKS
IN MATHEMATICS AND PHYSICS

Non-standard Analysis, *by Abraham Robinson*

General Theory of Relativity, *by P.A.M. Dirac*

Angular Momentum in Quantum Mechanics,
by A. R. Edmonds

Mathematical Foundations of Quantum Mechanics,
by John von Neumann

Introduction to Mathematical Logic, *by Alonzo Church*

Convex Analysis, *by R. Tyrrell Rockafellar*

Riemannian Geometry, *by Luther Pfahler Eisenhart*

The Classical Groups, *by Hermann Weyl*

Topology from the Differentiable Viewpoint,
by John W. Milnor

Algebraic Theory of Numbers, *by Hermann Weyl*

Continuous Geometry, *by John von Neumann*

Linear Programming and Extensions, *by George B. Dantzig*

Operator Techniques in Atomic Spectroscopy, *by Brian R. Judd*

THE TOPOLOGY OF FIBRE BUNDLES

By
NORMAN STEENROD

PRINCETON UNIVERSITY PRESS
PRINCETON, NEW JERSEY

Published by Princeton University Press, 41 William Street,
Princeton New Jersey 08540
In the United Kingdom: Princeton University Press,
Chichester, West Sussex

Second printing 1957, appendix added
Sixth printing 1967
Seventh printing, and first paperback printing, 1999

Steenrod, Norman Earl, 1910–1971.
The topology of fibre bundles / by Norman E. Steenrod.
p. cm. — (Princeton landmarks in mathematics
and physics) Originally published: Princeton :
Princeton University Press, 1951.
Includes bibliographical references and index.
ISBN 0-691-00548-6 (pbk. : alk. paper)
1. Fibre bundles (Mathematics) I. Title. II. Series.
QA 612.6.S74 1999
514'.224—dc21 99–17187

The paper used in this publication meets the minimum requirements
of ANSI/NISO Z39.48-1992 (R1997) (*Permanence of Paper*)

http://pup.princeton.edu

Printed in the United States of America

1 3 5 7 9 10 8 6 4 2

THE TOPOLOGY OF FIBRE BUNDLES

By
NORMAN STEENROD

PRINCETON UNIVERSITY PRESS
PRINCETON, NEW JERSEY

Published by Princeton University Press, 41 William Street,
Princeton New Jersey 08540
In the United Kingdom: Princeton University Press,
Chichester, West Sussex

Second printing 1957, appendix added
Sixth printing 1967
Seventh printing, and first paperback printing, 1999

Steenrod, Norman Earl, 1910–1971.
The topology of fibre bundles / by Norman E. Steenrod.
p. cm. — (Princeton landmarks in mathematics
and physics) Originally published: Princeton :
Princeton University Press, 1951.
Includes bibliographical references and index.
ISBN 0-691-00548-6 (pbk. : alk. paper)
1. Fibre bundles (Mathematics) I. Title. II. Series.
QA 612.6.S74 1999
514'.224—dc21 99–17187

The paper used in this publication meets the minimum requirements
of ANSI/NISO Z39.48-1992 (R1997) (*Permanence of Paper*)

http://pup.princeton.edu

Printed in the United States of America

1 3 5 7 9 10 8 6 4 2

Preface

The recognition of the domain of mathematics called fibre bundles took place in the period 1935–1940. The first general definitions were given by H. Whitney. His work and that of H. Hopf and E. Stiefel demonstrated the importance of the subject for the applications of topology to differential geometry. Since then, some seventy odd papers dealing with bundles have appeared. The subject has attracted general interest, for it contains some of the finest applications of topology to other fields, and gives promise of many more. It also marks a return of algebraic topology to its origin; and, after many years of introspective development, a revitalization of the subject from its roots in the study of classical manifolds.

No exposition of fibre bundles has appeared. The literature is in a state of partial confusion, due mainly to the experimentation with a variety of definitions of "fibre bundle." It has not been clear that any one definition would suffice for all results. The derivations of analogous conclusions from differing hypotheses have produced much overlapping. Many "known" results have not been published. It has been realized that certain standard theorems of topology are special cases of propositions about bundles, but the generalized forms have not been given.

The present treatment is an initial attempt at an organization. It grew out of lectures which I gave at the University of Michigan in 1947, and at Princeton University in 1948. The informed reader will find little here that is essentially new. Only such improvements and fresh applications are made as must accompany any reasonably successful organization.

The book is divided into three parts according to the demands made on the reader's knowledge of topology. The first part presupposes only a minimum of point set theory and closes with two articles dealing with covering spaces and the fundamental group. Part II makes extensive use of the homotopy groups of Hurewicz. Since no treatment of these has appeared in book form, Part II opens with a survey of the subject. Definitions and results are stated in detail; some proofs are given, and others are indicated. In Part III we make use of cohomology theory. Here, again, a survey is required because the standard treatments do not include the generalized form we must use. A reader who is familiar with the elements of homology theory will have little difficulty.

I must acknowledge my gratitude to Professor Sze-tsen Hu and Dr. R. L. Taylor who read the manuscript and suggested many improvements.

I wish to acknowledge also the aid of the National Academy of Sciences in support of publication of this volume.

Numbers enclosed in brackets refer to the bibliography.

NORMAN STEENROD

May, 1950
Princeton University

Contents

Part I. THE GENERAL THEORY OF BUNDLES

Part II. THE HOMOTOPY THEORY OF BUNDLES

Part III. THE COHOMOLOGY THEORY OF BUNDLES

THE TOPOLOGY OF FIBRE BUNDLES

Part I. The General Theory of Bundles

§1. Introduction

1.1. Provisional definition. A fibre bundle $\mathcal{B}$ consists, at least, of the following: (i) a topological space B called the *bundle space* (or, simply, bundle), (ii) a topological space X called the *base space*, (iii) a continuous map

$$p: \quad B \to X$$

of B onto X called the *projection*, and (iv) a space Y called the *fibre*. The set Y_x, defined by

$$Y_x = p^{-1}(x),$$

is called the *fibre over the point x of X*. It is required that each Y_x be homeomorphic to Y. Finally, for each x of X, there is a neighborhood V of x and a homeomorphism

$$\phi: \quad V \times Y \to p^{-1}(V)$$

such that

$$p\phi(x',y) = x' \qquad x' \in V,\ y \in Y.$$

A *cross-section* of a bundle is a continuous map $f: X \to B$ such that $pf(x) = x$ for each $x \in X$.

The above definition of bundle is not sufficiently restrictive. A bundle will be required to carry additional structure involving a group G of homeomorphisms of Y called the *group of the bundle*. Before imposing the additional requirements, consideration of a collection of examples will show the need for these. The discussion of these examples will be brief and intuitive; each will be treated later in detail.

1.2. The product bundle. The first example is the *product bundle* or product space $B = X \times Y$. In this case, the projection is given by $p(x,y) = x$. Taking $V = X$ and $\phi =$ the identity, the last condition is fulfilled. The cross-sections of B are just the graphs of maps $X \to Y$. The fibres are, of course, all homeomorphic, however there is a natural unique homeomorphism $Y_x \to Y$ given by $(x,y) \to y$. As will be seen, this is equivalent to the statement that the group G of the bundle consists of the identity alone.

1.3. The Möbius band. The second example is the Möbius band. The base space X is a circle obtained from a line segment L (as indicated in Fig. 1) by identifying its ends. The fibre Y is a line segment. The bundle B is obtained from the product $L \times Y$ by matching the two ends with a twist. The projection $L \times Y \to L$ carries over under this matching into a projection p: $B \to X$. There are numerous cross-sections; any curve as indicated with end points that match provides a cross-section. It is clear that any two cross-sections must agree on at least one point. There is no natural unique homeomorphism of Y_x

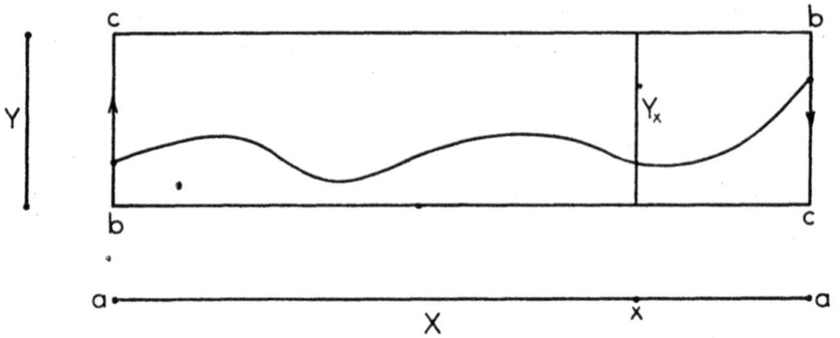

FIG. 1.

with Y. However there are two such which differ by the map g of Y on itself obtained by reflecting in its midpoint. In this case the group G is the cyclic group of order 2 generated by g.

1.4. The Klein bottle. The third example is the Klein bottle. The preceding construction is modified by replacing the fibre by a circle (Fig. 2). The ends of the cylinder $L \times Y$ are identified, as indicated, by reflecting in the diameter de. Again, the group G, is the cyclic group

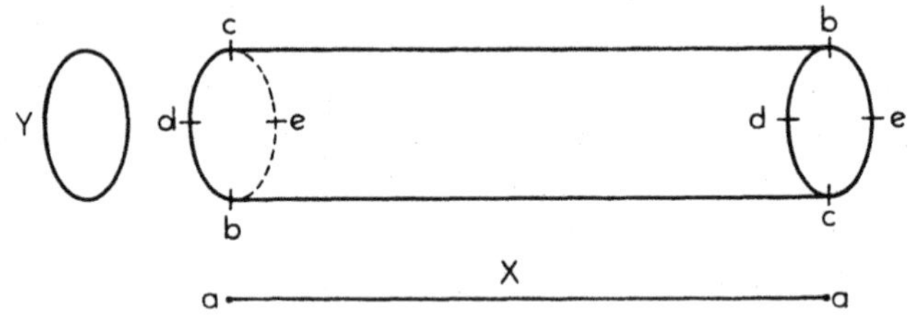

FIG. 2.

of order 2 generated by this reflection. (It is impossible to visualize this example in complete detail since the Klein bottle cannot be imbedded topologically in euclidean 3-space.)

1.5. The twisted torus. The fourth example, we will call the *twisted torus*. The construction is the same as for the Klein bottle except that reflection in the diameter de is replaced by reflection in the center of the circle (or rotation through 180°). As before, the group G is cyclic of order 2. In contrast to the preceding two examples, this bundle is homeomorphic to the product space $X \times Y$ and in such a way as to preserve fibres. However to achieve this one must use homeomorphisms $Y \to Y_x$ other than the two natural ones. But they need not differ from these by more than rotations of Y. This behavior is expressed by saying that the twisted torus is not a product bundle, but it is equivalent to one in the full group of rotations of Y.

1.6. Covering spaces. A covering space B of a space X is another example of a bundle. The projection p: $B \to X$ is the covering map. The usual definition of a covering space is the definition of bundle, in §1.1, modified by requiring that each Y_x is a discrete subspace of B, and that ϕ is a homeomorphism of $V \times Y_x$ with $p^{-1}(V)$ so that $\phi(x,y) = y$. If, in addition, it is supposed that X is arcwise connected, motion of a point x along a curve C in X from x_1 to x_2 can be covered by a continuous motion of Y_x in B from Y_{x_1} to Y_{x_2}. Choosing a base point x_0, each Y_x can be put in 1-1 correspondence with $Y = Y_{x_0}$ using a curve in X. This correspondence depends only on the homotopy class of the curve. Considering the action on Y of closed curves from x_0 to x_0, the fundamental group $\pi_1(X)$ appears as a group of permutations on Y. Any two correspondences of Y_x with Y differ by a permutation corresponding to an element of $\pi_1(X)$. Thus, *for covering spaces, the group of the bundle is a factor group of the fundamental group of the base space.*

1.7. Coset spaces. Another example of a bundle is a Lie group B operating as a transitive group of transformations on a manifold X. The projection is defined by selecting a point $x_0 \in X$ and defining $p(b) = b(x_0)$. If Y is the subgroup of B which leaves x_0 fixed, then the fibres are just the left cosets of Y in B. There are many natural correspondences $Y \to Y_x$, any $b \in Y_x$ defines one by $y \to b{\cdot}y$. However any two such $y \to b{\cdot}y$, $y \to b'{\cdot}y$ differ by the left translation of Y corresponding to $b^{-1}b'$. Thus the group G of the bundle coincides with the fibre Y and acts on Y by left translations. Finding a cross-section for such a bundle is just the problem of constructing in B a simply-transitive continuous family of transformations.

1.8. The tangent bundle of a manifold. As a final example let X be an n-dimensional differentiable manifold, let B be the set of all tangent vectors at all points of X, and let p assign to each vector its initial point. Then Y_x is the tangent plane at x. It is a linear space. Choosing a single representative Y, linear correspondences $Y_x \to Y$ can be constructed (using chains of coordinate neighborhoods in X), but not uniquely. In this case the group G of the bundle is the full linear group operating on Y. A cross-section here is just a vector field over X. The entire bundle is called the tangent bundle of X.

1.9. Generalizations of product spaces. It is to be observed that all the preceding examples of bundles are very much like product spaces. The language and notation has been designed to reflect this fact. A bundle is a generalization of a product space. The study of two spaces X and Y and maps f: $X \to Y$ is equivalent to the study of the product space $X \times Y$, its projections into X and Y, and graphs of maps f. This is broadened by replacing $X \times Y$ by a bundle space B, sacrificing the projection into Y, but replacing it, for each x, by a family of maps $Y_x \to Y$ any two of which differ by an element of a group G operating on Y. The graphs of continuous functions f: $X \to Y$ are replaced by cross-sections of the bundle.

This point of view would lead one to expect that most of the concepts of topology connected with pairs of spaces and their maps should generalize in some form. This is sustained in all that follows. For example, the Hopf theorem on the classification of maps of an n-complex into an n-sphere generalizes into the theory of the characteristic cohomology classes of a sphere-bundle.

The problems connected with bundles are of various types. The simplest question is the one of existence of a cross-section. This is of importance in differential geometry where a tensor field with prescribed algebraic properties is to be constructed. Is the bundle equivalent to a product bundle? If so, there exist many cross-sections. What are the relations connecting the homology and homotopy groups of the base space, bundle, fibre, and group? Can the bundle be simplified by replacing the group G by a smaller one? For given X, Y, G, what are the possible distinct bundles B? This last is the classification problem.

§2. COORDINATE BUNDLES AND FIBRE BUNDLES

2.1. The examples of §1 show that a bundle carries, as part of its structure, a group G of transformations of the fibre Y. In the last two examples, the group G has a topology. It is necessary to weave G and its topology into the definition of the bundle. This will be achieved through the intermediate notion of a *fibre bundle with coordinate systems* (briefly: "coordinate bundle"). The coordinate systems are eliminated by a notion of equivalence of coordinate bundles, and a passage to equivalence classes.

2.2. Transformation groups. A *topological group* G is a set which has a group structure and a topology such that (a) g^{-1} is continuous for g in G, and (b) g_1g_2 is continuous simultaneously in g_1 and g_2, i.e. the map $G \times G \to G$ given by $(g_1,g_2) \to g_1g_2$ is continuous when $G \times G$ has the usual topology of a product space.

If G is a topological group, and Y is a topological space, we say that G *is a topological transformation group of* Y *relative to a map* $\eta\colon\ G \times Y \to Y$ if (i) η is continuous, (ii) $\eta(e,y) = y$ where e is the identity of G, and (iii) $\eta(g_1g_2,y) = \eta(g_1,\eta(g_2,y))$ for all g_1,g_2 in G and y in Y.

As we shall rarely consider more than one such η, we shall abbreviate $\eta(g,y)$ by $g\cdot y$. Then (ii) becomes $e\cdot y = y$ and (iii) becomes $(g_1g_2)\cdot y = g_1\cdot(g_2\cdot y)$. For any fixed g, $y \to g\cdot y$ is a homeomorphism of Y onto itself; for it has the continuous inverse $y \to g^{-1}\cdot y$. In this way η provides a homomorphism of G into the group of homeomorphisms of Y.

We shall say that G is *effective* if $g\cdot y = y$, for all y, implies $g = e$. Then G is isomorphic to a group of homeomorphisms of Y. In this case one might identify G with the group of homeomorphisms, however we shall frequently allow the same G to operate on several spaces.

Unless otherwise stated, a topological transformation group will be assumed to be effective.

2.3. Definition of coordinate bundle. A *coordinate bundle* $\mathcal{B}$ is a collection as follows:

(1) A space B called the *bundle space*,
(2) a space X called the *base space*,
(3) a map $p\colon\ B \to X$ of B onto X called the *projection*,
(4) a space Y called the *fibre*,
(5) an effective topological transformation group G of Y called the *group of the bundle*,
(6) a family $\{V_j\}$ of open sets covering X indexed by a set J, the V_j's are called *coordinate neighborhoods*, and
(7) for each j in J, a homeomorphism

$$\phi_j\colon\ V_j \times Y \to p^{-1}(V_j)$$

called the *coordinate function*.

The coordinate functions are required to satisfy the following conditions:

(8) $$p\phi_j(x,y) = x, \qquad \text{for } x \in V_j,\ y \in Y,$$

(9) if the map $\phi_{j,x}\colon\ Y \to p^{-1}(x)$ is defined by setting

$$\phi_{j,x}(y) = \phi_j(x,y),$$

then, for each pair i,j in J, and each $x \in V_i \cap V_j$, the homeomorphism

$$\phi_{j,x}^{-1}\phi_{i,x}\colon\ Y \to Y$$

coincides with the operation of an element of G (it is unique since G is effective), and

(10) for each pair i,j in J, the map

$$g_{ji}: \quad V_i \cap V_j \to G$$

defined by $g_{ji}(x) = \phi_{j,x}^{-1}\phi_{i,x}$ is continuous.

It is to be observed that without (5), (9) and (10) the notion of bundle would be just that of §1.1. The condition (9) ties G essentially into the structure of the bundle, and (10) does the same for the topology of G.

As in §1, we denote $p^{-1}(x)$ by Y_x and call it the *fibre over* x.

The functions g_{ji} defined in (10) are called the *coordinate transformations* of the bundle. An immediate consequence of the definition is that, for any i,j,k in J,

$$g_{kj}(x)g_{ji}(x) = g_{ki}(x), \qquad x \in V_i \cap V_j \cap V_k. \tag{11}$$

If we specialize by setting $i = j = k$, then

$$g_{ii}(x) = \text{identity of } G, \qquad x \in V_i. \tag{12}$$

Now set $i = k$ in (11) and apply (12) to obtain

$$g_{jk}(x) = [g_{kj}(x)]^{-1}, \qquad x \in V_j \cap V_k. \tag{13}$$

It is convenient to introduce the map

$$p_j: \quad p^{-1}(V_j) \to Y \tag{14}$$

defined by

$$p_j(b) = \phi_{j,x}^{-1}(b) \qquad \text{where } x = p(b).$$

Then p_j satisfies the identities

$$p_j\phi_j(x,y) = y, \qquad \phi_j(p(b),p_j(b)) = b, \tag{14'}$$
$$g_{ji}(p(b))\cdot p_i(b) = p_j(b), \qquad p(b) \in V_i \cap V_j.$$

2.4. Definition of fibre bundle. Two coordinate bundles $\mathfrak{B}$ and $\mathfrak{B}'$ are said to be *equivalent in the strict sense* if they have the same bundle space, base space, projection, fibre, and group, and their coordinate functions $\{\phi_j\}$, $\{\phi'_k\}$ satisfy the conditions that

$$\bar{g}_{kj}(x) = \phi'^{-1}_{k,x}\phi_{j,x}, \qquad x \in V_j \cap V'_k \tag{15}$$

coincides with the operation of an element of G, and the map

$$\bar{g}_{kj}: \quad V_j \cap V'_k \to G$$

so obtained is continuous.

This can be stated briefly by saying that the union of the two sets of coordinate functions is a set of coordinate functions of a bundle.

That this is a proper equivalence relation follows quickly. Reflexivity is immediate. Symmetry follows from the continuity of $g \to g^{-1}$. Transitivity depends on the simultaneous continuity of $(g_1, g_2) \to g_1 g_2$.

With this notion of equivalence, a *fibre bundle* is defined to be an equivalence class of coordinate bundles.

One may regard a fibre bundle as a "maximal" coordinate bundle having all possible coordinate functions of an equivalence class. As our indexing sets are unrestricted, this involves the usual logical difficulty connected with the use of the word "all."

2.5. Mappings of bundles. Let $\mathfrak{B}$ and $\mathfrak{B}'$ be two coordinate bundles having the same fibre and the same group. By a *map* h: $\mathfrak{B} \to \mathfrak{B}'$ is meant a continuous map h: $B \to B'$ having the following properties

(16) h carries each fibre Y_x of B homeomorphically onto a fibre $Y_{x'}$ of B', thus inducing a continuous map $\bar{h}$: $X \to X'$ such that

$$p'h = \bar{h}p,$$

(17) if $x \in V_j \cap \bar{h}^{-1}(V'_k)$, and h_x: $Y_x \to Y_{x'}$ is the map induced by h $(x' = \bar{h}(x))$, then the map

$$\bar{g}_{kj}(x) = \phi'^{-1}_{k,x'} h_x \phi_{j,x} = p'_k h_x \phi_{j,x}$$

of Y into Y coincides with the operation of an element of G, and

(18) the map

$$\bar{g}_{kj}\colon \ V_j \cap \bar{h}^{-1}(V'_k) \to G$$

so obtained is continuous.

In the literature, the map h is called "fibre preserving." We shall use frequently the expression "bundle map" to emphasize that h is a map in the above sense.

It is readily proved that the identity map $B \to B$ is a map $\mathfrak{B} \to \mathfrak{B}$ in this sense. Likewise the composition of two maps $\mathfrak{B} \to \mathfrak{B}' \to \mathfrak{B}''$ is also a map $\mathfrak{B} \to \mathfrak{B}''$.

A map of frequent occurrence is an inclusion map $\mathfrak{B} \subset \mathfrak{B}'$ obtained as follows. Let $\mathfrak{B}'$ be a coordinate bundle over X', and let X be a subspace of X'. Let $B = p'^{-1}(X)$, $p = p'|B$, and define the coordinate functions of $\mathfrak{B}$ by $\phi_j = \phi'_j|(V'_j \cap X) \times Y$. Then $\mathfrak{B}$ is a coordinate bundle, and the inclusion map $B \to B'$ is a map $\mathfrak{B} \to \mathfrak{B}'$. We call $\mathfrak{B}$ *the portion of* $\mathfrak{B}'$ *over* X (or $\mathfrak{B}$ is $\mathfrak{B}'$ restricted to X), and we will use the notations

$$\mathcal{B} = \mathcal{B}'|X = \mathcal{B}'_X.$$

The functions $\bar{g}_{kj}$ of (17) and (18) are called the *mapping transformations.* There are two sets of relations which they satisfy:

$$\begin{aligned} \bar{g}_{kj}(x)g_{ji}(x) &= \bar{g}_{ki}(x), \quad x \in V_i \cap V_j \cap \bar{h}^{-1}(V'_k), \\ g'_{lk}(\bar{h}(x))\bar{g}_{kj}(x) &= \bar{g}_{lj}(x), \quad x \in V_j \cap \bar{h}^{-1}(V'_k \cap V'_l). \end{aligned} \tag{19}$$

These are verified by direct substitution using the definitions (10) and (17).

2.6. Lemma. *Let $\mathcal{B}$, $\mathcal{B}'$ be coordinate bundles having the same fibre Y and group G, and let $\bar{h}$: $X \to X'$ be a map of one base space into the other. Finally, let $\bar{g}_{kj}$: $V_j \cap \bar{h}^{-1}(V'_k) \to G$ be a set of continuous maps satisfying the conditions (19). Then there exists one and only one map h: $\mathcal{B} \to \mathcal{B}'$ inducing $\bar{h}$ and having $\{\bar{g}_{jk}\}$ as its mapping transformations.*

If $p(b) = x$ lies in $V_j \cap \bar{h}^{-1}(V'_k)$, define

$$h_{kj}(b) = \phi'_k(\bar{h}(x), \bar{g}_{kj}(x)\cdot p_j(b)). \tag{20}$$

Then h_{kj} is continuous in b, and $p'h_{kj}(b) = \bar{h}(p(b))$. Suppose $x \in V_i \cap V_j \cap \bar{h}^{-1}(V'_k \cap V'_l)$. Using the relations (14′) and (19), we have (with $x' = \bar{h}(x)$)

$$\begin{aligned} h_{kj}(b) &= \phi'_k(x', \bar{g}_{kj}(x)g_{ji}(x)\cdot p_i(b)) \\ &= \phi'_k(x', \bar{g}_{ki}(x)\cdot p_i(b)) = h_{ki}(b) \\ &= \phi'_l(x', g'_{lk}(x')\bar{g}_{ki}(x)\cdot p_i(b)) \\ &= \phi'_l(x', \bar{g}_{li}(x)\cdot p_i(b)) = h_{li}(b). \end{aligned}$$

It follows that any two functions of the collection $\{h_{kj}\}$ agree on their common domain. Since their domains are open and cover B, they define a single-valued continuous function h. Then $p'h = \bar{h}p$ follows from the same relation for h_{jk}. If, in (20), we replace b by $\phi_{j,x}(y)$, apply p'_k to both sides, and use the relations (14′), we obtain

$$\begin{aligned} p'_k h\phi_{j,x}(y) &= p'_k\phi'_k(x', \bar{g}_{kj}(x)\cdot p_j\phi_{j,x}(y)) \\ &= \bar{g}_{kj}(x)\cdot y \end{aligned}$$

which shows that h has the prescribed mapping transformations.

Conversely any h which has the prescribed mapping transformations must satisfy (20), and therefore h is unique.

2.7. Lemma. *Let $\mathcal{B}$, $\mathcal{B}'$ be coordinate bundles having the same fibre and group, and let h: $\mathcal{B} \to \mathcal{B}'$ be a map such that the induced map $\bar{h}$: $X \to X'$ is 1-1 and has a continuous inverse $\bar{h}^{-1}$: $X' \to X$. Then h has a continuous inverse h^{-1}: $B' \to B$, and h^{-1} is a map $\mathcal{B}' \to \mathcal{B}$.*

The fact that h is 1-1 in the large is evident. For any x' in $V'_k \cap \bar{h}(V_j)$, let $x = \bar{h}^{-1}(x')$, and, following (17), define

$$\bar{g}_{jk}(x') = \phi_{j,x}^{-1} h_x^{-1} \phi'_{k,x'}.$$

It follows that $\bar{g}_{jk}(x') = \bar{g}_{kj}(x)^{-1}$. Since $g \to g^{-1}$ is continuous in G, x is continuous in x', and $\bar{g}_{kj}(x)$ is continuous in x, it follows that $\bar{g}_{jk}(x')$ is continuous in x'. If $p'(b') = x'$ is in $V'_k \cap h(V_j)$, then h^{-1} is given by

$$h^{-1}(b') = \phi_j(\bar{h}^{-1}(x'), \bar{g}_{jk}(x') \cdot p'_k(b'))$$

which shows that h^{-1} is continuous on $p'^{-1}(V'_k \cap h(V_j))$. Since these sets are open and cover B', it follows that h^{-1} is continuous, and the lemma is proved.

Two coordinate bundles $\mathcal{B}$ and $\mathcal{B}'$ having the same base space, fibre and group are said to be *equivalent* if there exists a map $\mathcal{B} \to \mathcal{B}'$ which induces the identity map of the common base space.

The symmetry of this relation is provided by the above lemma. The reflexivity and transitivity are immediate. It is to be noted that strict equivalence, defined in §2.4, implies equivalence.

Two fibre bundles (see §2.4) having the same base space, fibre and group are said to be *equivalent* if they have representative coordinate bundles which are equivalent.

It is possible to define broader notions of equivalences of fibre bundles by allowing X or (Y,G) to vary by a topological equivalence. The effect of this is to reduce the number of equivalence classes. The definition chosen is the one most suitable for the classification theorems proved later.

2.8. Lemma. *Let $\mathcal{B},\mathcal{B}'$ be coordinate bundles having the same base space, fibre, and group, then they are equivalent if and only if there exist continuous maps*

$$\bar{g}_{kj}: \quad V_j \cap V'_k \to G \qquad j \in J,\ k \in J'$$

such that

$$\text{(19')} \qquad \begin{aligned} \bar{g}_{ki}(x) &= \bar{g}_{kj}(x) g_{ji}(x), & x \in V_i \cap V_j \cap V'_k \\ \bar{g}_{lj}(x) &= g'_{lk}(x) \bar{g}_{kj}(x), & x \in V_j \cap V'_k \cap V'_l. \end{aligned}$$

Suppose, first that $\mathcal{B},\mathcal{B}'$ are equivalent and h: $\mathcal{B} \to \mathcal{B}'$. Define $\bar{g}_{kj}$ by (17) (note that $x' = x$ since $\bar{h}$ is the identity). The relations (19) reduce to (19').

Conversely, suppose the $\bar{g}_{kj}$ are given. The relations (19') imply (19) in the case $\bar{h}$ = identity. The existence of h is provided by 2.6.

2.9. Let $\mathcal{B}$ be a coordinate bundle with neighborhoods $\{V_j\}$, and let $\{V'_k\}$ be a covering of X by an indexed family of open sets such that each V'_k is contained in some V_j (i.e., the second covering is a *refinement* of the first); then one constructs a strictly equivalent coordinate bundle $\mathcal{B}'$ with neighborhoods $\{V'_k\}$ by simply restricting ϕ_j to $V'_k \times Y$ where j is selected so that $V'_k \subset V_j$. When j,k are so related, the functions $\bar{g}_{kj}$ of (15) are constant and equal the identity of G.

Suppose now that $\mathcal{B},\mathcal{B}'$ are two coordinate bundles with the same base space, fibre, and group. The open sets $V_j \cap V'_k$, $j \varepsilon J$, $k \varepsilon J'$, cover X and form a refinement of $\{V_j\}$ and $\{V'_k\}$. It follows that $\mathcal{B},\mathcal{B}'$ are strictly equivalent to coordinate bundles $\mathcal{B}_1,\mathcal{B}'_1$, respectively, having the same set of coordinate neighborhoods. This observation lends weight to the following lemma.

2.10. Lemma. *Let $\mathcal{B},\mathcal{B}'$ be two coordinate bundles with the same base space, fibre, group and coordinate neighborhoods. Let g_{ji}, g'_{ji} denote their coordinate transformations. Then $\mathcal{B},\mathcal{B}'$ are equivalent if and only if there exist continuous functions λ_j: $V_j \to G$, defined for each j in J, and such that*

$$g'_{ji}(x) = \lambda_j(x)^{-1} g_{ji}(x) \lambda_i(x), \qquad x \varepsilon V_i \cap V_j. \tag{21}$$

If $\mathcal{B},\mathcal{B}'$ are equivalent, the functions $\bar{g}_{kj}$ provided by 2.8 enable us to define $\lambda_j = (\bar{g}_{jj})^{-1}$. Then the relations (19′) yield (21).

Conversely, suppose the λ's satisfying (21) are given. Define

$$\bar{g}_{kj}(x) = \lambda_k(x)^{-1} g_{kj}(x), \qquad x \varepsilon V_j \cap V_k.$$

Then the relations (19′) follow from (21) and (11), and the lemma is proved.

2.11. Lemma. *Let $\mathcal{B},\mathcal{B}'$ be coordinate bundles having the same fibre and group, and let h be a map $\mathcal{B} \to \mathcal{B}'$. Corresponding to each cross-section f': $X' \to B'$ there exists one and only one cross-section f: $X \to B$ such that*

$$hf(x) = f'\bar{h}(x), \qquad x \varepsilon X.$$

The cross-section f is said to be induced by h and f', and will be denoted by $h^{\#}f'$.

Let $x' = \bar{h}(x)$. Since $f(x)$ must lie in Y_x, and h_x: $Y_x \to Y_{x'}$ is a 1-1 map, it follows that $f(x) = h_x^{-1} f'(x')$. This defines f and proves its uniqueness. It remains to prove continuity. It suffices to show that f is continuous over any set of the form $V_j \cap \bar{h}^{-1}(V'_k)$ for these sets are open and cover X. Since $pf(x) = x$ is continuous, it remains to show that $p_j f(x)$ is continuous. By (18), $\bar{g}_{kj}(x)$ is continuous. Furthermore

$$\begin{aligned}\bar{g}_{kj}(x)\cdot[p_j f(x)] &= \phi'^{-1}_{k,x'} h_x \phi_{j,x} p_j f(x) = \phi'^{-1}_{k,x'} h_x f(x) \\ &= \phi'^{-1}_{k,x'} f'(\bar{h}(x)) = p'_k f'(\bar{h}(x)).\end{aligned}$$

Therefore

$$p_j f(x) = [\bar{g}_{kj}(x)]^{-1} \cdot p'_k f'(\bar{h}(x))$$

is also continuous.

The lemma shows that cross-sections behave contravariantly under mappings of bundles. In this respect they resemble covariant tensors.

2.12. Point set properties of B. It is well known that numerous topological properties of X and Y carry over to their product $X \times Y$. They also carry over to the bundle space B of any bundle with base space X and fibre Y. The argument given for the product space carries over to the bundle using the local product representations given by the coordinate functions.

As an example, suppose X and Y are Hausdorff spaces. Let b,b' be distinct points of B. If $p(b) \neq p(b')$, let U,V be neighborhoods of $p(b),p(b')$ such that $U \cap V = 0$. Then $p^{-1}(U)$ and $p^{-1}(V)$ are non-overlapping open sets containing b and b'. If $p(b) = p(b') = x$, then choose a j such that x is in V_j. Now $p_j(b) \neq p_j(b')$ since $b \neq b'$, therefore there exist neighborhoods U,U' of $p_j(b),p_j(b')$ such that $U \cap U' = 0$. Then $\phi_j(V_j \times U)$ and $\phi_j(V_j \times U')$ are non-overlapping open sets containing b and b'. Thus B is a Hausdorff space.

As a second example, suppose X and Y are compact Hausdorff spaces. For each point x in X, choose a j such that x is in V_j and choose an open set U_x such that x is in U_x and $\bar{U}_x \subset V_j$ (this can be done since any compact Hausdorff space is regular). The sets $\{U_x\}$ cover X; select a finite covering $U_1, \cdots, U_m$. Since $\bar{U}_r$ is compact, so is $\bar{U}_r \times Y$. Select j so that $\bar{U}_r \subset V_j$. Since ϕ_j is a topological map, it follows that $p^{-1}(\bar{U}_r)$ is compact. But these sets, for $r = 1, \cdots, m$, cover the space B. Therefore B is compact.

Among other common properties of X and Y which are also properties of B we mention (i) connectedness, (ii) the first axiom of countability, (iii) existence of a countable base for open sets, (iv) local compactness, (v) local connectedness, and (vi) arcwise connectedness.

2.13. In subsequent articles, the expression "bundle" will mean "coordinate bundle." Fibre bundles will not be the primary, but rather the ultimate, objects of study. They will be studied through their representatives. The various concepts introduced for coordinate bundles must behave properly under equivalence. The situation is similar to that in group theory when one studies groups given by generators and relations. Results which are not invariant under a change of base are of little interest.

The study of fibre bundles needs an invariant definition of bundle which is usable. A further discussion of this problem is given in §5.

§3. CONSTRUCTION OF A BUNDLE FROM COORDINATE TRANSFORMATIONS

3.1. Let G be a topological group, and X a space. By *a system of coordinate transformations in X with values in G* is meant an indexed covering $\{V_j\}$ of X by open sets and a collection of continuous maps

$$g_{ji}: \quad V_i \cap V_j \to G \qquad i,j \in J \tag{1}$$

such that

$$g_{kj}(x)g_{ji}(x) = g_{ki}(x), \qquad x \in V_i \cap V_j \cap V_k. \tag{2}$$

The relations $g_{ii}(x) = e$ and $g_{ij}(x) = (g_{ji}(x))^{-1}$ follow as in §2.3 (12), (13).

We have seen in §2.3 that any bundle over X with group G determines such a set of coordinate transformations. We shall prove a converse.

3.2. Existence theorem. *If G is a topological transformation group of Y, and $\{V_j\}$, $\{g_{ij}\}$ is a system of coordinate transformations in the space X, then there exists a bundle $\mathcal{B}$ with base space X, fibre Y, group G, and the coordinate transformations $\{g_{ij}\}$. Any two such bundles are equivalent.*

Let us regard the indexing set J for the covering $\{V_j\}$ as a topological space with the discrete topology. Let $T \subset X \times Y \times J$ be the set of those triples (x,y,j) such that $x \in V_j$. Then T is a topological space, and is the union of the disjoint open subsets $V_j \times Y \times j$. Define in T an equivalence relation:

$$(x,y,j) \sim (x',y',k)$$

if

$$x = x', \qquad g_{kj}(x)\cdot y = y'. \tag{3}$$

That this is a proper equivalence follows immediately from (2). Define B to be the set of equivalence classes of this relation in T. Let

$$q: \quad T \to B$$

assign to each (x,y,j) its equivalence class $\{(x,y,j)\}$. A set U in B is called open if $q^{-1}(U)$ is an open set of T. Then B is a topological space and q is continuous.

Define p: $B \to X$ by

$$p(\{(x,y,j)\}) = x. \tag{4}$$

By (3), p is uniquely defined. If W is an open set of X, then $(pq)^{-1}(W) = q^{-1}(p^{-1}(W))$ is the intersection of T with the open set $W \times Y \times J$.

It is therefore an open set of T. Then, by definition, $p^{-1}(W)$ is an open set of B; so p is continuous.

Define the coordinate function ϕ_j by

$$\phi_j(x,y) = q(x,y,j), \qquad x \in V_j,\ y \in Y. \tag{5}$$

Since q is continuous, so is ϕ_j. By (4), $pq(x,y,j) = x$, and therefore $p\phi_j(x,y) = x$. Thus ϕ_j maps $V_j \times Y$ into $p^{-1}(V_j)$.

If $b = \{(x,y,k)\}$ is in $p^{-1}(V_j)$, then $x \in V_j \cap V_k$, and $(x,y,k) \sim (x,g_{jk}(x)\cdot y,j)$. Therefore $b = \phi_j(x,g_{jk}(x)\cdot y)$. Thus ϕ_j maps $V_j \times Y$ onto $p^{-1}(V_j)$.

If $(x,y,j) \sim (x',y',j)$, then $x = x'$, and $g_{jj}(x)\cdot y = y'$. Since $g_{jj}(x) = e$, we have $y = y'$. Therefore ϕ_j is a continuous 1-1 map of $V_j \times Y$ onto $p^{-1}(V_j)$.

To prove that ϕ_j^{-1} is continuous, we must show that W open in $V_j \times Y$ implies $\phi_j(W)$ is open in B, i.e. $q^{-1}\phi_j(W)$ is open in T. Since the sets $V_k \times Y \times k$ are open and cover T, it is enough to show that $q^{-1}\phi_j(W)$ meets $V_k \times Y \times k$ in an open set. This intersection is contained in $(V_j \cap V_k) \times Y \times k$ which is itself open in T. The function q restricted to the latter set can be factored into a composition

$$(V_j \cap V_k) \times Y \times k \xrightarrow{\ r\ } V_j \times Y \xrightarrow{\ \phi_j\ } B$$

where

$$r(x,y,k) = (x,g_{jk}(x)\cdot y).$$

Then r is continuous; so $r^{-1}(W)$ is an open set as required.

Consider now the map $\phi_{j,x}^{-1}\phi_{i,x}$ of Y on itself $(x \in V_i \cap V_j)$. If $y' = \phi_{j,x}^{-1}\phi_{i,x}(y)$; by definition, we have $\phi_j(x,y') = \phi_i(x,y)$, or $q(x,y',j) = q(x,y,i)$; which means $(x,y',j) \sim (x,y,i)$, and therefore $y' = g_{ji}(x)\cdot y$. Thus, for each $y \in Y$,

$$\phi_{j,x}^{-1}\phi_{i,x}(y) = g_{ji}(x)\cdot y.$$

This proves that the $\{g_{ji}\}$ are the coordinate transformations of the constructed bundle.

If, in §2.10, we choose the λ's to be constant and equal to the identity element e in G, then the conclusion asserts that any two bundles having the same coordinate transformations are equivalent. Thus the bundle constructed in §3.2 is unique up to an equivalence.

In §2.8 we have a necessary and sufficient condition for the equivalence of two bundles expressed solely in terms of their coordinate transformations. If we take these as defining an equivalence relation among the systems of coordinate transformations in X with group G, we obtain the following result:

3.3. THEOREM. *The operation of assigning to each bundle with base space X, fibre Y, and group G the system of its coordinate transformations sets up a 1-1 correspondence between equivalence classes of bundles and equivalence classes of systems of coordinate transformations.*

It is to be observed that this result reduces the problem of classifying bundles to that of classifying coordinate transformations. In the latter problem, only the space X and the topological group G are involved; the fibre Y plays no role.

3.4. Examples. The construction of some of the examples of §1 can now be clarified in terms of coordinate transformations. In all of the examples §1.3, §1.4, and §1.5, the base space X is a circle and G is a cyclic group of order 2. Cover X by two open sets V_1 and V_2 each of which is an open arc. Then $V_1 \cap V_2$ is the union of two disjoint open arcs U and W. Define $g_{12}(x)$ to be e in G if $x \in U$, and to be the non-trivial element of G if $x \in W$. Defining $g_{11} = g_{22} = e$, and $g_{21} = (g_{12})^{-1}$, we have a system of coordinate transformations in X. By allowing G to operate on various fibres, we obtain from §3.2 corresponding bundles. The examples §1.3, §1.4 and §1.5 are three such.

3.5. The definition of coordinate transformations in §3.1 and the bundle construction of §3.2 is nothing more than a clarification of the definition of bundle given by Whitney in [103] for the special case of sphere bundles over complexes. In Whitney's scheme, the coordinate transformations are defined for incident simplexes. The bundle is constructed by forming the product of each simplex with the fibre and then assembling these products according to the coordinate transformations.

§4. THE PRODUCT BUNDLE

4.1. A coordinate bundle is called *a product bundle* if there is just one coordinate neighborhood $V = X$, and the group G consists of the identity element e alone.

THEOREM. *If the group of a bundle consists of the identity element alone, then the bundle is equivalent to a product bundle.*

This is a trivial application of §2.8. One defines the functions $\bar{g}_{kj} = e$, and then (19) and (20) must hold.

4.2. Enlarging the group of a bundle. Let H be a closed subgroup of the topological group G. If $\mathfrak{B}$ is a bundle with group H, the same coordinate neighborhoods, and the same coordinate transformations, altered only by regarding their values as belonging to G, define a new bundle called the *G-image of* $\mathfrak{B}$.

In this definition a bundle is regarded as just the collection consisting of the base space, group, and coördinate transformations (see §3.3). If H operates on a fibre Y, it may or may not occur that G operates on Y or even that such operations can be defined.

If two H-bundles are equivalent, it is clear that their G-images are also equivalent. The operation therefore maps equivalences classes into such.

Let H and K be two closed subgroups of G, and let $\mathfrak{B},\mathfrak{B}'$ be bundles having the same base space and the groups H, K respectively. We say that $\mathfrak{B},\mathfrak{B}'$ are *equivalent in* G (or G-equivalent) if the G-images of $\mathfrak{B}$ and $\mathfrak{B}'$ are equivalent.

As a special case, let K be the subgroup of G consisting of the identity element alone. Then $\mathfrak{B}'$ is equivalent to the product bundle. In this case we say that *the H-bundle $\mathfrak{B}$ is G-equivalent to the product bundle.*

4.3. Equivalence theorem. *Let $\mathfrak{B}$ be a bundle with group H and coordinate transformations $\{g_{ji}\}$. Let H be a subgroup of G. Then $\mathfrak{B}$ is G-equivalent to the product bundle if and only if there exist maps λ_j: $V_j \to G$ such that*

$$g_{ji}(x) = \lambda_j(x)\,\lambda_i(x)^{-1}, \qquad x \in V_i \cap V_j.$$

Let $\mathfrak{B}'$ be the bundle having the same coordinate neighborhoods as $\mathfrak{B}$, but with group consisting of the identity element so that all $g'_{ji} = 1$. The result follows now from §2.10.

The statement that a bundle is equivalent to the product bundle means, as stated in §4.1, that its group consists of the identity. We shall allow ourselves to say that *a bundle with group G (not the identity) is equivalent to the product bundle,* and mean thereby that it is G-equivalent to the product bundle. The expression "simple bundle" is used for this in the literature.

4.4. An example may clarify the definitions. Let $\mathfrak{B}$ denote the twisted torus, §1.5, defined explicitly in §3.4. Denote the group of $\mathfrak{B}$ by H (a cyclic group of order 2). *This bundle is not H-equivalent to the product bundle.* This is proved using §4.3. For suppose λ_1,λ_2 exist as in §4.3. Since V_1,V_2 are connected sets, and H consists of 2 elements, the λ's must each be constant. Taking x in U, we find $\lambda_1\lambda_2^{-1} = e$, taking x in W, we find $\lambda_1\lambda_2^{-1} \neq e$.

Now let G be the full group of rotations of the circle Y. Then H is a subgroup of G. We assert that *$\mathfrak{B}$ is G-equivalent to the product bundle.* Define $\lambda_1(W)$ to be the 180° rotation of Y, $\lambda_1(U) = e$, and extend continuously over the rest of V_1 to obtain an arc in G joining these two rotations. Define $\lambda_2(V_2) = e$. Then $g_{12} = \lambda_1\lambda_2^{-1}$.

Thus the twisted torus is not a product bundle, but is equivalent to the product bundle in the full group of rotations.

Consider now the Möbius band or Klein bottle (§1.3, §1.4). The base space, group, and coordinate transformations are the same as for the twisted torus. If we ignore the fibre, and use the above imbedding of H in the connected group G, then this bundle is G-equivalent to the product. However the imbedding of H in G does not conform to any prescription of the operations of G on the fibre.

In general, it is not very significant to consider equivalence of bundles in a larger group unless this group is also a group of homeomorphisms of the common fibre.

§5. The Ehresmann-Feldbau Definition of Bundle

5.1. We have remarked earlier the need for a direct definition of fibre bundle—one which avoids coordinate functions and equivalence classes. Ehresmann and Feldbau [22] have proposed such a definition. It does not agree with the one we are using. In this article we discuss their definition and the differences.

5.2. Definition. A bundle, in the sense of Ehresmann-Feldbau (briefly, an E-F bundle), consists of B, p, X, and Y as usual. In addition there is given a group G (not topologized) of homeomorphisms of Y, and for each x a family of homeomorphisms G_x of Y into Y_x such that (i) $\xi,\xi' \in G_x$ implies $\xi^{-1}\xi' \in G$, and (ii) $\xi \in G_x$, $g \in G$ implies $\xi g \in G_x$. Finally, for each x, there exists a neighborhood V of x and a homeomorphism ϕ: $V \times Y \rightarrow p^{-1}(V)$ such that $p\phi(x,y) = x$, and $\phi_x \in G_x$ for each x.

Two such bundles $\mathfrak{B},\mathfrak{B}'$, having the same base space, fibre, and group, are *equivalent* if there is a homeomorphism h: $B \rightarrow B'$ which maps Y_x into Y'_x in such a way that $\xi \in G_x$ implies $h\xi \in G'_x$.

5.3. Relations with coordinate bundles. If $\xi \in G_x$, the operation $g \rightarrow \xi g$ clearly defines a 1-1 map of G onto G_x. Thus, if one knows G and a single element of G_x the entire family can be constructed. In a coordinate bundle, such an element of G_x is provided by $\phi_{j,x}$ if $x \in V_j$. Thus a coordinate bundle determines uniquely an E-F bundle. It is easily verified that equivalent coordinate bundles determine equivalent E-F bundles.

The essential feature of this passage from coordinate bundles to E-F bundles is the dropping of the topology of G. There may be various topologies of G under which a given set of coordinate transformations are continuous. Two such topologies provide inequivalent coordinate bundles. But all such yield the same E-F bundle. It follows that our fibre bundle is a somewhat more definitive notion than the E-F bundle. The inclusion of the topology of G in the structure of the bundle is not just excess baggage. It will play a very important role in later work, especially in the classification theorems.

It would appear from this that the E-F bundle is inadequate. This is far from true. To see this, consider the reverse process of trying to assign a coordinate bundle to an E-F bundle. The existence of product representations enables us to choose coordinate neighborhoods and coordinate functions. Then the coordinate transformations are defined. It remains to select a topology in G in which they are continuous. Also G must be a topological transformation group of Y. For suitably restricted Y and G this can always be done by assigning to G the "compact-open" topology defined as follows (see [33]).

5.4. The compact-open topology. If $U \subset Y$ is open and $C \subset Y$ is compact, let $W(C,U)$ be the set of $g \in G$ such that $g \cdot C \subset U$. The totality of sets $W(C,U)$ are taken as a subbase for the open sets of G (form arbitrary unions of finite intersections of sets $W(C,U)$). The resulting topology in G is called the *compact-open topology* (abbreviated: C-O topology). It is easily proved that, if Y is a Hausdorff space, so also is G.

If G has the C-O topology, then the continuity of the coordinate functions ϕ_i, ϕ_j implies the continuity of $g_{ji}(x) = \phi_{j,x}^{-1}\phi_{i,x}$.

To prove this, suppose $g_{ji}(x_0)$ is in $W(C,U)$. The set $Z = \phi_i^{-1}\phi_j(V_j \times U)$ in $V_i \times Y$ is clearly open. If $y \in C$, then $g_{ji}(x_0)\cdot y \in U$. Since

$$\phi_i(x_0,y) = \phi_j(x_0,g_{ji}(x_0)\cdot y),$$

it follows that $(x_0,y) \in Z$. Thus the open set Z contains the compact set $x_0 \times C$. By a standard argument, there exists a neighborhood N of x_0 such that $N \times C \subset Z$. We can suppose that $N \subset V_j$. If $x \in N$ and $y \in C$, then

$$\phi_i(x,y) = \phi_j(x,g_{ji}(x)\cdot y) \in \phi_j(V_j \times U).$$

But this implies that $g_{ji}(x)\cdot y \in U$. Therefore $g_{ji}(x) \in W(C,U)$, and continuity of g_{ji} at x_0 follows.

If G has the C-O topology, and Y is regular and locally-compact, then the natural maps $G \times G \to G$ and $G \times Y \to Y$ are continuous.

Suppose $g_1g_2 \in W(C,U)$. Then $g_1g_2\cdot C \subset U$, or $g_2\cdot C \subset g_1^{-1}\cdot U$, and the latter set is open. Since Y is regular and locally-compact, there exists an open set V of Y such that $g_2\cdot C \subset V$, $\bar{V} \subset g_1^{-1}\cdot U$, and $\bar{V}$ is compact. If $g_1' \in W(\bar{V},U)$ and $g_2' \in W(C,V)$, it follows that $g_1'g_2' \in W(C,U)$. Thus $W(\bar{V},U)$ and $W(C,V)$ are neighborhoods of g_1 and g_2

whose product lies in $W(C,U)$. This implies that $G \times G \to G$ is continuous.

Suppose now that $g_0 \cdot y_0 \varepsilon U$ (open). Since Y is regular and locally-compact, there is a neighborhood V of y_0 such that $\bar{V}$ is compact and $\bar{V} \subset g_0^{-1} \cdot U$. Hence $g_0 \varepsilon W(\bar{V},U)$. If $g \varepsilon W(\bar{V},U)$ and $y \varepsilon V$, it follows that $g \cdot y \varepsilon U$. This implies that $G \times Y \to Y$ is continuous.

If G has the C-O topology and Y is compact Hausdorff, then the map $G \to G$ which sends g into g^{-1} is continuous.

Suppose $g_0^{-1} \varepsilon W(C,U)$. This implies that $C \subset g_0 \cdot U$ or $Y - C \supset g_0 \cdot (Y - U)$. Then $Y - U$ is compact, $Y - C$ is open, and $g_0 \varepsilon W(Y - U, Y - C)$. If $g \varepsilon W(Y - U, Y - C)$, it follows quickly that $g^{-1} \varepsilon W(C,U)$; and the result is proved.

The restriction that Y be compact, for this last result, can be relaxed if the group G is restricted to being an equi-continuous family of homeomorphisms.

5.5. It appears from these results that the C-O topology has all the desired features. Given an E-F bundle in which Y is compact-Hausdorff, we have only to assign to G the C-O topology, select a family of coordinate functions, and then we have a coordinate bundle. The same is true if Y is regular and locally-compact, and G is suitably restricted. Furthermore this mapping of E-F bundles into coordinate bundles preserves equivalence.

For most bundles that arise in practice, such as sphere-bundles, tensor bundles over manifolds, and coset spaces of Lie groups, the conditions on Y,G are satisfied, and G has a natural topology which coincides with the C-O topology. For all such, the E-F definition of bundle is essentially equivalent to the one we are using.

When Y is locally compact (but not compact) and G is unrestricted, the C-O topology fails only in the continutiy of g^{-1}. It is to be remarked that a modification of the C-O topology given by Arens [2] eliminates this defect.

Recently, Ehresmann [29] has introduced a definition of fibre bundle in which the topology of G plays a role. This definition is equivalent to ours. It has the advantage of being invariant in form. The associated principal bundle (§8) is regarded as a part of the structure of the original bundle.

§6. Differentiable Manifolds and Tensor Bundles

6.1. Coordinate systems in a manifold. An *n-dimensional manifold* (or n-manifold) is a topological space in which each point has a neighborhood homeomorphic to some open set in cartesian n-space. We shall restrict attention to manifolds which are separable, metric, and connected.

A system S of differentiable coordinates in an n-manifold X is an indexed family $\{V_j, j \in J\}$ of open sets covering X, and, for each j, a homeomorphism

$$\psi_j: \quad E_j \to V_j,$$

where E_j is an open set in cartesian n-space, such that the map

$$\psi_j^{-1}\psi_i: \quad \psi_i^{-1}(V_i \cap V_j) \to \psi_j^{-1}(V_i \cap V_j), \qquad i,j \in J, \tag{1}$$

is differentiable. If each such map has continuous derivatives of order r, then S is said to be of class r. If S is of class r for each r, then S is said to be of class ∞. If each such map is analytic, then S is said to be analytic or of class ω.

If S,S' are two systems of coordinates in X of class r they are said to be *r-equivalent* if the composite families $\{V_j, V'_k\}$, $\{\psi_j, \psi'_k\}$ form a system of class r. *A differentiable n-manifold X of class r* is an n-manifold X together with an r-equivalence class of systems of coordinates in X.

Let X,X' be differentiable manifolds of dimensions n,n', and of classes $\geqq r$. A map f: $X \to X'$ is said to be of class r if there exist representative systems of coordinates S and S' such that, for $j \in J$, $k \in J'$, the map

$$\psi_k'^{-1}f\psi_j: \quad \psi_j^{-1}(V_j \cap f^{-1}V'_k) \to E'_k$$

has continuous derivatives to the order r. Clearly, if this is true of one such pair S,S', it will be true of any other.

6.2. The linear group. Denote by L_n the group of non-singular, real $n \times n$-matrices. If the elements of a matrix are regarded as the coordinates of a point in an n^2-dimensional cartesian space, then L_n is an open subset. The identity map ψ defines an analytic system S of coordinates in L_n—just one neighborhood $V = L_n$. Then L_n together with the ω-equivalence class of S form an analytic differentiable manifold. The operations in L_n of multiplication and inverse are analytic in terms of S.

6.3. Jacobian matrices and orientability. Let S be a system of coordinates of class r in the n-manifold X. If $x \in V_i \cap V_j$, denote by $a_{ji}(x)$ the $n \times n$-matrix of first partial derivatives of the functions (1) evaluated at $\psi_i^{-1}(x)$, i.e. the Jacobian matrix of (1). The equation

$$a_{kj}(x)\cdot a_{ji}(x) = a_{ki}(x), \qquad x \in V_i \cap V_j \cap V_k \tag{2}$$

follows immediately from the function of a function rule for derivatives. If we set $k = i$, it follows that $a_{ji}(x)$ has an inverse, and therefore it lies in L_n. Thus a_{ji}: $V_i \cap V_j \to L_n$. Since the functions (1) are of class r, their derivatives are of class $r - 1$. This implies that a_{ji} is a function of class $r - 1$.

A system S of coordinates is called *oriented* if the determinant of $a_{ij}(x)$ is positive for all i,j and $x \varepsilon V_i \cap V_j$. If S,S' are oriented systems, it is easy to prove that the Jacobian matrices of $\psi_j'^{-1}\psi_i$ have determinants which are either positive for all i,j and $x \varepsilon V_i \cap V_j'$, or negative for all i,j and $x \varepsilon V_i \cap V_j'$. We say that S and S' are *positively* or *negatively* related accordingly. It follows that the oriented systems divide into two classes, those within the same class are positively related, two in different classes are negatively related. Each class is called an *orientation* of X. If X admits an oriented system, it is said to be *orientable;* in this case it has two orientations.

The simplest example of a non-orientable manifold is the Möbius band (1.3) with its edge removed.

6.4. Tensor bundles. Let G be a topological transformation group of a space Y, and let h: $L_n \rightarrow G$ be a continuous homomorphism. Define $g_{ji}(x) = ha_{ji}(x)$. It follows from (2) that the set $\{g_{ji}\}$ is a system of coordinate transformations in X as defined in §3.1. By §3.2, there is a bundle $\mathcal{B}$ with base space X, fibre Y, group G, and these coordinate transformations; and by §3.3, $\mathcal{B}$ is unique up to an equivalence. This bundle is called *the tensor bundle of type h over the differentiable manifold X*.

A cross-section of a tensor bundle of type h is called a *tensor field* over X of type h.

If G is a Lie group it is well known that h: $L_n \rightarrow G$ is analytic [12]. This implies that the g_{ji} are of class $r - 1$. Suppose moreover that Y is a differentiable manifold of class $\geqq r - 1$ and that $G \times Y \rightarrow Y$ has class $\geqq r - 1$. Then $g_{ji}(x)\cdot y$ is a function of class $r - 1$ in (x,y). In this case, the tensor bundle space B becomes a differentiable manifold of class $r - 1$ using, as coordinates in B, maps ψ': $E_j \times E \rightarrow B$ of the form $\psi'(u,v) = \phi_j(\psi_j(u),\psi(v))$ where ψ: $E \rightarrow W$ is a coordinate neighborhood in Y. It follows quickly that p: $B \rightarrow X$ and p_j: $p^{-1}(V_j) \rightarrow Y$ are functions of class $r - 1$.

The preceding definitions of tensor and tensor field are essentially equivalent to the classical definitions. The novelty of our treatment lies in the assignment of a topology to the set of tensors (of a prescribed type) at the various points of X. This is done in such a way as to form a bundle space under the natural projection into X. In most applications, Y is a linear space and G is a linear group; hence B is a differentiable manifold. The advantage of our approach is that a tensor field becomes a function in the ordinary sense. Its continuity and differentiability need not be given special definitions. A full discussion of the classical point of view is given in the books of Veblen and Whitehead [95, 96].

6.5. Examples of tensor bundles. It is perhaps worthwhile to consider some of the standard examples of tensor bundles. First, let Y be the real number system and G equal the identity. Then $\mathfrak{B}$ is called *the bundle of absolute scalars over X*. Of course $B = X \times Y$.

Again, let Y be the real numbers, and G the 1-dimensional linear group. Let w be a positive integer, and define $h(a)$ to be the determinant of a raised to the power w. Then h: $L_n \to G$. The resulting bundle is called *the bundle of relative scalars over X of weight w*. When $w = 1$ they are called *scalar densities*. It will follow from the results of §12 that: *if w is even, $\mathfrak{B}$ is equivalent to the product $X \times Y$; if w is odd, it is equivalent to the product if and only if X is an orientable manifold.* In particular, the scalar densities over the projective plane do not form a product space.

Let Y be n-dimensional real linear space: $y = (y^1, \cdots, y^n)$. For any $a \in L_n$, $a = \|a^\alpha_\beta\|$, define $(a \cdot y)^\alpha = a^\alpha_\beta y^\beta$ (summation on repeated indices). This is the standard representation of L_n as the group of linear transformations of n-space. The resulting bundle is called *the tangent bundle of X* or bundle of contravariant vectors on X. The system of coordinates constructed for B in §6.4 is made up of maps ψ'_j: $E_j \times Y \to B$ defined by $\psi'_j(u,y) = \phi_j(\psi_j(u),y)$. If $b \in p^{-1}(V_i \cap V_j)$, then the Jacobian matrix $g'_{ji}(b)$ of $\psi'^{-1}_j\psi'_i$ is defined. Direct computation shows that the determinant of the $2n \times 2n$-matrix $g'_{ji}(b)$ is the square of the determinant of $g_{ji}(p(b))$. Thus, the former is always positive. But this means that *the space B of the tangent bundle is always orientable* even though X is not. It is an easy exercise to verify that, if the product $X \times Y$ of two manifolds is orientable, then X and Y are orientable. This implies: *the tangent bundle of a non-orientable manifold is never a product.*

The reader may wish to verify that L_n, considered as a transformation group in Y, has the compact-open topology (see §5.4).

Let h: $L_n \to L_n$ send each matrix into the transpose of its inverse. Let Y and the operations of L_n in Y be as above. This choice of h yields *the bundle of covariant tangent vectors of X*. It is shown in §12.11 that the covariant and contravariant tangent bundles are equivalent.

For the general tensor bundle, contravariant of order s, covariant of order t, and weight w, the fibre Y is a linear space of dimension n^{s+t}. An element $y \in Y$ has coordinates $(y^{\alpha_1 \cdots \alpha_s}_{\beta_1 \cdots \beta_t})$ where $\alpha_i, \beta_i = 1, \cdots, n$. If $a \in L_n$, then $h(a)$ is the linear transformation of Y given by

$$(h(a) \cdot y)^{\alpha_1 \cdots \alpha_s}_{\beta_1 \cdots \beta_t} = |a|^w a^{\alpha_1}_{\gamma_1} \cdots a^{\alpha_s}_{\gamma_s} y^{\gamma_1 \cdots \gamma_s}_{\delta_1 \cdots \delta_t} \bar{a}^{\delta_1}_{\beta_1} \cdots \bar{a}^{\delta_t}_{\beta_t}$$

where $\bar{a}$ is the transpose inverse of a, and $|a|$ is its determinant. It is easily verified that h maps L_n isomorphically onto a subgroup G of the full linear group of Y, and G is the group of the bundle.

Let $\mathcal{B}$ be a bundle with fibre Y, and let Y' be a subspace of Y mapped on itself by every transformation of the group G of $\mathcal{B}$ (i.e. Y' is *invariant* under G). It follows that, for any i,j and $x \in V_i \cap V_j$, we have

$$\phi_{i,x}(Y') = \phi_{j,x}(Y').$$

Let $B' \subset B$ denote the union of the subspaces $\phi_{i,x}(Y')$ for all i and $x \in V_i$. It is easily proved that the functions ϕ_i restricted to $V_i \times Y'$ define a bundle structure $\mathcal{B}'$ in B' having the fibre Y', and the same coordinate neighborhoods and coordinate transformations as $\mathcal{B}$. We refer to $\mathcal{B}'$ as the *subbundle* of $\mathcal{B}$ determined by the invariant subspace Y' of Y. If G does not operate effectively in Y', the group of $\mathcal{B}'$ is a factor group of G.

For example, if Y is the space of covariant tensors of order t, and $h\colon\ L_n \to G$ is as above, then the subspace Y' of tensors which are skew-symmetric is invariant under G. The same is true of the space of symmetric tensors.

In the case of contravariant vectors, Y' = the zero vector in Y is invariant under L_n. Then B' is homeomorphic to X under the projection p. Thus $B' \subset B$ provides a cross-section of the tangent bundle. The set Y'' of non-zero vectors in Y is likewise invariant, and B'' is the complement in B of B'.

Consider now the fibre Y of covariant tensors of order 2. The subspace Y' of symmetric, positive-definite elements is invariant under G, and determines a subbundle of the bundle of covariant tensors of order 2. A cross-section of the subbundle is just a Riemannian metric tensor in X.

6.6. Bundles of linear spaces. Let $\mathcal{B}$ be a bundle in which the fibre Y is a vector space over the real numbers, and G is a group of linear transformations. Such a bundle we call a *bundle of linear spaces*. If $b_1,b_2 \in Y_x$, $x \in V_j$, and u,v are real numbers, define

$$ub_1 + vb_2 = \phi_{j,x}(up_j(b_1) + vp_j(b_2)). \tag{3}$$

Since

$$\begin{aligned}\phi_{i,x}(up_i(b_1) + vp_i(b_2)) &= \phi_{j,x}g_{ji}(x)\cdot(up_i(b_1) + vp_i(b_2))\\ &= \phi_{j,x}(ug_{ji}(x)p_i(b_1) + vg_{ji}(x)p_i(b_2))\\ &= \phi_{j,x}(up_j(b_1) + vp_j(b_2)),\end{aligned}$$

the definition is independent of the choice of the coordinate function ϕ_j.

Thus each Y_x becomes a linear space. If f_1 and f_2 are two cross-sections and u,v are real numbers define $uf_1 + vf_2$ by

$$(uf_1 + vf_2)(x) = uf_1(x) + vf_2(x).$$

From (3) above it follows that $uf_1 + vf_2$ is continuous in each V_j, and therefore continuous over X. We conclude that the set of all cross-sections forms a vector space.

It is to be noted that most of the examples of tensor bundles are of this kind.

6.7. Differentiable approximations to a cross-section. The most important applications of the theory of bundles, developed in the sequel, are to differential geometry. In particular the existence or non-existence of cross-sections are determined for particular tensor bundles. The results are always in terms of the existence or non-existence of *continuous* cross-sections. The differential geometer is interested in *differentiable* tensor fields. The following theorem is a general justification for the consideration of continuity alone in subsequent work.

Let $\mathfrak{B}$ be a bundle over the differentiable manifold X such that B is a differentiable manifold, p is differentiable, and B, p, X, ϕ_j, and p_j have a class $\geqq r$ $(r = 1, 2, \cdots, \infty)$. Let f: $X \to B$ be a continuous cross-section. We shall suppose that f is differentiable of class $\geqq r$ on a closed subset A of X. This means that f is of class $\geqq r$ in some open set U of X containing A. (The case of A being vacuous is not excluded.) Finally, let ρ be a metric on B, and let ϵ be a positive number.

Theorem. *Under the above hypotheses, there exists a differentiable cross-section f': $X \to B$ of class $\geqq r$ such that $\rho(f(x),f'(x)) < \epsilon$ for each $x \in X$, and $f'(x) = f(x)$ for $x \in A$.*

It should be emphasized that $r \leqq \infty$. The analytic case is unsolved. The main proof is preceded by several lemmas.

Lemma. *Let D,D' be two rectangular domains in cartesian n-space R defined by $a_i < x < b_i$, $a'_i < x < b'_i$ $(i = 1, \cdots, n)$ respectively and such that D' contains the closure of D. Then there exists a real-valued function g defined in R of class ∞, and such that $0 \leqq g(x) \leqq 1$ for all x, $g(x) = 1$ for $x \in D$, and $g(x) = 0$ for $x \in R - D'$.*

For any interval $[c,d]$ of real numbers let

$$\psi_{cd}(x) = \begin{cases} \exp\left(-\dfrac{1}{x-c} + \dfrac{1}{x-d}\right), & x \in [c,d], \\ 0 & x \text{ not } \in [c,d]. \end{cases}$$

Then ψ is of class ∞, and $\psi \geqq 0$. Define

$$\phi_{cd}(x) = \int_c^x \psi_{cd}(x)dx \Big/ \int_c^d \psi_{cd}(x)dx.$$

Then ϕ is of class ∞, $0 \leqq \phi(x) \leqq 1$, $\phi(x) = 0$ for $x \leqq c$, and $\phi(x) = 1$ for $x \geqq d$. If $[a,b]$, $[a',b']$ are two intervals with $a' < a$, $b < b'$, then, by piecing together two such functions as ϕ, we define

$$h(x) = \begin{cases} \phi_{a'a}(x), & x \leqq b, \\ 1 - \phi_{bb'}(x), & x > b, \end{cases}$$

and obtain a function h of class ∞, $0 \leqq h(x) \leqq 1$, $h(x) = 1$ for $x \in [a,b]$ and $h(x) = 0$ for x outside $[a',b']$. Let h_i be such a function for the intervals $[a_i,b_i]$, $[a'_i,b'_i]$. Then the product function

$$g(x_1, \cdots, x_n) = h_1(x_1) \cdots h_n(x_n)$$

has the properties asserted in the lemma.

LEMMA. *Let U be an open set in R with compact $\bar{U}$, and let V be an open set containing $\bar{U}$. Then there exists a real-valued function g defined in R of class ∞ such that $0 \leqq g(x) \leqq 1$ for all x, $g(x) = 1$ for $x \in \bar{U}$, and $g(x) = 0$ for $x \in R - V$.*

As $\bar{U}$ is compact, we can choose a finite number $D_1, \cdots, D_n$ of rectangular domains covering $\bar{U}$ such that the closure of each is in V. Let D'_i be a rectangular domain containing $\bar{D}_i$ and contained in V. Let g_i be a function for the pair D_i, D'_i as asserted in the preceding lemma. Define the function g by

$$1 - g = (1 - g_1)(1 - g_2) \cdots (1 - g_m).$$

Then g is of class ∞, $0 \leqq g \leqq 1$, some $g_i(x) = 1$ implies $g(x) = 1$, and every $g_i(x) = 0$ implies $g(x) = 0$. Thus $g(x) = 1$ for $x \in \bigcup D_i$, and $g(x) = 0$ for x outside $\bigcup D'_i$.

LEMMA. *Let F be a real-valued continuous function defined in an open set W' in R, and of class $\geqq r$ in an open set $U \subset W'$. Let U',V' be open sets such that $\bar{U}' \subset V'$, $\bar{V}'$ is compact and $\subset W'$. Finally, let δ be a positive number. Then there exists a real-valued continuous function F' defined in W' such that $|F'(x) - F(x)| < \delta$ for all $x \in W'$, F' is of class $\geqq r$ in $U \cup U'$, and $F'(x) = F(x)$ for x in $W' - \bar{V}'$.*

By the Weierstrass approximation theorem, there is a polynomial $G(x)$ such that $|G(x) - F(x)| < \delta$ for $x \in \bar{V}'$. By the preceding lemma there exists a function g of class ∞ such that $0 \leqq g \leqq 1$, $g = 1$ on $\bar{U}'$

and $g = 0$ outside V'. Define

$$F'(x) = g(x)G(x) + (1 - g(x))F(x), \qquad x \in W'.$$

Then $F' = G$ on U' and $F' = F$ on $W' - V'$. On $\bar{V}'$,

$$|F'(x) - F(x)| = |g(x)||G(x) - F(x)| < \delta.$$

Furthermore F' is of class $\geqq r$ wherever F is of class $\geqq r$, in particular, in U. Hence F' is of class $\geqq r$ in $U \cup U'$.

We return now to the proof of the theorem. For each $x \in X$, we can choose a j_x such that $x \in V_{j_x}$, and a coordinate neighborhood $E_x \subset Y$ such that $p_{j_x}f(x) \in E_x$. Since f is continuous, there is a neighborhood $C_x \subset V_{j_x}$ of x mapped into E_x by $p_{j_x}f$. Let D_x, contained in C_x, be the compact closure of a neighborhood of x.

Since X is separable, metric, and locally-compact, there is a countable sequence of open sets $\{P_i\}$ with compact closures whose union is X. Let $Q_j = \bigcup \bar{P}_i$ for $i \leqq j$. Then $\{Q_i\}$ is a monotone increasing sequence of compact sets whose union is X. Construct now a third sequence of compact sets $\{R_i\}$ such that $Q_i \subset R_i$ and $R_i \subset$ the interior of R_{i+1}. This is done inductively, suppose $R_1, \cdots, R_k$ are defined. Since $R_k \cup Q_{k+1}$ is compact, there is a finite number of open sets covering it with compact closures. Let R_{k+1} be the union of their closures. Let S_i be the closure of $R_i - R_{i-1}$. Then S_i is compact, $X = \bigcup S_i$ and $S_i \cap S_j = 0$ if $j \neq i - 1, i$ or $i + 1$. For each $x \in S_i$ there is a neighborhood D_x as above. By reducing the size of D_x we can insure that it does not meet S_j for $j \neq i - 1, i$ or $i + 1$. Choose a finite number of such D's with interiors covering S_i. Do this for each i, and arrange the totality of these D's in a simple sequence $\{D_{x_i}\}$. Then the interiors of the D's cover X, and any D intersects only a finite number of the other D's of the sequence. Abbreviate $D_{x_i}, C_{x_i}, \phi_{x_i}, E_{x_i}$ by D_i, C_i, etc.

Define the sequence of compact sets $A_0, A_1, \cdots$ inductively by $A_0 = A$, and $A_i = A_{i-1} \cup D_i$. Then X is the union of the interiors of the sets A_i. We shall define a sequence of functions $f_0, f_1, \cdots$ such that

(i) $f_j(x) = f_i(x)$ for $x \in A_i$ if $i < j$,
(ii) $\rho(f_j(x), f(x)) < \epsilon$, $x \in X$,
(iii) f_j is of class $\geqq r$ on A_j,
(iv) $p_i f_j$ maps D_i into E_i for all i and j.

The sequence is constructed inductively. Define $f_0 = f$. Suppose f_i defined for $i \leqq k$ satisfying these conditions.

Since f_k is of class $\geqq r$ on A_k, by definition it is of class $\geqq r$ in some open set $U \supset A_k$. Let $D = D_{k+1} - U \cap D_{k+1}$. Then, by (iv),

$p_{k+1}f_k = F$ maps D into E_{k+1}. Choose an open set C such that $D \subset C$ and $F(C)$ lies in E_{k+1}. Since $D \cap A_k = 0$, there are open sets U', V', W' such that

$$D \subset U', \quad \bar{U}' \subset V', \quad \bar{V}' \subset W', \quad W' \subset C, \quad W' \cap A_k = 0,$$

and $\bar{V}'$ is compact, and W' meets only a finite number of the sets D_i. Since E_{k+1} is a coordinate neighborhood the function F, restricted to W', is given by real-valued components $F^\alpha(\alpha = 1, \cdots, \dim Y)$. We may apply the last lemma above to each component and obtain a function F': $W' \to E_{k+1}$ such that $|F'^\alpha - F^\alpha| < \delta$ for each α, F' is of class $\geqq r$ in $(U \cap W') \cup U'$ and $F' = F$ in $W' - \bar{V}'$. Define $f_{k+1}(x) = f_k(x)$ for x not in W' and $= \phi_{k+1}(x,F'(x))$ for x in W'. Since $W' \cap A_k = 0$, condition (i) holds. Since ϕ_{k+1} has class $\geqq r$, and F' has class $\geqq r$ in $(U \cap W') \cup U'$, it follows that f_{k+1} has class $\geqq r$ in $U \cup U'$, therefore (iii) holds. Condition (ii) holds for f_k; by restricting the size of δ it will clearly hold for f_{k+1}. The same is true for condition (iv); but here we must remark that, since f_{k+1} differs from f_k only in V', and V' meets only a finite number of the sets D_i, we need impose only a finite number of restrictions on δ to achieve (iv).

Assuming the sequence $\{f_k\}$ constructed, define $f'(x) = f_i(x)$ for $x \in A_i$. Then, by (i), f' is uniquely defined for each x. By (ii), $\rho(f'(x), f(x)) < \epsilon$. By (iii), f_i is of class $\geqq r$ on the interior of A_i, the same therefore holds for f'. But X is the union of the interiors of the sets A_i. So f' is of class $\geqq r$ over all of X. Finally $f'(x) = f_0(x) = f(x)$ for $x \in A_0 = A$.

6.8. Much that has been said in this article carries over to the case of complex analytic manifolds. One is naturally restricted to fibres Y and groups G which are complex analytic. The preceding approximation theorem is of no interest in the complex case since analyticity is not established.

§7. Factor Spaces of Groups

7.1. Definition of factor space. Let B be a topological group and let G be a closed subgroup of B. A left coset of G in B is a set of the form $b \cdot G$. Any such set is closed, and any two such either coincide or have no point in common. Let B/G denote the set whose elements are the left cosets of G in B. Define the *natural map*

$$p: \quad B \to B/G \quad \text{by} \quad p(b) = b \cdot G.$$

A subset U of B/G is said to be open if $p^{-1}(U)$ is an open set of B. It is readily verified that these open sets define a topology in B/G. The set B/G with this topology is called *the factor space* (*or coset space*) *of* B

by G. Clearly p is continuous by definition, and the topology of B/G is maximal with respect to this property.

If U is an open set in B, then $p^{-1}p(U) = U{\cdot}G$ (i.e. the set of products ug). But this set is open in B. Therefore $p(U)$ is open in B/G; and *p is an interior map.* If x,x' are distinct points of B/G, choose $b \in p^{-1}(x)$, $b' \in p^{-1}(x')$. Then $b^{-1}b'$ is not in G. Let W be a neighborhood of $b^{-1}b'$ with $W \cap G = 0$. Let U,V be neighborhoods of b,b' respectively such that $U^{-1}{\cdot}V \subset W$. Then $p(U)$, $p(V)$ are neighborhoods of x,x'. They have no common point. For if x'' is such a point and $p(b'') = x''$, there are elements $g,g' \in G$ such that $b''g \in U$ and $b''g' \in V$. This implies $(b''g)^{-1}(b''g') = g^{-1}g' \in W$ which is impossible. Thus *B/G is a Hausdorff space.*

Notice that the transformation $B \to B$, sending each b into its inverse, maps each left coset of G into a right coset and conversely. This induces a homeomorphism between the left and right coset spaces, so all results for left coset spaces hold equally for right coset spaces.

7.2. Translations of factor spaces. If $x \in B/G$ and $b \in B$, define the *left translation of x by b* by

$$b{\cdot}x = p(b{\cdot}p^{-1}(x)). \tag{1}$$

It is readily proved that $(b_1b_2){\cdot}x = b_1{\cdot}(b_2{\cdot}x)$ so that B is a group of transformations of B/G under the operation (1). Clearly, B is transitive (i.e. for any pair x,x', there is a $b \in B$ such that $b{\cdot}x = x'$).

If U is open in B/G, then $p^{-1}(U)$, $b{\cdot}p^{-1}(U)$, and $p(b{\cdot}p^{-1}(U))$ are also open. Therefore $b{\cdot}U$ is open. Thus B is a group of homeomorphisms of B/G.

Define G_0 to be the intersection of all the subgroups bGb^{-1} conjugate to G in B. Then G_0 is a closed invariant subgroup of B, and it is the largest subgroup of G which is invariant in B. If $g \in G_0$, then

$$gbG = b(b^{-1}gb)G = bG.$$

Thus each element of G_0 acts as the identity transformation in B/G. Conversely, if $cbG = bG$ for every b, then $cb \in bG$ or $c \in bGb^{-1}$ for every b. Hence $c \in G_0$. Thus the factor group B/G_0 acts effectively in B/G. Let p_0: $B \to B/G_0$ be the natural map.

Suppose now that $h \in B/G_0$, $x \in B/G$ and $h{\cdot}x$ lies in the open set U. Choose $b_1 \in p_0^{-1}(h)$, $b \in p^{-1}(x)$. Then $b_1b \in p^{-1}(U) = U'$. Choose neighborhoods V' and W' of b_1,b respectively such that $V'{\cdot}W' \subset U'$. Since p,p_0 are interior maps, $V = p_0(V')$, $W = p(W')$ are neighborhoods of h,x respectively. It follows quickly that $h' \in V$, $x' \in W$ implies $h'{\cdot}x' \in U$. This proves that *B/G_0 is a topological transformation group of B/G.*

7.3. Transitive groups. Conversely, suppose we are given that B is a transitive topological transformation group of X. Choose a base point $x_0 \varepsilon X$. Define p': $B \to X$ by $p'(b) = b \cdot x_0$. It is clear that p' is continuous. Let G be the subgroup of elements of B which map x_0 into itself. Then G is a closed subgroup, and, for each $x \varepsilon X$, $p'^{-1}(x)$ is a left coset of G in B. This defines a unique 1-1 map q: $B/G \to X$ such that $qp(b) = p'(b)$ for all b. If U is open in X, $p'^{-1}(U)$ is open in B. This latter set coincides with $p^{-1}q^{-1}(U)$. Hence $q^{-1}(U)$ is open in B/G. It follows that q is continuous.

In general, q^{-1} is not continuous. There are circumstances under which q^{-1} is continuous and which occur frequently. For example, if B is compact, so also is B/G; and one can apply the well-known result that a continuous 1-1 map of a compact Hausdorff space onto a Hausdorff space is a homeomorphism.

Suppose q^{-1} is continuous. If U is open in B, it follows that $p'(U) = qp(U)$ is open in X. Hence p' is an interior map. Conversely, if p' is interior, and V is open in B/G, we have $q(V) = p'p^{-1}(V)$ is open in X. This means that q^{-1} is continuous. Summarizing we have the

THEOREM. *If B is compact, or if p': $B \to X$ is an interior map, then the natural map q: $B/G \to X$ is a homeomorphism, and the maps p' and p: $B \to B/G$ are topologically equivalent.*

If p' maps a neighborhood of e onto a neighborhood of x_0, it follows from the homogeneity of p' that it is an interior map.

7.4. The bundle structure theorem. We desire to prove that B is a bundle over B/G with respect to the projection p. Or, more generally, if H is a closed subgroup of G and p: $B/H \to B/G$ assigns to each coset of H the coset of G which contains it, then B/H is a bundle over B/G with projection p. It is an unsolved problem whether this is always the case. Some mild restriction seems to be necessary.

Let G be a closed subgroup of B. Then G is a point $x_0 \varepsilon B/G$. A *local cross-section of G in B* is a function f mapping a neighborhood V of x_0 continuously into B and such that $pf(x) = x$ for each $x \varepsilon V$. If B is a bundle over B/G, it is clear that such an f must exist.

THEOREM. *If the closed subgroup G of B admits a local cross-section f, if H is a closed subgroup of G, and p: $B/H \to B/G$, is the map induced by the inclusion of cosets, then we can assign a bundle structure to B/H relative to p. The fibre of the bundle is G/H, and the group of the bundle is G/H_0 acting in G/H as left translations where H_0 is the largest subgroup of H invariant in G. Furthermore, any two cross-sections lead to strictly equivalent bundles. Finally, the left translations of B/H by elements of B are bundle mappings of this bundle onto itself.*

Taking $H = e$, we have the

COROLLARY. *If G has a local cross-section in B, then B is a fibre bundle over B/G relative to the projection p which assigns to each b the coset bG. The fibre of the bundle is G and the group is G acting on the fibre by left translations.*

Introduce the natural maps:

$$\begin{array}{ccc} & B & \\ p_1 \swarrow & & \searrow p_2 \\ B/H & \xrightarrow[p]{} & B/G \end{array}$$

As observed in §7.1, p_1 and p_2 are continuous maps. If U is open in B/G, then, by definition, $p_2^{-1}(U)$ is open in B. Since $pp_1 = p_2$, we have $p_2^{-1}(U) = p_1^{-1}p^{-1}(U)$. This means that $p^{-1}(U)$ is open in B/H, and, therefore, p is continuous. It is clear that $G/H \subset B/H$ and $p(G/H) = x_0$. We will denote elements of G/H by y and elements of B/H by z.

We construct the coordinate bundle as follows. The indexing set J is just the set B. For each $b \in B$, define the coordinate neighborhood V_b in B/G by $V_b = b \cdot V$ (f is defined on V). Define f_b: $V_b \to B$ by $f_b(x) = bf(b^{-1} \cdot x)$. Then f_b is continuous, and $p_2 f_b(x) = x$. For any $x \in V_b$ and $y \in G/H$ define the coordinate function ϕ_b by

$$\phi_b(x,y) = f_b(x) \cdot y. \tag{1}$$

As proved in §7.2, left translation of B/H by an element of B is continuous in both variables. Therefore ϕ_b is continuous in (x,y). Since p_1 maps G onto G/H, we can choose $g \in G$ so that $p_1(g) = y$. Then $p_1(f_b(x)g) = f_b(x) \cdot y$ and $p_2(f_b(x)g) = p_2 f_b(x) = x$. Since $pp_1 = p_2$, it follows that $p\phi_b(x,y) = x$.

Define p_b: $p^{-1}(V_b) \to G/H$ by

$$p_b(z) = [f_b(p(z))]^{-1} \cdot z. \tag{2}$$

Clearly p_b is continuous, $p_b\phi_b(x,y) = y$, and $\phi_b(p(z),p_b(z)) = z$. The existence of the continuous maps p and p_b with these properties shows that ϕ_b maps $V_b \times G/H$ homeomorphically onto $p^{-1}(V_b)$.

Now suppose $x \in V_b \cap V_c$, then

$$\begin{aligned} p_c\phi_b(x,y) &= f_c(x)^{-1} \cdot [f_b(x) \cdot y] \\ &= [f_c(x)^{-1} f_b(x)] \cdot y \end{aligned}$$

is a left translation of y by the element

$$g_{cb}(x) = f_c(x)^{-1} f_b(x). \tag{3}$$

Since $p_2f_c = p_2f_b$, $g_{cb}(x)$ lies in G. The continuity of f_c, f_b and of inverses implies the continuity of g_{cb}.

As observed in §7.2, the group which operates effectively in G/H is G/H_0. The image of g_{cb} under the natural map $G \to G/H_0$ is the coordinate transformation in $V_b \cap V_c$. This completes the construction of the coordinate bundle.

Now let f, f' be two local cross-sections defined in neighborhoods V, V' of x_0. Define V'_b, f'_b, ϕ'_b, p'_b as above using f', V' instead of f, V. Then

$$\begin{aligned} \bar{g}_{cb}(x)\cdot y &= \phi'^{-1}_{c,x}\phi_{b,x}(y) = p'_c\phi_b(x,y) \\ &= [f'_c(x)]^{-1}\cdot f_b(x)\cdot y, \qquad x \in V_b \cap V'_c. \end{aligned}$$

Since $f'_c(x)$ and $f_b(x)$ both lie in the left coset of G over x, $\bar{g}_{cb}(x) = f'_c(x)^{-1}f_b(x)$ is in G. It is clearly continuous. Therefore, by §2.4, the two bundle structures based on f and f' are equivalent.

To prove the last statement, let $b_1 \in B$. Left translation of B by b_1 does not disturb the inclusion relations among left cosets of G and H. Therefore $b_1\cdot p(z) = p(b_1\cdot z)$ for each $z \in B/H$. Let

$$x \in V_b, \qquad x' = b_1\cdot x \in V_c, \qquad y \in G/H.$$

Then the mapping transformation $\bar{g}_{cb}(x)$ is given by

$$\bar{g}_{cb}(x)\cdot y = \phi^{-1}_{c,x'}(b_1\cdot\phi_{b,x}(y)) = f_c(x')^{-1}b_1f_b(x)\cdot y.$$

Hence $\bar{g}_{cb}(x)$ is the image in G/H_0 of the element $f_c(x')^{-1}b_1f_b(x)$ of G. Since the latter is continuous in x, so also is the former. Having verified the conditions of §2.5, b_1 is a bundle mapping.

REMARK. In the special case $H = e$, we have, by §4.3 and the form of (3), that the B-image of the constructed bundle is a product bundle. Note that B is not a transformation group of G. However we do have that G operates on B by left translations, and the bundle over B/G with fibre B associated with $B \to B/G$ (see §9.1) is B-equivalent to a product.

7.5. Lie groups. A *Lie group* B is a topological group and a differentiable manifold of class 1 in which the operation $B \times B \to B$ given by $(b,b') \to bb'$ and the operation $B \to B$ given by $b \to b^{-1}$ are differentiable maps of class 1. It is a standard theorem of Lie theory that B is differentiably equivalent to an analytic manifold in which the two operations are analytic.

A Lie group may have more than one connected component. But each component is an open set.

It is proved also [Chevalley, 12, p. 135] that any closed subgroup G of B is itself a Lie group and the inclusion map $G \subset B$ is analytic and non-singular. Furthermore, an analytic structure is defined in the left coset space B/G in such a way that the projection p: $B \to B/G$ is analytic and of maximum rank at each point of B. A central step in this process is the construction of a local cross-section of G in B [Chevalley, 12, Proposition 1, p. 110]. Consequently, *the bundle structure theorem of §7.4 applies to any Lie group B and any closed subgroup G of B.*

Since all of the examples of topological groups considered in subsequent sections are Lie groups, the bundle structure theorem will be used without further comment. In every case, though, the construction of an explicit local cross-section is a simple matter.

The problem has not been solved of determining the most general conditions on B and G for the existence of a local cross-section. Gleason [37] has shown the existence when G is a compact Lie group and B is an arbitrary group. An unpublished example of Hanner provides a compact abelian group of infinite dimension and a closed 0-dimensional subgroup without a local cross-section. It seems probable that the local cross-section will always exist when B is compact and finite dimensional. (See App. sect. 1.)

7.6. Orthogonal groups. We shall consider a number of examples of factor spaces of groups.

Let O_n denote the real orthogonal group of transformations in euclidean n-space E^n. It is a transitive group on the unit $(n - 1)$-sphere S^{n-1}. If $x_0 \in S^{n-1}$, the subgroup leaving x_0 fixed is just an orthogonal group O_{n-1}. By §7.3, we may make the identification

$$S^{n-1} = O_n/O_{n-1},$$

and, by §7.4, O_n is a bundle over S^{n-1} with fibre and group O_{n-1}.

7.7. Stiefel manifolds. A *k-frame*, v^k, in E^n is an ordered set of k independent vectors. Let L_n be the full linear group. Any fixed k-frame v_0^k can be transformed into any other v^k by an element of L_n. Let $V'_{n,k}$ denote the set of all k-frames, and let $L_{n,k}$ be the subgroup of L_n leaving fixed each vector of v_0^k. Then we may identify

$$V'_{n,k} = L_n/L_{n,k}.$$

The coset space on the right is a manifold with an analytic structure. We assign this structure to $V'_{n,k}$. The space $V'_{n,k}$ is called *the Stiefel manifold [91] of k-frames in n-space.*

If we restrict attention to k-frames in which the vectors are of unit length and pairwise orthogonal (briefly: *an orthogonal k-frame*), the set of these, $V_{n,k}$, is a subspace of $V'_{n,k}$. The group O_n maps $V_{n,k}$ on itself, and is transitive. The subgroup leaving fixed a v_0^k is just the orthogonal group O_{n-k} operating in the space orthogonal to all the vectors of v_0^k. Thus

$$V_{n,k} = O_n/O_{n-k}.$$

If we translate any v^k along its first vector to its end point on S^{n-1} we obtain a $(k-1)$-frame of vectors tangent at a point of S^{n-1}. The process is clearly reversible. Thus we may interpret $V_{n,k}$ as *the manifold of orthogonal $(k-1)$-frames tangent to S^{n-1}*. In particular, when $k = 2$, *$V_{n,2}$ is the manifold of unit tangent vectors on S^{n-1}*. For another interpretation, let S^{k-1} be the unit sphere in the plane of the vectors v_0^k. An orthogonal map of S^{k-1} into S^{n-1} corresponds exactly to a map of v_0^k into another v^k. Thus $V_{n,k}$ is the manifold of orthogonal maps of S^{k-1} into S^{n-1}.

7.8. Let v_0^n be a fixed orthogonal n-frame in E^n, and let v_0^k denote the first k vectors of v_0^n. Let O_{n-k} be the subgroup leaving v_0^k fixed. Then $O_{n-k} \supset O_{n-k-1}$. Passing to the coset spaces by these subgroups and introducing the natural projections (inclusion of cosets) we obtain a chain of Stiefel manifolds and projections

$$O_n = V_{n,n} \to V_{n,n-1} \to \cdots \to V_{n,2} \to V_{n,1} = S^{n-1}.$$

Each projection or any composition of them is a bundle mapping. By the theorem 7.4, the fibre of $V_{n,n-k+1} \to V_{n,n-k}$ is the coset space $O_k/O_{k-1} = S^{k-1}$ and the group of the bundle is O_k.

Any bundle in which the fibre is a k-sphere and the group is the orthogonal group is called a *k-sphere bundle*. Thus the Stiefel manifolds provide a chain of sphere bundles connecting O_n and S^{n-1}.

Any orthogonal $(n-1)$-frame in n-space can be completed to an orthogonal n-frame in just two ways by the addition of the nth vector. This corresponds to the fact that $V_{n,n} \to V_{n,n-1}$ has a 0-sphere as fibre. This is not a double covering in the strict sense, $V_{n,n} = O_n$ is a space having two connected components—the subgroup R_n of matrices of determinant $+1$ (the rotation group of S^{n-1}), and a second component of matrices of determinant -1. Now O_1 is a group of two elements and the determinant of the non-trivial element is -1. Therefore $V_{n,n} \to V_{n,n-1}$ maps each component of O_n topologically onto $V_{n,n-1}$. Thus we may identify $V_{n,n-1}$ with the rotation group R_n of S^{n-1}. Therefore the projection $O_n \to V_{n,k}$ maps R_n onto $V_{n,k}$. This leads to the identification

$$V_{n,k} = R_n/R_{n-k}, \qquad k < n.$$

A bundle in which the fibre is a sphere and the group is the rotation group is called an *orientable sphere bundle*. It follows from the above remarks that $V_{n,n-k} \to V_{r,n-k-1}$ is an orientable k-sphere bundle.

7.9. Grassmann manifolds. Let $M_{n,k}$ denote the set of k-dimensional linear subspaces (k-planes through the origin) of E^n. Any element of O_n carries a k-plane into a k-plane, and, in fact, O_n is transitive on $M_{n,k}$. If E^k is a fixed k-plane and E^{n-k} is its orthogonal complement, the subgroup of O_n mapping E^k on itself splits up into the direct product $O_k \times O'_{n-k}$ of two orthogonal subgroups the first of which leaves E^{n-k} pointwise fixed and the second leaves E^k pointwise fixed. It follows that we may identify

$$M_{n,k} = O_n/O_k \times O'_{n-k}.$$

The set $M_{n,k}$ with this structure as an analytic manifold is called the *Grassmann manifold of k-planes in n-space.*

One of the subgroups O_k, O'_{n-k} contains an element of determinant -1, therefore the projection $O_n \rightarrow M_{n,k}$ maps the rotation group R_n onto $M_{n,k}$. Let R_k and R'_{n-k} be the rotation subgroups of O_k, O'_{n-k}. Define

$$\tilde{M}_{n,k} = R_n/R_k \times R'_{n-k}.$$

Then $M_{n,k}$ is called *the manifold of oriented k-planes of n-space.* The natural projection $\tilde{M}_{n,k} \rightarrow M_{n,k}$ is a 2-fold covering (both spaces are connected and the fibre is a 0-sphere).

If we identify the Stiefel manifold $V_{n,k}$ with O_n/O'_{n-k} it follows that *$V_{n,k}$ is a bundle over $M_{n,k}$ with fibre and group O_k.* Passing to rotation groups, we obtain that *$V_{n,k}$ is a bundle over $\tilde{M}_{n,k}$ with fibre R_k.*

The correspondence between any k-plane and its orthogonal $(n - k)$-plane sets up a 1-1 correspondence $M_{n,k} \leftrightarrow M_{n,n-k}$. The space of lines through the origin, $M_{n,1}$, or pairs of antipodal points on S^{n-1}, is just projective $(n - 1)$-space; and $\tilde{M}_{n,1} = S^{n-1}$.

7.10. Unitary groups. The unitary group U_n operating in complex n-space is also transitive on the unit $(2n - 1)$-sphere. As in the real case,

$$S^{2n-1} = U_n/U_{n-1}.$$

A series of bundles, analogous to those formed from O_n, can be constructed for U_n. A similar construction can be given for the symplectic group (see §20).

§8. The Principal Bundle and the Principal Map

8.1. The associated principal bundle. A bundle $\mathfrak{B} = \{B,p,X,Y,G\}$ is called a *principal bundle* if $Y = G$ and G operates on Y by left translations.

A slightly broader definition is that G is simply-transitive on Y and the mapping $G \rightarrow Y$ given by $g \rightarrow g \cdot y_0$ (y_0 fixed) is an interior mapping. Then G is homeomorphic to Y, and the operations of G in Y correspond to left translations in G.

If B is a Lie group and G is a closed subgroup, the bundle structure given, in §7.4, to p: $B \to B/G$ is that of a principal bundle.

Let $\mathcal{B} = \{B,p,X,Y,G\}$ be an arbitrary bundle. The *associated principal bundle* $\tilde{\mathcal{B}}$ of $\mathcal{B}$ is the bundle given by the construction theorem 3.2 using the same base space X, the same $\{V_j\}$ the same $\{g_{ji}\}$, and the same group G as for $\mathcal{B}$ but replacing Y by G and allowing G to operate on itself by left translations.

The concept of the associated principal bundle is due to Ehresmann [21], and also the general notion of associated bundles (§9).

8.2. Equivalence theorem. *Two bundles having the same base space, fibre and group are equivalent if and only if their associated principal bundles are equivalent.*

This is an immediate consequence of §2.8 which states that equivalence is purely a property of the coordinate transformations; for a bundle and its associated principal bundle have the same coordinate transformations.

8.3. The cross-section theorem. *A principal bundle with group G is equivalent in G to the product bundle (see §4.3) if and only if it admits a cross-section.*

Suppose a cross-section f: $X \to B$ is given. Define $\lambda_i(x) = p_i(f(x))$ for $x \in V_i$. From the relation

$$g_{ji}(p(b)) \cdot p_i(b) = p_j(b), \qquad p(b) \in V_i \cap V_j,$$

(see §2.3), we obtain immediately that

$$g_{ji}(x) \cdot \lambda_i(x) = \lambda_j(x), \qquad x \in V_i \cap V_j. \tag{1}$$

By §4.3, the bundle is equivalent to a product.

Conversely, suppose $\mathcal{B}$ is equivalent to a product bundle. By §4.3, there exist functions λ_i satisfying (1). Define

$$f_i(x) = \phi_i(x,\lambda_i(x)), \qquad x \in V_i.$$

Then f_i is continuous. From (1) we obtain $f_i(x) = f_j(x)$ for $x \in V_i \cap V_j$. It follows that $f(x) = f_i(x)$ for $x \in V_i$ defines a continuous single-valued cross-section.

Combining §8.2 and §8.3, we have

8.4. COROLLARY. *A bundle with group G is equivalent in G to a product bundle if and only if the associated principal bundle admits a cross-section.*

8.5. Examples. One advantage of passing to the principal bundle is that its structure is often simpler than that of the given bundle.

Consider as examples the Möbius band, Klein bottle and twisted torus as bundles over the circle (§1.3, §1.4, §1.5). All these bundles have the same group and coordinate transformations (§3.4); hence the same principal bundle $\tilde{\mathcal{B}}$. It is easily seen that $\tilde{\mathcal{B}}$ is a circle and p: $\tilde{B} \to X$ is a double covering. Simple considerations of connectedness show that $\tilde{\mathcal{B}}$ does not admit a cross-section.

As another example, consider the 4-dimensional real space of quaternions

$$q = x_1 + ix_2 + jx_3 + kx_4.$$

The usual multiplication rule satisfies the norm condition $|q \cdot q'| = |q| \cdot |q'|$ where $|q|^2 = \Sigma x_i^2$. Then the unit 3-sphere S^3 ($|q| = 1$) is a subgroup. If $q \varepsilon S^3$, the transformation of 4-space given by $q' \to qq'$ preserves the norm. Thus to each $q \varepsilon S^3$ is assigned an orthogonal transformation $f(q)$ in O_4 (see §7.6). Denoting by $e \varepsilon S^3$ the unit quaternion, define p: $O_4 \to S^3$ by $p(o) = o(e)$. By §7.6, this is a principal bundle mapping. Clearly $pf(q) = q$; so f is a cross section. It follows from §8.3, that O_4 is a product bundle over S^3.

Exactly the same argument may be carried through using Cayley numbers (an algebra on 8 units, see §20.5) in place of the quaternions.

In both cases the image of the unit element is the identity transformation. Since the sphere is connected, its image must lie in the rotation subgroup R_4 (R_8) of O_4 (O_8) (see §7.8) which is itself a bundle over S^3 (S^7) with fibre and group R_3 (R_7).

Summarizing, we have

8.6. Theorem. *For $n = 3$ and $n = 7$, the rotation group R_{n+1} of the n-sphere S^n, as a bundle over S^n with group and fibre R_n, is equivalent to the product bundle $S^n \times R_n$.*

It will be shown later (§§22–24) that this is not true for most values of n. It is conjectured that it holds for integers n of the form $2^k - 1$. If one could construct a division algebra in a real vector space of dimension 2^k, the conjecture could be proved. It is not known whether this can be done in a space of 16 dimensions.

8.7. The principal map. Let $\mathcal{B} = \{B,p,X,Y,G\}$ be a bundle, and let $\tilde{\mathcal{B}} = \{\tilde{B},\tilde{p},X,G,G\}$ be its associated principal bundle. Form now the product bundle

$$\tilde{\mathcal{B}} \times Y = \{\tilde{B} \times Y, q, \tilde{B}, Y, G\}, \quad q(\tilde{b},y) = \tilde{b}$$

treated as a bundle with group G. We define the *principal map*

$$P: \quad \tilde{\mathcal{B}} \times Y \to \mathcal{B}$$

as follows: if $x = \tilde{p}(\tilde{b}) \in V_i$, set

$$P(\tilde{b},y) = \phi_i(x,\tilde{p}_i(\tilde{b})\cdot y). \tag{2}$$

This formula, of course, defines only a set of functions $\{P_i\}$. However, for $x \in V_i \cap V_j$,

$$\phi_{i,x}(\tilde{p}_i(\tilde{b})\cdot y) = \phi_{j,x}(g_{ji}(x)\cdot \tilde{p}_i(\tilde{b})\cdot y) = \phi_{j,x}(\tilde{p}_j(\tilde{b})\cdot y).$$

Therefore $P_i = P_j$ on $\tilde{p}^{-1}(V_i \cap V_j) \times Y$, and a unique P is defined. Since each P_i is continuous, so also is P.

Clearly,

$$pP(\tilde{b},y) = x = \tilde{p}(\tilde{b}) = \tilde{p}q(\tilde{b},y).$$

Therefore commutativity holds in the diagram

$$\begin{array}{ccc} \tilde{B} \times Y & \xrightarrow{P} & B \\ \downarrow q & & \downarrow p \\ \tilde{B} & \xrightarrow{\tilde{p}} & X \end{array}$$

This means that P carries fibres into fibres and induces the map $\tilde{p}$ of the base spaces.

To prove that P is a bundle mapping (as defined in §2.5), recall first that $\tilde{\mathcal{B}} \times Y$ has a single coordinate neighborhood $V'_j = \tilde{B}$ and the coordinate function ϕ'_j is the identity map. If $x = \tilde{p}(\tilde{b}) \in V_i$ and we compute $\bar{g}_{ij}$ by (17) of §2.5, we obtain

$$\begin{aligned} \bar{g}_{ij}(\tilde{b})\cdot y &= \phi_{i,x}^{-1} P_{\tilde{b}} \phi'_{j,\tilde{b}}(y) = \phi_{i,x}^{-1} P(\tilde{b},y) \\ &= \phi_{i,x}^{-1}\phi_{i,x}(\tilde{p}_i(\tilde{b})\cdot y) = \tilde{p}_i(\tilde{b})\cdot y. \end{aligned}$$

But $\tilde{p}_i(\tilde{b}) \in G$ and is continuous in $\tilde{b}$; therefore $\bar{g}_{ij}$ is a continuous map of $p^{-1}(V_i)$ into G, and P is a bundle mapping. We have proved

8.8. THEOREM. *If $\tilde{\mathcal{B}}$ is the associated principal bundle of $\mathcal{B}$, then the principal map P: $\tilde{\mathcal{B}} \times Y \to \mathcal{B}$ is a bundle mapping and P induces the projection $\tilde{p}$: $\tilde{B} \to X$ of the base spaces.*

8.9. Admissible maps. There are several interpretations to be given of the principal map. For the first of these, let us say that a map ξ: $Y \to Y_x$ ($Y_x = p^{-1}(x)$) is *admissible* if the map $p_i\xi$: $Y \to Y$ ($x \in V_i$) is in G. If $x \in V_i \cap V_j$, then $p_j\xi = g_{ji}(x)p_i\xi$ is also in G, so that admissibility is independent of the coordinate neighborhood.

If Y is regarded as the bundle space of the trivial bundle in which the base space is a point, the fibre is Y, and the group is G, then ξ: $Y \to Y_x$ is admissible if and only if ξ: $Y \to \mathcal{B}$ is a bundle map.

For any $\tilde{b} \varepsilon \tilde{B}$, define the map

$$\tilde{b}: \quad Y \to Y_x \quad (x = \tilde{p}(\tilde{b})) \quad \text{by} \quad \tilde{b}(y) = P(\tilde{b},y).$$

Then

$$p_i\tilde{b}(y) = p_i\phi_i(x,\tilde{p}_i(\tilde{b})\cdot y) = \tilde{p}_i(\tilde{b})\cdot y,$$

and $\tilde{b}$ is an admissible map. Let $G_x = \tilde{p}^{-1}(x)$. Since $\tilde{p}_i$ maps G_x homeomorphically onto G, it follows that distinct elements of G_x give distinct admissible maps $Y \to Y_x$.

Let ξ: $Y \to Y_x$ be an admissible map. Let $\tilde{b} = \tilde{\phi}_i(x,p_i\xi)$ if $x \varepsilon V_i$. Then $\tilde{b} \varepsilon G_x$, and

$$\tilde{b}(y) = \phi_i(x,\tilde{p}_i(\tilde{b})\cdot y) = \phi_i(x,p_i\xi(y)) = \xi(y).$$

Thus *G_x is the set of all admissible maps of Y into Y_x.*

It follows that $\tilde{B}$ (*the space of the principal bundle*) *can be interpreted as the set of all admissible maps of the fibre Y into the bundle space B.*

With this interpretation the cross-section theorem 8.4 takes on intuitive content. A cross-section f of the principal bundle gives, for each x, an admissible map $Y \to Y_x$, and thereby a map $X \times Y \to B$.

The Ehresmann-Feldbau definition of bundle (§5.2) is in terms of admissible maps of Y into B. We see, in retrospect, that their definition of bundle involves directly the principal bundle. In their invariant approach the space $\tilde{B}$ would be defined directly as the set of admissible maps and assigned the compact-open topology. We leave it as an exercise for the reader to verify: *If G has the compact-open topology, then the topology assigned above to $\tilde{B}$ coincides with the compact-open topology assigned to $\tilde{B}$ as the function space of admissible maps.*

8.10. Right translations of B. In the preceding section, the effect of fixing the variable $\tilde{b}$ in $P(\tilde{b},y)$ was considered. If we fix instead the variable y we obtain a map y: $\tilde{B} \to B$ called a *principal map* of $\tilde{B}$ into B: $y(\tilde{b}) = P(\tilde{b},y)$. Clearly $py(\tilde{b}) = \tilde{p}(\tilde{b})$ so that y maps G_x into Y_x for each x.

Let us specialize further to the case where $\mathfrak{B}$ is itself a principal bundle so that $\tilde{\mathfrak{B}}$ is equivalent to $\mathfrak{B}$. Choose the natural equivalence h: $\tilde{\mathfrak{B}} \to \mathfrak{B}$ given by $h(\tilde{b}) = \phi_i(x,\tilde{p}_i(\tilde{b}))$ for $x = \tilde{p}(\tilde{b}) \varepsilon V_i$. With this identification any element g of G gives a principal map of B on itself carrying each G_x onto G_x. Using the identification h and formula (2) above, it follows that g: $B \to B$ is given by

$$g(b) = \phi_i(x,p_i(b)g), \qquad x = p(b) \varepsilon V_i. \tag{3}$$

Therefore, for any $g' \varepsilon G$ and $x \varepsilon V_i$,

$$p_i g\phi_{i,x}(g') = p_i\phi_i(x,g'g) = g'g.$$

This means that *the map g of G_x on itself is equivalent under $\phi_{i,x}$ to the right translation of G by g.* For this reason the principal map g: $B \to B$ is called a *right translation* of B. It is to be noted that (3) provides a direct definition of the right translation g.

If ξ: $G \to G_x$ is admissible, $b \in G_x$ and $g \in G$, then

$$g(b) = \xi([\xi^{-1}(b)]\cdot g). \tag{3'}$$

To prove this, suppose $x \in V_i$. Then $p_i\xi$: $G \to G$ is a left translation by g_1, say. From (3) we have

$$g(b) = \xi\xi^{-1}\phi_{i,x}([p_i\xi\xi^{-1}(b)]\cdot g) = \xi(g_1^{-1}([g_1\cdot\xi^{-1}(b)]\cdot g)),$$

and (3′) follows from the associative law in G.

An immediate consequence of (3) is

$$p_i(g(b)) = p_i(b)g, \qquad p(b) \in V_i. \tag{4}$$

Using this we have

$$\begin{aligned}(g_1g_2)(b) &= \phi_i(x,p_i(b)g_1g_2) \qquad (x = p(b))\\ &= \phi_i(x,p_i(g_1(b))g_2) = g_2(g_1(b)).\end{aligned}$$

Taking $g_2 = g_1^{-1}$, it follows that a right translation is a homeomorphism. Summarizing, we have

8.11. THEOREM. *If $\mathfrak{B}$ is a principal bundle with group G, then the right translations of B by elements of G map each fibre on itself, and provide an anti-representation of G as a topological transformation group of B.*

In general, a right translation g: $B \to B$ of a principal bundle does not provide a bundle mapping (see §2.5). For suppose $x \in V_j$. Computing $\bar{g}_{jj}(x)$ by (17) of §2.5, we obtain

$$\bar{g}_{jj}(x)g' = g'g, \qquad \text{all } g' \in G.$$

Taking $g' = 1$, we have $\bar{g}_{jj}(x) = g$. Therefore $gg' = g'g$, for all $g' \in G$, is the precise condition for g: $B \to B$ to be a bundle map, i.e. g is in the center of G. In particular, *if G is abelian, a right translation is a bundle mapping.*

8.12. THEOREM. *The principal map P: $\tilde{B} \times Y \to B$ is the projection of a bundle structure having the fibre G, group G, coordinate neighborhoods $p^{-1}(V_j)$, and the coordinate transformations $g_{ji}(p(b))$. Furthermore, the map q: $\tilde{B} \times Y \to \tilde{B}$ ($q(\tilde{b},y) = \tilde{b}$) is a bundle mapping of this bundle into $\tilde{\mathfrak{B}}$.*

From $pP = \tilde{p}q$, we obtain

$$P^{-1}p^{-1}(V_j) = q^{-1}\tilde{p}^{-1}(V_j) = \tilde{p}^{-1}(V_j) \times Y.$$

Define

$$\psi_j\colon\ p^{-1}(V_j) \times G \to P^{-1}p^{-1}(V_j)$$

by

$$\psi_j(b,g) = (\tilde{\phi}_j(p(b),g),\ g^{-1}\cdot p_j(b)).$$

Letting $x = p(b)$, and using (2) of §8.7, we have

$$P\psi_j(b,g) = P(\cdot\ \cdot\ \cdot\ ,\ \cdot\ \cdot\ \cdot) = \phi_j(x,[\tilde{p}_j\tilde{\phi}_j(x,g)]\cdot g^{-1}\cdot p_j(b))$$
$$= \phi_j(x,gg^{-1}\cdot p_j(b)) = b.$$

Define r_j: $P^{-1}p^{-1}(V_j) \to G$ by $r_j = \tilde{p}_j q$. Then

$$r_j\psi_j(b,g) = \tilde{p}_j q(\cdot\ \cdot\ \cdot\ ,\ \cdot\ \cdot\ \cdot) = \tilde{p}_j\tilde{\phi}_j(x,g) = g.$$

This proves that ψ_j is a product representation. A similar calculation shows that

$$r_j\psi_i(b.g) = g_{ji}(x)\cdot g,$$

and therefore the coordinate transformations are $g_{ji}(p(b))$. The same calculation yields

$$\tilde{p}_j q\psi_{i,b}(g) = g_{ji}(x)\cdot g,$$

and therefore q is a bundle mapping.

8.13. Associated maps. Let $\mathfrak{B},\mathfrak{B}'$ be bundles having the same fibre and group and let h be a map $\mathfrak{B} \to \mathfrak{B}'$. The mapping transformations $\{\bar{g}_{kj}\}$ of h are as defined in §2.5. Let $\tilde{\mathfrak{B}},\tilde{\mathfrak{B}}'$ be the associated principal bundles. According to §2.6, there is a unique map

$$\tilde{h}\colon\ \tilde{\mathfrak{B}} \to \tilde{\mathfrak{B}}'$$

having the mapping transformations $\{\bar{g}_{kj}\}$. We call $\tilde{h}$ *the associated map of the principal bundles.*

THEOREM. *If h is a map $\mathfrak{B} \to \mathfrak{B}'$ and $\tilde{h}$ is the associated map $\tilde{\mathfrak{B}} \to \tilde{\mathfrak{B}}'$, then*

$$P'(\tilde{h}(\tilde{b}),y) = hP(\tilde{b},y), \qquad \tilde{b} \,\varepsilon\, \tilde{B},\ y \,\varepsilon\, Y,$$

where P,P' are the related principal maps.

If $x = \tilde{p}(\tilde{b}) \,\varepsilon\, V_j$, and $x' = \bar{h}(x) \,\varepsilon\, V'_k$, then, by (20) of §2.6, we have

$$\tilde{p}'_k\tilde{h}(\tilde{b}) = \tilde{p}'_k\tilde{\phi}'_k(x',\bar{g}_{kj}(x)\cdot\tilde{p}_j(\tilde{b})) = \bar{g}_{kj}(x)\cdot\tilde{p}_j(\tilde{b}).$$

Applying the definitions of P,P' we obtain the result:

$$P'(\tilde{h}(\tilde{b}),y) = \phi'_k(x',[\tilde{p}'_k\tilde{h}(\tilde{b})]\cdot y) = \phi'_k(x',\bar{g}_{kj}(x)\tilde{p}_j(\tilde{b})\cdot y)$$
$$= h\phi_j(x,\tilde{p}_j(\tilde{b})\cdot y) = hP(\tilde{b},y).$$

The intuitive content of the theorem is based on the interpretation of $\tilde{b}$ as an admissible map $Y \to Y_x$. Then $\tilde{b}$ followed by h_x: $Y_x \to Y_{x'}$ is an admissible map $Y \to Y_{x'}$ and is therefore an element $\tilde{h}(\tilde{b})$ in $G_{x'}$.

This naturally defined function from $\mathfrak{B}$ to $\mathfrak{B}'$ is the associated map, and it is a bundle map as might be expected.

8.14. COROLLARY. *If ξ: $Y \to \mathfrak{B}$ is admissible $\tilde{\xi}$: $G \to \tilde{\mathfrak{B}}$ is the associated map, and e is the identity element of G, then*

$$P(\tilde{\xi}(g),y) = P(\tilde{\xi}(e),g\cdot y).$$

It is to be understood that Y is treated as a G-bundle having a single point as base space. Its principal bundle is G. Then ξ is a bundle map, and $\tilde{\xi}$ is defined. Let P' be the principal map for the bundle Y. Clearly $P'(g,y) = g\cdot y$. Applying §8.13, we have

$$P(\tilde{\xi}(g),y) = \xi P'(g,y) = \xi(g\cdot y)$$
$$P(\tilde{\xi}(e),g\cdot y) = \xi(e\cdot(g\cdot y)) = \xi(g\cdot y).$$

8.15. The principal bundle of a coset space.

THEOREM. *If B is a topological group, G is a closed subgroup which has a local cross-section, and H is a closed subgroup of G, then the bundle*

$$B/H \to B/G \qquad \text{(see §7.4)}$$

has, as its principal bundle, the bundle

$$B/H_0 \to B/G$$

where H_0 is the largest subgroup of H invariant in G.

By §7.4, the second bundle has G/H_0 as fibre and group; hence it is principal. If the proof of §7.4 is examined, it is found that the coordinate transformations of the two bundles coincide—they are the images in G/H_0 of the functions $f_c(x)^{-1}f_b(x)$ defined on $V_b \cap V_c$.

To apply the preceding result to coset spaces of the orthogonal group O_n we need the following:

8.16. LEMMA. *The largest subgroup of O_k which is invariant in O_{k+1} consists of e alone.*

As a subgroup of O_{k+1}, O_k operates on the first k variables. If v_0 is the unit vector for the $(k+1)$st variable, then v_0 is fixed under O_k. If $a \in O_{k+1}$, then av_0 is fixed under aO_ka^{-1}. Hence, if $b \in aO_ka^{-1}$ for all $a \in O_{k+1}$, then $bav_0 = av_0$ for every a. Since O_{k+1} is transitive on the unit sphere, it follows that $b = e$.

We apply these results to the Stiefel manifolds (§7.7, §7.8).

8.17. THEOREM. *If $j > k$, the bundle $V_{n,j} \to V_{n,k}$ has $O_n \to V_{n,k}$ as its principal bundle.*

8.18. COROLLARY. *If $n = 2, 4,$ or 8, then the bundle $V_{n,k} \to S^{n-1}$ is a product bundle.*

By §8.6, the associated principal bundle $O_n \to S^{n-1}$ is a product; hence $V_{n,k} \to S^{n-1}$ is a product.

§9. Associated Bundles and Relative Bundles

9.1. Associated bundles. Two bundles, having the same base space X and the same group G, are said to be *associated* if their associated principal bundles (§8.1) are equivalent.

In particular, a bundle and its associated principal bundle are associated.

From §8.2, it follows that, if two bundles are equivalent, they are also associated. In addition, if two associated bundles have the same fibre, and the same action of the group on the fibre, then they are equivalent.

It is easily checked that the relation of being associated is reflexive, symmetric, and transitive. In this way all bundles over X with group G are divided into non-overlapping classes; within each class is just one equivalence class of principal bundles.

If $\mathcal{B} = \{B,p,X,Y,G\}$, and G is also a topological transformation group of a space Y', then the construction theorem 3.2 provides a new bundle $\mathcal{B}' = \{B',p',X,Y',G\}$ having the same coordinate transformations as $\mathcal{B}$. It follows from §8.1 that $\mathcal{B}$ and $\mathcal{B}'$ have the same principal bundle. This establishes the existence of an associated bundle having any prescribed fibre (on which G operates).

There are, of course, a large number of bundles associated with a given bundle $\mathcal{B}$. One need only choose a space Y' and a continuous isomorphism of G onto a group of homeomorphisms of Y'. Many such arise naturally. Most of the tensor bundles over a differentiable manifold X constructed in §6.5 are associated bundles. The tangent bundle is regarded as the central bundle. A tensor bundle is associated with the tangent bundle if and only if $h\colon\ L_n \rightarrow G$ is an isomorphism onto. This is the case for all the standard tensor bundles of order > 0. In the case of a bundle of scalars the group L_n is mapped into L_1 and is not an isomorphism for $n > 1$. In this case we say that the bundle is *weakly associated* with the tangent bundle.

A second class of examples arises by using fibres Y' of the form G/H where H is a closed subgroup of G. If G operates effectively on G/H (§7.2), we obtain an associated bundle; otherwise, a weakly associated bundle. The Stiefel manifolds (§7.8) $V_{n,k}$ are all bundles over S^{n-1} and have the common principal bundle O_n over S^{n-1}.

9.2. Relative bundles. Let $\mathcal{B}$ be a bundle $\{B,p,X,Y,G\}$. Let A be a closed subspace of X and H a closed subgroup of G. If, for every i,j and $x \in V_i \cap V_j \cap A$, the coordinate transformation $g_{ji}(x)$ is an element of H, then the portion of the bundle over A may be regarded as a bundle with group H. One has only to restrict the coordinate neighborhoods and functions to A. Whenever this occurs we shall say that $\mathcal{B}$ *is a relative* (G,H)*-bundle over the base space* (X,A).

The related notions of strict equivalence, of mapping, and of equivalence are defined for relative bundles just as for absolute bundles with the exception that cutting down to A restricts the group to H. For example, let $\mathcal{B}$ be a (G,H)-bundle over (X,A), and let $\mathcal{B}'$ be an (H,H)-bundle over (X,A). *A* (G,H)*-equivalence of* $\mathcal{B}$ *and* $\mathcal{B}'$ is a map h: $\mathcal{B} \rightarrow \mathcal{B}'$ which is, first, a G-equivalence of the two absolute bundles over X, and, second, an H-equivalence when restricted to the portions of $\mathcal{B}$, $\mathcal{B}'$ lying over A.

A general point of view which will prevail is that the smaller the group of a bundle, the simpler the bundle. We shall always attempt to simplify a bundle with group G by seeking a G-equivalent bundle with a group $H \subset G$. Such an attempt will usually be a step-wise procedure of simplifying the bundle over successively larger portions of the space X. At each stage we will have a relative bundle in the above sense.

9.3. The canonical cross-section of a relative bundle. Let $\mathcal{B}$ be a (G,H)-bundle over (X,A). Let $\mathcal{B}'$ denote the associated bundle over X having G/H as fibre and group G acting as left translations. As remarked in §7.2, the group of $\mathcal{B}'$ is actually G/H_0 where H_0 is the largest subgroup of H invariant in G. If H_0 is bigger than the identity then $\mathcal{B}'$ is only weakly associated. Let e_0 denote the coset H treated as an element of G/H. We define a cross-section over A of the bundle $\mathcal{B}'$ by

$$f_0(x) = \phi_j'(x,e_0), \qquad x \in V_j \cap A. \tag{1}$$

If $x \in V_i \cap V_j \cap A$, then

$$\phi_j'(x,e_0) = \phi_i'(x,g_{ij}(x)\cdot e_0) = \phi_i'(x,e_0)$$

since $g_{ij}(x)$ is in H. Thus (1) defines a unique continuous function over A. *We call* f_0 *the canonical cross-section of the* (G,H)*-bundle.*

9.4. Reduction of the group of a bundle.

THEOREM. *Let H be a closed subgroup of G which has a local cross-section. A (G,H)-bundle over (X,A) is (G,H)-equivalent to an (H,H)-bundle over (X,A) if and only if the canonical cross-section (defined only over A) can be extended to a full cross-section of the weakly associated bundle with fibre G/H.*

Suppose $\mathcal{B}$ is (G,H)-equivalent to an (H,H)-bundle. By §2.10, there exist maps λ_i: $V_i \rightarrow G$ such that $\lambda_i(x) \in H$ if $x \in V_i \cap A$, and the functions $g_{ji}'(x) = \lambda_j(x)^{-1}g_{ji}(x)\lambda_i(x)$ have values in H for all x,i,j. Let q: $G \rightarrow G/H$ be the natural map. Define

$$f(x) = \phi_j'(x,q\lambda_j(x)), \qquad x \in V_j. \tag{2}$$

If $x \in V_i \cap V_j$, then

$$\begin{aligned}\phi_j'(x,q\lambda_j(x)) &= \phi_i'(x,g_{ij}(x)\cdot q\lambda_j(x)) \\ &= \phi_i'(x,q[g_{ij}(x)\lambda_j(x)]) = \phi_i'(x,q[\lambda_i(x)g_{ij}'(x)]) \\ &= \phi_i'(x,q\lambda_i(x))\end{aligned}$$

since $g_{ij}'(x)$ lies in H. Therefore (2) defines a unique continuous function over all of X. If $x \in A$, then $\lambda_j(x) \in H$, and $q\lambda_j(x) = e_0$. Therefore $f(x) = f_0(x)$ for x in A.

Conversely, let f be a cross-section of $\mathcal{B}'$ defined over all of X and which is an extension of f_0. Let W be a neighborhood of e_0 in G/H and d a local cross-section of H in G defined over W. For each $g \in G$, let $W_g = g\cdot W$ and $d_g(y) = gd(g^{-1}\cdot y)$ for $y \in W_g$. We can suppose that each coordinate neighborhood V_j of $\mathcal{B}$ is so small that $p_j'f\colon\ V_j \to G/H$ is such that $p_j'f(V_j)$ lies in some set W_g. (If this were not the case we would pass to the refinement of $\{V_j\}$ consisting of all open sets of the form $(p_j'f)^{-1}(W_g)$, see §2.9.) For each j, select such a g and define $\lambda_j(x) = d_g p_j' f(x)$. Then $q\lambda_j(x) = p_j'f(x)$. Define $g_{ji}'' = \lambda_j^{-1}g_{ji}\lambda_i$. Then

$$\begin{aligned}qg_{ji}''(x) &= \lambda_j^{-1}(x)g_{ji}(x)\cdot q(\lambda_i(x)) = \lambda_j^{-1}(x)g_{ji}(x)\cdot p_i'f(x) \\ &= \lambda_j^{-1}(x)\cdot p_j'f(x) = e_0,\end{aligned}$$

and $g_{ji}''\colon\ V_i \cap V_j \to H$. It follows from §2.10 that $\mathcal{B}$ is G-equivalent to a bundle in H. If $x \in A$, then $q\lambda_j(x) = p_j'f_0(x) = p_j'\phi_j'(x,e_0) = e_0$; and $\lambda_j(x)$ is in H. Thus the equivalence is a (G,H)-equivalence, and the proof is complete.

Taking A to be vacuous gives

9.5. Corollary. *If H has a local cross-section in G, then a G-bundle over X is G-equivalent to an H-bundle if and only if the weakly associated bundle with fibre G/H has a cross-section.*

9.6. A bundle of bundles. Let H be a closed subgroup of G with a local cross-section, and K a closed subgroup of H. Let

$$Y = G/H, \qquad Y' = G/K,$$

and let $\eta\colon\ Y' \to Y$ be the natural map. According to §7.4, η is the projection of a bundle structure with fibre H/K and group H/K_0 where K_0 is the largest subgroup of K invariant in H.

Let $\mathcal{B}_0$ be a bundle over X with group G, and let $\mathcal{B},\mathcal{B}'$ be the (possibly weakly) associated bundles over X with fibres Y and Y' respectively. We may suppose that $\mathcal{B}_0,\mathcal{B},\mathcal{B}'$ have the same coordinate neighborhoods $\{V_j\}$ and the coordinate transformations of $\mathcal{B},\mathcal{B}'$ are the images in their respective groups of the coordinate transformations of $\mathcal{B}_0$. Define the *natural map*

(3) $$\nu\colon\ B' \to B$$

by

$$\nu(b') = \phi_j(p'(b'),\eta p'_j(b')), \quad p'(b') \in V_j. \tag{4}$$

If $x = p'(b') \in V_i \cap V_j$, then

$$\begin{aligned}\phi_i(x,\eta p'_i(b')) &= \phi_j(x,g_{ji}(x)\cdot\eta p'_i(b')) \\ &= \phi_j(x,\eta(g_{ji}(x)\cdot p'_i(b'))) = \phi_j(x,\eta p'_j(b')).\end{aligned}$$

Therefore (4) is independent of the choice of j. Clearly the relation $p\nu = p'$ holds in the diagram

$$\begin{array}{ccc} & \nu & \\ B' & \rightarrow & B \\ & p' \searrow \swarrow p & \\ & X & \end{array}$$

THEOREM. *With respect to the natural map ν, B' is a bundle over B with the fibre H/K and the group H/K_0.*

Referring to the proof of §7.4, to each $\alpha \in G$ corresponds a coordinate neighborhood U_α of the bundle η: $Y' \rightarrow Y$, and a coordinate function

$$\psi_\alpha(y,z) = f_\alpha(y)\cdot z$$

where $z \in H/K$ and f_α is a cross-section of the part of G over U_α.

The indexing set for the bundle $B' \rightarrow B$ will be the product $J \times G$. For $j \in J$ and $\alpha \in G$, we take

$$W_{j\alpha} = \phi_j(V_j \times U_\alpha) \tag{5}$$

as the coordinate neighborhood, and

$$\omega_{j\alpha}(b,z) = \phi'_j(p(b),\psi_\alpha(p_j(b),z))$$

as the coordinate function. Then

$$\begin{aligned}\nu\omega_{j\alpha}(b,z) &= \nu\phi'_j(\,\cdot\cdot\cdot\,) = \phi_j(p'\phi'_j(\,\cdot\cdot\cdot\,),\eta p'_j\phi'_j(\,\cdot\cdot\cdot\,)) \\ &= \phi_j(p(b),\eta\psi_\alpha(p_j(b),z)) = \phi_j(p(b),p_j(b)) = b\end{aligned}$$

as required.

If $b' \in \nu^{-1}(W_{j\alpha})$, then $p\nu(b') = p'(b')$ is in V_j, and $p'_j(b')$ is defined. Referring to (4) and (5) we see that $\eta p'_j(b')$ lies in U_α. Hence η_α: $\eta^{-1}(U_\alpha) \rightarrow H/K$ is defined. Let

$$\nu_{j\alpha}(b') = \eta_\alpha p'_j(b'), \qquad b' \in \nu^{-1}(W_{j\alpha}).$$

Then

$$\begin{aligned}\nu_{j\alpha}\omega_{j\alpha}(b,z) &= \eta_\alpha p'_j\phi'_j(p(b),\psi_\alpha(p_j(b),z)) \\ &= \eta_\alpha\psi_\alpha(p_j(b),z) = z.\end{aligned}$$

This proves that $\omega_{j\alpha}$ maps $W_{j\alpha} \times (H/K)$ topologically onto $\nu^{-1}(W_{j\alpha})$.

Assuming that b lies in $W_{i\alpha} \cap W_{j\beta}$, let $x = p(b)$, and $y = p_i(b)$. Then the coordinate transformation $\gamma_{j\beta,i\alpha}(b)$ of $B' \to B$ is given by

$$\begin{aligned}\gamma_{j\beta,i\alpha}(b)\cdot z &= \nu_{j\beta}\omega_{i\alpha}(b,z) = \eta_\beta p'_j\phi'_i(x,\psi_\alpha(y,z))\\ &= \eta_\beta(g_{ji}(x)\cdot\psi_\alpha(y,z)) = f_\beta(g_{ji}(x)\cdot y)^{-1}\cdot(g_{ji}(x)\cdot(f_\alpha(y)\cdot z))\\ &= [f_\beta(g_{ji}(x)\cdot y)^{-1}g_{ji}(x)f_\alpha(y)]\cdot z.\end{aligned}$$

Therefore $\gamma_{j\beta,i\alpha}(b)$ is the image in H/K_0 of the element of H represented by the expression in brackets. This implies the continuity of $\gamma_{j\beta,i\alpha}$; and the proof is complete.

§10. The Induced Bundle

10.1. First definition. Let $\mathcal{B}'$ be a bundle with base space X', fibre Y, and group G, and let η: $X \to X'$ be a continuous map. The *induced bundle* $\eta^{-1}\mathcal{B}'$ having base space X, fibre Y, and group G is defined as follows. The coordinate neighborhoods are the inverse images of those of $\mathcal{B}'$: $V_j = \eta^{-1}V'_j$. The coordinate transformations are given by

(1) $$g_{ji}(x) = g'_{ji}(\eta(x)), \qquad x \in V_i \cap V_j.$$

The bundle $\eta^{-1}\mathcal{B}'$ is the one provided by the construction theorem 3.2 with these coordinate transformations.

The *induced map* h: $\eta^{-1}\mathcal{B}' \to \mathcal{B}'$ is defined by

(2) $$h(b) = \phi'_j(\eta p(b),p_j(b)), \qquad p(b) \in V_j.$$

Referring back to §3.2, if b is the equivalence class of (x,y,j), then $h(b) = \phi'_j(\eta(x),y)$. If $x = p(b)$ lies in $V_i \cap V_j$, then

$$\begin{aligned}\phi'_j(\eta(x),p_j(b)) &= \phi'_i(\eta(x),g'_{ij}(\eta(x))\cdot p_j(b))\\ &= \phi'_i(\eta(x),g_{ij}(x)\cdot p_j(b)) = \phi'_i(\eta(x),p_i(b)).\end{aligned}$$

Therefore (2) defines a unique continuous function. Clearly $p'h(b) = \eta p(b)$. This means that *the map* $X \to X'$ *induced by* h *coincides with* η.

To show that h is a bundle map, we compute $\bar{g}_{kj}$ according to (17) of §2.5:

$$\begin{aligned}\bar{g}_{kj}(x)\cdot y &= \phi'^{-1}_{k,x'}h_x\phi_{j,x}(y)\\ &= \phi'^{-1}_{k,x'}\phi'_{i,x'}(y) = g'_{kj}(x')\cdot y.\end{aligned}$$

Therefore $\bar{g}_{kj} = g_{kj}$, and it is a continuous map of $V_k \cap V_j$ into G.

10.2. Second definition. There is a second definition of induced bundle as follows. Let $\mathcal{B}',X,\eta$ be as above. Form the product space $X \times B'$, and let p: $X \times B' \to X$, h: $X \times B' \to B'$ be the natural projections. Define B to be the subspace of $X \times B'$ of those pairs (x,b') such that $\eta(x) = p'(b')$. We have, therefore, commutativity in the diagram

$$\begin{array}{ccc} B & \xrightarrow{h} & B' \\ \downarrow{\scriptstyle p} & & \downarrow{\scriptstyle p'} \\ X & \xrightarrow{\eta} & X' \end{array}$$

Define $V_j = \eta^{-1}(V_j')$, and set

(3) $$\phi_j(x,y) = (x,\phi_j'(\eta(x),y)).$$

Then $p\phi_j(x,y) = x$. Set $p_j(x,b') = p_j'(b')$ whenever $\eta(x) = p'(b')$ is in V_j. Then $p_j\phi_j(x,y) = y$; and ϕ_j maps $V_j \times Y$ topologically onto $p^{-1}(V_j) \cap B$. Finally,

$$\begin{aligned} g_{ji}(x)\cdot y &= p_j\phi_i(x,y) = p_j(x,\phi_i'(\eta(x),y)) \\ &= p_j'\phi_i'(\eta(x),y) = g_{ji}'(\eta(x))\cdot y. \end{aligned}$$

This proves that (3) provides a bundle structure, and that the coordinate transformations of this bundle coincide with those of the induced bundle defined in §10.1. It follows from §2.10 that the two bundles are equivalent. The equivalence is established directly by assigning to the element $\{(x,y,j)\}$, in the first definition, the element $(x,\phi_j'(\eta(x),y))$ in the second definition. Under this correspondence the two definitions of h likewise correspond.

10.3. Equivalence theorem. *Let $\mathcal{B},\mathcal{B}'$ be two bundles having the same fibre and group, let h: $\mathcal{B} \to \mathcal{B}'$ be a bundle map (see §2.5), and let η: $X \to X'$ be the induced map of the base spaces. Then the induced bundle $\check{\mathcal{B}} = \eta^{-1}\mathcal{B}'$ is equivalent to $\mathcal{B}$, and there is an equivalence h_0: $\mathcal{B} \to \check{\mathcal{B}}$ such that h is the composition of h_0 and the induced map $\check{h}$: $\mathcal{B} \to \mathcal{B}'$.*

Using the second definition of the induced bundle $\check{\mathcal{B}}$, define h_0: $B \to X \times B'$ by

(4) $$h_0(b) = (p(b),h(b)).$$

Note that $\check{h}h_0(b) = h(b)$ follows immediately. It remains to prove that h_0 gives a bundle map $\mathcal{B} \to \check{\mathcal{B}}$. Now $\eta p(b) = p'h(b)$ is the definition of the induced map η. Therefore h_0 maps B into $\check{B}$. Furthermore $ph_0(b) = p(p(b),h(b)) = p(b)$, so h_0 maps the fibre over x in B into the fibre over x in $\check{B}$. Finally, computing the functions $\bar{g}_{ji}$ for h_0 as in (17) of 2.5, we obtain

$$\begin{aligned} \bar{g}_{kj}(x)\cdot y &= \check{p}_k h_0\phi_j(x,y) = \check{p}_k(x,h\phi_j(x,y)) \\ &= p_k'h\phi_j(x,y). \end{aligned}$$

This means that the functions $\bar{g}_{kj}$ for h_0 are the same as for h. Hence h_0 is a bundle map.

10.4. The square of a bundle. If $\mathcal{B} = \{B,p,X,Y,G\}$ is a bundle, the map p: $B \to X$ induces a bundle $p^{-1}\mathcal{B}$ having B as base space. This bundle is called the *square* of the original bundle and is denoted by $\mathcal{B}^2 = \{\check{B},\check{p},B,Y,G\}$.

Let h: $\mathcal{B}^2 \to \mathcal{B}$ be the induced map. Using the second definition of induced bundle, we find that $\check{B}$ is the subspace of $B \times B$ consisting of pairs (b_1,b_2) such that $p(b_1) = p(b_2)$ (i.e. b_1 and b_2 belong to the same fibre). Then $h(b_1,b_2) = b_2$ and $\check{p}(b_1,b_2) = b_1$. Thus h is topologically equivalent to $\check{p}$ under the involution $(b_1,b_2) \to (b_2,b_1)$ of $\check{B}$.

The *natural cross-section* f: $B \to \check{B}$ *of* $\mathcal{B}^2$ is defined by $f(b) = (b,b)$. This means in particular that $\mathcal{B}^2$ *admits a cross-section.*

If $\mathcal{B}$ is a principal bundle, the existence of f and §8.3 imply that $\mathcal{B}^2$ is equivalent to a product bundle. Using the principal map P of §8.7, the correspondence ψ: $B \times G \to \check{B}$ is given directly by $\psi(b,g) = (b,P(b,g))$, from which it follows that $h\psi(b,g) = P(b,g)$.

10.5. Properties of induced bundles. The proofs of the following properties of induced bundles are omitted since they are entirely straightforward.

If $\mathcal{B}'_1$ and $\mathcal{B}'_2$ are equivalent bundles over X', and η: $X \to X'$, then the induced bundles are equivalent.

If $\mathcal{B}'_1$ and $\mathcal{B}'_2$ are associated (weakly associated) bundles over X' (see §9.1), and η: $X \to X'$, then the induced bundles are associated (weakly associated).

If $\mathcal{B}'$ is a principal bundle, so also is any induced bundle $\eta^{-1}\mathcal{B}'$.

If $\mathcal{B}$ is a bundle over X, and η: $X \to X$ is the identity map, then the induced map $\eta^{-1}\mathcal{B} \to \mathcal{B}$ is an equivalence.

If $\mathcal{B}'$ is a bundle over X', and η: $X \to X'$ is a constant (i.e. $\eta(X)$ is a point of X'), then $\eta^{-1}\mathcal{B}'$ is equivalent to the product bundle.

If η: $X \to X'$, η': $X' \to X''$, and $\mathcal{B}''$ is a bundle over X'', then $(\eta'\eta)^{-1}\mathcal{B}''$ is equivalent to $\eta^{-1}\eta'^{-1}\mathcal{B}''$.

§11. Homotopies of Maps of Bundles

11.1. The bundle $\mathcal{B} \times I$. Let $I = [0,1]$ be the closed unit interval of real numbers, and let

$$\mathcal{B} = \{B,p,X,Y,G,V_j,\phi_j\}$$

be a bundle. The bundle $\mathcal{B} \times I$ is defined by

$$\mathcal{B} \times I = \{B \times I,q,X \times I,Y,G,V_j \times I,\psi_j\}$$

where

$$q(b,t) = (p(b),t), \qquad \psi_j(x,t,y) = (\phi_j(x,y),t).$$

Then the coordinate transformations of $\mathcal{B} \times I$ satisfy

$$g_{ji}(x,t) = g_{ji}(x).$$

If π: $X \times I \to X$ is the projection $\pi(x,t) = x$, then an equivalent definition of $\mathcal{B} \times I$ is that it is the induced bundle $\pi^{-1}\mathcal{B}$.

The projection $\bar{\pi}$: $B \times I \to B$, given by $\bar{\pi}(b,t) = b$ is obviously a bundle mapping $\bar{\pi}$: $\mathcal{B} \times I \to \mathcal{B}$. For any t in I, the map μ_t: $B \to B \times I$ given by $\mu_t(b) = (b,t)$ is also a bundle map μ_t: $\mathcal{B} \to \mathcal{B} \times I$.

11.2. Definition of homotopy. Let $\mathcal{B},\mathcal{B}'$ be two bundles having the same fibre and group, and let h_0 and h_1 be two bundle maps of $\mathcal{B}$ into $\mathcal{B}'$. We shall say that h_0 and h_1 are *homotopic* (as bundle maps), written $h_0 \simeq h_1$, if there exists a bundle map h: $\mathcal{B} \times I \to \mathcal{B}'$ such that $h(b,0) = h_0(b)$ and $h(b,1) = h_1(b)$. The map h is called a *homotopy.*

It is clear that the induced map $\bar{h}$: $X \times I \to X'$ is a homotopy (in the ordinary sense) connecting the induced maps $\bar{h}_0$ and $\bar{h}_1$.

It is a trivial matter to verify that the relation $h_0 \simeq h_1$ is reflexive, symmetric and transitive. Thus the bundle maps $\mathcal{B} \to \mathcal{B}'$ divide naturally into *homotopy classes.*

A homotopy h is said to be *stationary with the induced map* $\bar{h}$ if, for each $b \in B$ and each t-interval $[t_1,t_2]$ such that $\bar{h}(p(b),t)$ is constant for $t_1 \leqq t \leqq t_2$, then $h(b,t)$ is constant for $t_1 \leqq t \leqq t_2$. Expressed roughly, if the image of $p(b)$ fails to move during part of the homotopy, then the same is true of the image of b.

One of the most important tools in the study of bundles is provided by the following result.

11.3. First covering homotopy theorem. *Let $\mathcal{B},\mathcal{B}'$ be two bundles having the same fibre and group. Let the base space X of $\mathcal{B}$ be normal, locally-compact and such that any covering of X by open sets is reducible to a countable covering (such a space we call a C_σ-space). Let h_0: $\mathcal{B} \to \mathcal{B}'$ be a bundle map, and let $\bar{h}$: $X \times I \to X'$ be a homotopy of the induced map $\bar{h}_0$: $X \to X'$. Then there is a homotopy h: $\mathcal{B} \times I \to \mathcal{B}'$ of h_0 whose induced homotopy is $\bar{h}$, and h is stationary with $\bar{h}$.**

The homotopy h is said to *cover* $\bar{h}$. The geometric idea behind the proof is to decompose the homotopy h into a succession of homotopies each of which is small in the sense that all motion which takes place does so inside a single coordinate neighborhood V'_j of $\mathcal{B}'$. The part of $\mathcal{B}'$ over V'_j is a product space. But the theorem is trivial for a product space. The complete covering homotopy is built up in pieces.

The theorem is proved first in the special case of a compact X. Since $X \times I$ is compact, the covering $\{\bar{h}^{-1}V'_j\}$ has a refinement of the form $\{U_\lambda \times I_\nu\}$ where $\{U_\lambda\}$ is an open covering of X, and $I_1, \cdots, I_r$ is a finite sequence of open intervals covering I such that I_ν meets only

* See App. sect. 2.

$I_{\nu-1}$ and $I_{\nu+1}$ for each $\nu = 2, \cdots, r-1$. Choose numbers

$$0 = t_0 < t_1 < \cdots < t_r = 1$$

so that $t_\nu \in I_\nu \cap I_{\nu+1}$ $(\nu = 1, \cdots, r-1)$. We shall suppose, inductively, that $h(b,t)$ has been defined over $\mathfrak{B} \times [0,t_\nu]$ so as to satisfy the conclusion of the theorem over this part, and proceed to extend h over $\mathfrak{B} \times [t_\nu,t_{\nu+1}]$.

For each $x \in X$, there is a pair of neighborhoods (W,W') such that $x \in W$, $\bar{W} \subset W'$, and $\bar{W}'$ lies in some U_λ. Choose a finite number of such pairs (W_α,W'_α) $(\alpha = 1, \cdots, s)$ covering X. According to the Urysohn lemma, there is a map u_α: $X \rightarrow [t_\nu,t_{\nu+1}]$ such that $u_\alpha(\bar{W}_\alpha) = t_{\nu+1}$ and $u_\alpha(X - W'_\alpha) = t_\nu$. Define $\tau_0(x) = t_\nu$ for $x \in X$, and

$$\tau_\alpha(x) = \max\,(u_1(x), \cdots, u_\alpha(x)), \qquad x \in X, \qquad \alpha = 1, \cdots, s.$$

Then

$$t_\nu = \tau_0(x) \leqq \tau_1(x) \leqq \cdots \leqq \tau_s(x) = t_{\nu+1}.$$

Define X_α to be the set of pairs (x,t) such that $t_\nu \leqq t \leqq \tau_\alpha(x)$. Let $\mathfrak{B}_\alpha$ be the part of $\mathfrak{B} \times I$ over X_α. Then

$$\mathfrak{B} \times t_\nu = \mathfrak{B}_0 \subset \mathfrak{B}_1 \subset \cdots \subset \mathfrak{B}_s = \mathfrak{B} \times [t_\nu,t_{\nu+1}].$$

We suppose, inductively, that h has been properly extended over $\mathfrak{B}_{\alpha-1}$ and shall proceed to define its extension over $\mathfrak{B}_\alpha$.

By definition of the τ's, the set $X_\alpha - X_{\alpha-1}$ is contained in $W'_\alpha \times [t_\nu,t_{\nu+1}]$; and by definition of the W's, $\bar{W}'_\alpha \times [t_\nu,t_{\nu+1}] \subset U_\lambda \times I_\nu$ and is mapped by $\bar{h}$ into some V'_j. With j fixed in this manner, define

$$h(b,t) = \phi'_j(\bar{h}(x,t),p'_jh(b,\tau_{\alpha-1}(x))), \tag{1}$$
$$\text{for } x = p(b), \quad (x,t) \in X_\alpha - X_{\alpha-1}.$$

Notice that the right side is defined and continuous for (x,t) in the closure of $X_\alpha - X_{\alpha-1}$, and, for $t = \tau_{\alpha-1}(x)$, it reduces to $h(b,\tau_{\alpha-1}(x))$. Hence h on $B_{\alpha-1}$ has been continuously extended over B_α. Clearly $p'h(b,t) = \bar{h}(x,t)$, so h maps fibres into fibres and covers $\bar{h}$. The variable t occurs in just one place on the right side of (1). If $\bar{h}(x,t) = \bar{h}(x,t')$ for $\tau_{\alpha-1}(x) \leqq t < t' \leqq \tau_\alpha(x)$, then $h(b,t) = h(b,t')$. This shows that h is stationary with $\bar{h}$.

To show that h is a bundle map of $\mathfrak{B}_\alpha$ it suffices to prove this on the closure of $B_\alpha - B_{\alpha-1}$ since h is a bundle map on $\mathfrak{B}_{\alpha-1}$. Suppose $x \in V_k$ and (x,t) lies in the closure of $X_\alpha - X_{\alpha-1}$. Abbreviate $\tau_{\alpha-1}(x)$ by τ, then

$$\begin{aligned}\bar{g}_{jk}(x,t)\cdot y &= p'_jh_{(x,t)}\psi_k(x,t,y) = p'_jh(\phi_k(x,y),t)\\ &= p'_j\phi'_j(\bar{h}(x,t),p'_jh(\phi_k(x,y),\tau))\\ &= p'_jh(\phi_k(x,y),\tau)\\ &= p'_jh_{(x,\tau)}\psi_k(x,\tau,y)\\ &= \bar{g}_{jk}(x,\tau_{\alpha-1}(x))\cdot y.\end{aligned}$$

The hypothesis that h is a bundle map of $\mathcal{B}_{\alpha-1}$ in $\mathcal{B}'$ means that $\bar{g}_{jk}(x,\tau)$ yields a continuous map into G of the intersection of $V_k \times I$ with the boundary of $X_\alpha - X_{\alpha-1}$. It follows that $\bar{g}_{jk}(x,t)$ provides a continuous map into G of the intersection of $V_k \times I$ with the closure of $X_\alpha - X_{\alpha-1}$. Therefore h is a bundle map of $\mathcal{B}_\alpha$ into $\mathcal{B}'$. This completes the general step of the two-fold induction, and proves the theorem for a compact space X.

In the general case, the hypothesis on X implies that X is the union of an expanding sequence of open sets $\{W_n\}$ such that $\bar{W}_n$ is compact and contained in W_{n+1} for each n. By the Urysohn lemma, there is a map τ_n: $X \to [0,1]$ such that $\tau_n(\bar{W}_n) = 1$ and $\tau_n(X - W_{n+1}) = 0$. Set $\tau_0(x) = 0$. Then $\tau_n(x) \leqq \tau_{n+1}(x)$, and, for each x, there is an n such that $\tau_n(x) = 1$. Define X_n to be the set of pairs (x,t) such that $0 \leqq t \leqq \tau_n(x)$. Let $\mathcal{B}_n$ be the part of $\mathcal{B} \times I$ lying over X_n. Then h_0 provides a map of $\mathcal{B}_0$ into $\mathcal{B}'$. Suppose, inductively, that h has been defined on $\mathcal{B}_{n-1}$ so as to satisfy the conditions of the theorem.

The set $A = \bar{W}_{n+1} - W_{n-1}$ is compact and contains all points x for which $\tau_{n-1}(x) < \tau_n(x)$. Let λ_x: $[0,1] \to [\tau_{n-1}(x), \tau_n(x)]$ be the linear map

$$\lambda_x(s) = s\tau_n(x) + (1 - s)\tau_{n-1}(x).$$

The inverse map λ_x^{-1} is given by

$$\lambda_x^{-1}(t) = (t - \tau_{n-1}(x))/(\tau_n(x) - \tau_{n-1}(x)).$$

Define $h_0'(b) = h(b,\tau_{n-1}(p(b)))$, and $\bar{h}'(x,s) = \bar{h}(x,\lambda_x(s))$ restricting x to be in A, and b to lie in $p^{-1}(A)$. Then h_0' is a bundle map, $\bar{h}'$ is a homotopy of $\bar{h}_0'$, and the base space A is compact. The theorem, having been proved for this case, provides a covering homotopy h'. Define

$$h(b,t) = h'(b,\lambda_x^{-1}(t)) \qquad \text{for } x = p(b), \quad \tau_{n-1}(x) < t \leqq \tau_n(x). \tag{2}$$

This extends h over B_n. To prove continuity, notice first that the right side of (2) is defined for $t = \tau_{n-1}(x)$ and gives $h(b,\tau_{n-1}(x))$. This shows that h is continuous except possibly at a point (b,t) such that $t = \tau_{n-1}(x) = \tau_n(x)$; for, in this case, λ_x^{-1} is not defined. But then $\bar{h}'(x,s)$ is constant in s. The stationary property implies that $h'(b,s)$ is likewise constant. So, if N' is a neighborhood of $h'(b,s)$, $h'^{-1}(N')$ contains $b \times I$. Since $b \times I$ is compact, there is a neighborhood N of b such that $h'(N \times I) \subset N'$. If $b_1 \varepsilon N$ and $\tau_{n-1}(p(b_1)) < t \leqq \tau_n(p(b_1))$, it follows that $h(b,t)$ is in N'. This proves the continuity of h.

The relation $p'h(b,t) = \bar{h}(p(b),t)$ is an immediate consequence of (2). That the extended h is stationary with $\bar{h}$ follows from the same property of h'. It remains to check that h is a bundle map. Let $x \varepsilon V_j$,

$\tau_{n-1}(x) \leqq t \leqq \tau_n(x)$, and $\bar{h}(x,t) \in V'_k$. If we compute the function $\bar{g}_{kj}$ defined in (17) of §2.5 for h we obtain

$$\bar{g}_{kj}(x,t) = \bar{g}'_{kj}(x,\lambda_x^{-1}(t))$$

where $\bar{g}'_{kj}$ is the analogous function for h'. This shows that $\bar{g}_{kj}(x,t)$ is in G and is continuous except possibly at points (x,t) where $\tau_{n-1}(x) = t = \tau_n(x)$. An argument similar to that of the preceding paragraph establishes continuity in this case.

This completes the inductive step showing that h can be extended from $\mathfrak{B}_{n-1}$ to $\mathfrak{B}_n$. The complete function h: $B \times I \to B'$ is the result of the successive extensions. Its continuity and bundle mapping properties follow from the fact that B_{n-1} lies in the interior of B_n relative to $B \times I$. This completes the proof.

11.4. Theorem. *If X is a C_σ-space, then any bundle $\mathfrak{B}'$ over the base space $X \times I$ is equivalent to a bundle of the form $\mathfrak{B} \times I$.*

Define $\bar{h}_0(x) = (x,0)$. Set $\mathfrak{B} = \bar{h}_0^{-1}\mathfrak{B}'$, and let h_0: $\mathfrak{B} \to \mathfrak{B}'$ be the induced map. The function $\bar{h}(x,t) = (x,t)$ is a homotopy of $\bar{h}_0$. The covering homotopy provides a bundle map h: $\mathfrak{B} \times I \to \mathfrak{B}'$. But $\bar{h}$ is the identity map of $X \times I$. Hence h is a bundle equivalence.

11.5. Theorem. *Let $\mathfrak{B}'$ be a bundle over X', let X be a C_σ-space, and let $\bar{h}_0,\bar{h}_1$ be homotopic maps of X into X', then the induced bundles $\bar{h}_0^{-1}\mathfrak{B}'$ and $\bar{h}_1^{-1}\mathfrak{B}'$ are equivalent.*

Let $\bar{h}$: $X \times I \to X'$ be a homotopy of $\bar{h}_0$ into $\bar{h}_1$. Let $\bar{\mu}_t(x) = (x,t)$. Then $\bar{h}_t = \bar{h}\bar{\mu}_t$ $(t = 0,1)$. By §11.4, we have a bundle equivalence

$$\bar{h}^{-1}\mathfrak{B}' \sim \mathfrak{B} \times I.$$

This implies

$$\bar{h}_t^{-1}\mathfrak{B}' = \bar{\mu}_t^{-1}\bar{h}^{-1}\mathfrak{B}' \sim \bar{\mu}_t^{-1}(\mathfrak{B} \times I), \qquad \text{for } t = 0,1. \tag{3}$$

As observed in §11.1, the map $\mu_t(b) = (b,t)$ is a bundle map $\mathfrak{B} \to \mathfrak{B} \times I$ for each t. It follows from §10.3 that $\mathfrak{B}$ is equivalent to $\bar{\mu}_t^{-1}(\mathfrak{B} \times I)$ for each value of t. Using the two equivalences for $t = 0$ and 1 and the equivalences in (3) we obtain the desired result.

11.6. Corollary. *If X is a C_σ-space and is contractible on itself to a point, then any bundle over X is equivalent to a product bundle.*

The identity map of X induces a bundle equivalent to the given bundle. A constant map induces a product bundle. The result follows now from §11.5.

The result can be paraphrased by saying that contractible spaces X admit only trivial bundles. Bundles over closed or open n-cells are therefore of little interest. The simplest base spaces that provide nontrivial bundles are the spheres of various dimensions.

In the literature the name "covering homotopy theorem" has been applied to the following consequence of §11.3:

11.7. Second covering homotopy theorem. *Let $\mathfrak{B}'$ be a bundle over X'. Let X be a C_σ-space, let f_0: $X \to B'$ be a map, and let $\bar{f}$: $X \times I \to X'$ be a homotopy of $p'f_0 = \bar{f}_0$. Then there is a homotopy $f : X \times I \to B'$ of f_0 covering $\bar{f}$ (i.e. $p'f = \bar{f}$), and f is stationary with $\bar{f}$.**

Let $\mathfrak{B} = \bar{f}_0^{-1}\mathfrak{B}'$, and let h_0: $\mathfrak{B} \to \mathfrak{B}'$ be the induced map. Let h: $\mathfrak{B} \times I \to \mathfrak{B}'$ be a covering homotopy (§11.3) of $\bar{f}$ which is stationary with $\bar{f}$. Define the cross-section ϕ of $\mathfrak{B}$, using the second definition of induced bundle (§10.2), as follows: $\phi(x) = (x, f_0(x))$. Then $h_0\phi = f_0$. Define $f(x,t) = h(\phi(x),t)$. It follows immediately that f is the desired homotopy.

11.8. A map f: $X \to Y$ is called *reducible* if it is homotopic to a map of X into a proper subset of Y; otherwise, it is called *irreducible.* It is called *inessential* if it is homotopic to a constant map; otherwise, *essential.* A space is called *reducible* (*irreducible*) if its identity map is reducible (irreducible).

THEOREM. *Let $\mathfrak{B}$ be a bundle in which B is a C_σ-space and p: $B \to X$ is reducible. Then B is reducible.*

We have only to apply §11.7 to the identity map of B.

COROLLARY. *If B is irreducible so also are p and X. In particular, if X has more than one point, then p is essential.*

Since X reducible implies p reducible, the corollary is immediate.

It is a well known result of homology theory that any compact n-manifold is irreducible (i.e. it is the irreducible carrier of an n-cycle mod 2). As a consequence, the above results apply to the bundles constructed (§7) out of coset spaces of compact Lie groups.

§12. CONSTRUCTION OF CROSS-SECTIONS

12.1. Solid spaces. A space Y will be called *solid* if, for any normal space X, closed subset A of X, and map f: $A \to Y$, there exists a map f': $X \to Y$ such that $f'|A = f$.

The Tietze extension theorem [1, p. 73] asserts that any interval of real numbers (open or closed) is a solid space. It is easy to see that the topological product of a collection of solid spaces is likewise solid—one simply extends each component function. It follows that euclidean n-space, and a closed n-dimensional cube are solid. Their homeomorphs, open and closed n-cells, are likewise solid. The Hilbert parallelotope is solid.

There are several properties which are closely related to solidity. A subspace Y of Z is called a *retract* of Z if there exists a map f: $Z \to Y$

* See App. sect. 2.

such that $f(y) = y$ for $y \varepsilon Y$. A compact metric space Y is called an *absolute retract* if it is a retract of any metric separable space which contains Y (see [3]).

For a compact metric space Y, the properties absolute retract and solid are euqivalent. Since a retraction is just an extension of the identity map, it follows that solid implies absolute retract. Conversely, let Y be an absolute retract. Imbed Y topologically in the Hilbert parallelotope Z. Let r: $Z \to Y$ be a retraction. If A is closed in X and f: $A \to Y$, we regard f as a map $A \to Z$. Since Z is solid, f extends to a map f': $X \to Z$. Then rf' is an extension of f to a map $X \to Y$. Hence Y is solid.

Let Y be solid and such that $Y \times I$ is normal. Let y_0 be a point of Y. Let $A = (Y \times 0) \cup (y_0 \times I) \cup (Y \times 1)$. Define f: $A \to Y$ by $f(y,0) = y$, $f(y,1) = y_0$ and $f(y_0,t) = y_0$. Then solidity implies that f extends to f': $Y \times I \to Y$. It follows that Y is contractible to a point. Since y_0 is arbitrary, it also follows that Y is locally contractible.

12.2. Existence theorem. *Let X be a normal space with the property that every covering of X by open sets is reducible to a countable covering (e.g. X is compact, or has a countable base, or is a union of countably many compact subsets). Let $\mathcal{B}$ be a bundle over X with a fibre Y which is solid. Let f be a cross-section of $\mathcal{B}$ defined on a closed subset A of X. Then f can be extended to a cross-section over all of X. (Taking $A = 0$, it follows that $\mathcal{B}$ has a cross-section.)* (See App. sect. 3.)

For each point x choose a neighborhood U_x of x such that $\bar{U}_x$ is contained in some coordinate neighborhood V_j of $\mathcal{B}$. Choose a countable number of these $U_1, U_2, \cdots$ covering X. Set $A_0 = A$ and define A_n inductively by $A_n = \bar{U}_n \cup A_{n-1}$. Set $f_0 = f$. Suppose cross-sections f_i, defined on A_i, are given for $i < n$, and are such that $f_i|A_{i-1} = f_{i-1}$. Choose a V_j which contains $\bar{U}_n$. Let $C_n = \bar{U}_n \cap A_{n-1}$. Define h: $C_n \to Y$ by $h(x) = p_j f_{n-1}(x)$. Since $\bar{U}_n$ is closed in X, it is a normal space, and C_n is a closed subset of $\bar{U}_n$. Since Y is solid, h extends to a map h': $\bar{U}_n \to Y$. Set $h''(x) = \phi_j(x,h'(x))$ for $x \varepsilon \bar{U}_n$. Then h'' is continuous, $ph''(x) = x$, and $h''|C_n = f_{n-1}|C_n$. If we define $f_n(x) = f_{n-1}(x)$ for $x \varepsilon A_{n-1}$ and $f_n(x) = h''(x)$ for $x \varepsilon A_n - A_{n-1}$, it follows that f_n is a continuous cross-section over A_n extending f_{n-1}. We construct thus a sequence $\{f_n\}$ where for each n, f_n is a cross-section of $\mathcal{B}|A_n$, and f_n extends f_{n-1}. Now define f' by $f'(x) = f_n(x)$ for $x \varepsilon A_n - A_{n-1}$. Since X is the union of the interiors of the sets A_n, it follows that f' is everywhere continuous. Then f' is the desired cross-section.

12.3. Corollary. *If X is a space as in §12.2, and the topological group G is solid, then any bundle over X with group G is equivalent to the product bundle.*

It follows from §12.2 that the associated principal bundle (§8.1) has a cross-section. The result follows from §8.4.

This result should be compared with §11.6. The two together can be paraphrased: if either X or G is homotopically trivial then every bundle over X with group G is trivial.

12.4. COROLLARY. *If X is a differentiable manifold, the bundles of scalars over X of even weight (see §6.5) are all equivalent to product bundles. If X is orientable, this is true of bundles of scalars of all weights. If X is non-orientable, the bundles of scalars of odd weights are not product bundles.*

In the first case, the determinant of the Jacobian at each point of $V_i \cap V_j$, raised to the power w, is a positive real number. The group G of positive real numbers is solid. In the second case the coordinate systems can be chosen so that all determinants of Jacobians are positive; and the same argument applies. For the last statement, note that the product bundle is non-orientable. Just as in the case of the tangent bundle (see §6.5), one proves that the bundle of scalars of an odd weight is an orientable manifold; hence, not equivalent to the product.

12.5. Reducing the group of a bundle to a subgroup.

THEOREM. *Let X be a space as in §12.2, and let A be closed in X. Let G be a Lie group and H a closed subgroup such that G/H is solid. Then any (G,H)-bundle over (X,A) (see §9.2) is (G,H)-equivalent to an (H,H)-bundle.*

This is an immediate consequence of §9.4 and §12.2. An important special case is

12.6. COROLLARY. *If X,G and H are as in §12.5, then any bundle over X with group G is equivalent in G to a bundle with group H.*

12.7. THEOREM. *Let X be a C_σ-space (§11.3), let G be a Lie group. and H a closed subgroup such that G/H is solid, then any two H-bundles over X which are G-equivalent are also H-equivalent.*

To prove this, form $X \times I$ where $I = [0,1]$ is the closed unit interval. Let $\mathcal{B}_0,\mathcal{B}_1$ be two H-bundles over X which are G-equivalent. As remarked in §2.9, we can suppose they have the same coordinate neighborhoods $\{V_j\}$. Denote their coordinate transformations by ${}_0h_{ij}$ and ${}_1h_{ij}$ respectively. Using half open intervals $[0,1)$ and $(0,1]$, form the open sets ${}_0V_j = V_j \times [0,1)$, and ${}_1V_j = V_j \times (0,1]$. These cover $X \times I$. Since $\mathcal{B}_0,\mathcal{B}_1$ are G-equivalent there are maps $\lambda_j\colon\ V_j \to G$ such that ${}_0h_{ji} = \lambda_j^{-1}\,{}_1h_{ji}\lambda_i$ (see §2.10). Define

$$
\begin{aligned}
{}_0g_{ji}(x,t) &= {}_0h_{ji}(x), && (x,t) \in {}_0V_i \cap {}_0V_j,\\
{}_1g_{ji}(x,t) &= {}_1h_{ji}(x), && (x,t) \in {}_1V_i \cap {}_1V_j,\\
{}_{01}g_{ji}(x,t) &= {}_0h_{ji}(x)\lambda_i(x)^{-1} && \\
&= \lambda_j(x)^{-1}\,{}_1h_{ji}(x), && (x,t) \in {}_0V_j \cap {}_1V_i.
\end{aligned}
$$

Note, in particular, that ${}_{01}g_{jj}(x,t) = \lambda_j(x)^{-1}$. It is easily checked that these functions satisfy the transitivity law for coordinate transformations. By §3.2, there exists a bundle over $X \times I$ with these coordinate transformations. The portion over $X \times 0$ ($X \times 1$) is essentially the same bundle as $\mathfrak{B}_0$ ($\mathfrak{B}_1$). We have therefore a relative (G,H)-bundle over the pair $(X \times I,\ X \times 0 \cup X \times 1)$: By §12.5, it is (G,H)-equivalent to an (H,H)-bundle. By §11.4, the resulting H-bundle over $X \times I$ is H-equivalent to a bundle of the form $\mathfrak{B} \times I$. Then $\mathfrak{B}_0$ and $\mathfrak{B}_1$ are both H-equivalent to $\mathfrak{B}$.

Combining §12.6 and §12.7 we have

12.8. Corollary. *If X, G and H are as in §12.7, then the equivalence classes of bundles over X with group G are in a natural 1-1 correspondence with the equivalence classes of bundles over X with group H.*

12.9. Reduction of tensor bundles to the orthogonal group. The preceding results have important consequences in simplifying a bundle and in reducing the classification problem to a simpler one. An example is the following

Theorem. *Let L_n be the full linear group on n real variables, and O_n the orthogonal subgroup. Let X be a C_σ-space. Then any bundle over X with group L_n is equivalent in L_n to a bundle with group O_n. If two O_n-bundles over X are L_n-equivalent, then they are O_n-equivalent.*

Let H_n be the space of positive definite, symmetric matrices in L_n. As shown in [12, p. 16], L_n is the product space $O_n \times H_n$, the correspondence being given by the multiplication of matrices. It follows that the natural projection $L_n \rightarrow L_n/O_n$ maps H_n homeomorphically on L_n/O_n. But H_n is an open cell of dimension $n(n+1)/2$ (see proof of §12.12). Therefore L_n/O_n is solid. The theorem follows now from §12.6 and §12.7.

A second decomposition of L_n is $L_n = O_n \times T_n$ where T_n is the *subgroup* of triangular matrices $||a_{ij}||$ ($a_{ij} = 0$ for $i < j$, and $a_{ii} > 0$). This is proved by applying the Gram-Schmidt orthogonalization process to the n vectors which are the rows of a matrix of L_n. It is evident that T_n is an open cell.

12.10. A differentiable n-manifold X is a C_σ-space and any tensor bundle over X has, for its group, L_n or a factor group of L_n. Then §12.9 implies that all tensor bundles (see §6.4) and their equivalences are reducible to the group O_n or to a factor group of O_n. A single reduction of the tangent bundle provides a simultaneous reduction of all the associated tensor bundles.

Having reduced the group of the tangent bundle to O_n, the unit $(n-1)$-sphere S^{n-1} in real n-space Y^n becomes invariant under the group. We are led therefore to the associated subbundle of the tangent bundle having S^{n-1} as fibre. This bundle is called *the tangent sphere bundle of the differentiable manifold.* It is somewhat simpler to study than the tangent bundle. It is of one less dimension. If X is a compact manifold, so is the tangent sphere bundle.

We have noted in §6.5 that the tangent bundle admits the trivial cross-section of zero vectors. An important problem is to decide the question of existence of a vector field which is non-zero at each point. This is equivalent to the problem of finding a cross-section of the tangent sphere bundle. In one direction this is clear, a cross-section of the tangent sphere bundle is a non-singular vector field. For the other direction, let Z be n-space with the origin deleted. By radial deformation, Z can be retracted into S^{n-1}. This retraction is invariant under O_n, and provides thereby a deformation retraction of the subbundle of the tangent bundle with fibre Z into the tangent sphere bundle. The latter deformation retracts any non-singular vector field into a cross-section of the tangent sphere bundle.

12.11. THEOREM. *The contravariant and covariant tangent bundles of a differentiable manifold are equivalent.*

Recall (§6.5) that the passage $h\colon\ L_n \to L_n$ from contravariant to covariant replaces each matrix by its transpose inverse. In O_n, this operation is the identity. Having reduced the tangent bundle to O_n, the operation becomes the identity.

12.12. Metrizability of manifolds.

THEOREM. *A differentiable manifold always admits a Riemannian metric.*†

The problem here is to prove the existence of a cross-section of the tensor bundle whose fibre Y consists of symmetric, positive-definite matrices. Observe first that the symmetric matrices Y' form a linear subspace of all matrices. The condition of being positive-definite ($a_{ij}v^iv^j > 0$ for all vectors v of length 1) is obviously preserved under small variations of the a_{ij}. This means that Y is an open subspace of Y'. If $a = ||a_{ij}||$ and $b = ||b_{ij}||$ are positive-definite matrices, then, for $0 \leqq t \leqq 1$,

$$(ta_{ij} + (1 - t)b_{ij})v^iv^j = ta_{ij}v^iv^j + (1 - t)b_{ij}v^iv^j > 0.$$

Hence $ta + (1 - t)b$ is also positive-definite. Therefore Y is a convex open set of the linear space Y'. It follows that Y is an open cell (of dimension $n(n + 1)/2$). But this means that Y is solid. The existence of the cross-section follows from §12.2.

If X is of class r, by the result of §6.7, the metric tensor can be constructed to be of class r-1.

† In §6.1 we assumed that manifolds are separable.

12.13. Reduction to the unitary group. The reduction of bundles with the group L_n to the group O_n is paralleled by a similar reduction for bundles having as group the complex linear group CL_n to the unitary subgroup U_n. As in §12.9, we have $CL_n = U_n \times H_n$ where H_n is the set of positive definite Hermitian manifolds. The latter is also an open cell.

The preceding remark applies immediately to tensor bundles over complex analytic manifolds. Much that has been said for real manifolds carries over to the complex analytic case. It should be noticed however that U_n is not a complex analytic manifold. Neither is the tangent sphere bundle. Similarly the existence of a continuous cross-section of the bundle of symmetric Hermitian tensors does not provide an *analytic* Hermitian metric.

12.14. The reduction from L_n to O_n appears to be a special case of a general result for bundles with Lie groups. According to E. Cartan, any connected semi-simple Lie Group G is topologically a product space $H \times E$ where H is a compact subgroup of G and E is a euclidean space. Proofs of this result have been given by Malcev and Mostow [70]. A more recent result of Iwasawa [59] and Mostow is that the theorem holds without the condition of semi-simplicity.

The general conclusion appears to be that *a bundle whose group is a connected Lie group is equivalent in its group to a bundle whose group is a compact subgroup.*

§13. Bundles Having a Totally Disconnected Group

13.1. Translating a fibre along a curve. Let X be an arcwise connected space, and let $\mathfrak{B}$ be a bundle over X with fibre Y and group G. Let C be a curve in X from x_0 to x_1 (i.e. $C\colon\ I \to X$ with $C(0) = x_0$, $C(1) = x_1$). Let Y_t denote the fibre over $C(t)$. Let $h_0\colon\ Y \to Y_0$ be an admissible map. We may regard Y as a bundle with base space consisting of a single point 0. Then C is a homotopy of $\bar{h}_0$. The covering homotopy theorem (§11.3) provides a bundle map $h\colon\ I \times Y \to \mathfrak{B}$ such that $h(0,y) = h_0(y)$ and $ph(t,y) = C(t)$. By fixing t, h defines a map $h_t\colon\ Y \to Y_t$. Then

$$h_t h_0^{-1}\colon\ Y_0 \to Y_t \tag{1}$$

is a 1-parameter family of maps beginning with the identity and ending with a map $Y_0 \to Y_1$. We call (1) *a translation of Y_0 along C into Y_1.*

Let $\tilde{\mathfrak{B}}$ denote the associated principal bundle of $\mathfrak{B}$ (§8.1). Now $I \times G$ is the associated principal bundle of $I \times Y$. According to

§8.13, to the map h: $I \times Y \to B$ corresponds an associated map $\tilde{h}$: $I \times G \to \tilde{\mathfrak{B}}$. Then

$$\tilde{h}_t \tilde{h}_0^{-1}: \; G_0 \to G_t, \qquad 0 \leqq t \leqq 1, \tag{2}$$

is called the *associated* translation of G_0 along C into G_1.

Let P: $\tilde{B} \times Y \to B$ denote the principal map of $\mathfrak{B}$ (§8.7), and let P': $I \times G \times Y \to I \times Y$ be the same for $I \times Y$. Direct calculation, see (2) of §8.7, yields, for special case of the product $I \times Y$, that

$$P'(t,g,y) = (t,g\cdot y).$$

By the result of 8.13, we have $hP'(t,g,y) = P(\tilde{h}(t,g),y)$, and therefore

$$h(t,g\cdot y) = P(\tilde{h}(t,g),y). \tag{3}$$

13.2. LEMMA. *If the topology of the group G is totally disconnected (i.e. no connected subsets except single points), then the translation of fibres along curves is a unique operation.*

It follows from (3) that h is determined by $\tilde{h}$, so it suffices to prove the uniqueness of the associated translation of G_0 along C. Using the compactness of I, we may divide it into a finite number of subintervals $I_k = [t_k, t_{k+1}]$ such that $C(I_k)$ is contained in a single coordinate neighborhood. Suppose inductively that uniqueness has been proved for $t \leqq t_k$; and suppose V_j contains $C(I_k)$. Let $\tilde{h}, \tilde{h}'$ be two maps $I_k \times G \to \tilde{\mathfrak{B}}$ which cover $C|I_k$ and which agree for $t = t_k$. Let g be fixed in G. Then $\tilde{p}_k\tilde{h}(t,g)$ and $\tilde{p}_k\tilde{h}'(t,g)$ are two maps $I_k \to G$ which coincide for $t = t_k$. Since G is totally disconnected, both must be constant. Hence $\tilde{p}_k\tilde{h} = \tilde{p}_k\tilde{h}'$. Since also $\tilde{p}\tilde{h} = \tilde{p}\tilde{h}' = C$, it follows that $\tilde{h} = \tilde{h}'$. Therefore uniqueness holds for $t \leqq t_{k+1}$, and the lemma is proved.

13.3. Composition of translations. We shall assume in the remainder of this article that G is totally disconnected. Although this is a strong restriction, it includes the special case of a *discrete* group G (i.e. every subset of G is open, e.g. a finite group) which will provide important applications of the results to be obtained.

Let C be a curve from x_0 to x_1. Define

$$C^{\#}: \; Y_1 \to Y_0 \tag{4}$$

to be the result of translating Y_1 back along C into Y_0. That is, $C^{\#}$ is the result of translating Y_1 along the curve C^{-1} given by $C^{-1}(t) = C(1 - t)$. Since G is totally disconnected, $C^{\#}$ depends on C alone.

If C_1 is a curve from x_0 to x_1, and C_2 is a curve from x_1 to x_2, define their *composition* C_1C_2 to be the curve

$$C_1C_2(t) = \begin{cases} C_1(2t) & 0 \leqq t \leqq 1/2, \\ C_2(2t - 1) & 1/2 \leqq t \leqq 1, \end{cases} \tag{5}$$

from x_0 to x_2. As an immediate consequence of the uniqueness, we have

$$(C_1C_2)^{\#} = C_1^{\#}C_2^{\#}. \tag{6}$$

(Note: if we had defined $C^{\#}$ to be the translation of Y_0 along C into Y_1, a reversal of order would occur on the right side.)

13.4. Invariance under a homotopy of the curve. Now, let C,C' be two curves from x_0 to x_1 which are homotopic leaving their end points fixed. Let h be a covering homotopy of C. We may apply the covering homotopy theorem to the map h: $I \times Y \to \mathfrak{B}$ and the homotopy of C into C'. The result will be a homotopy of h into a map h': $I \times Y \to \mathfrak{B}$ which covers C'. Since x_0,x_1 are stationary during the homotopy, it follows that h' and h coincide for $t = 0$ and $t = 1$. Therefore $C^{\#} = C'^{\#}$: *the map $C^{\#}$ depends only on the homotopy class of C.*

13.5. The characteristic class χ. Consider now *closed* curves beginning and ending at x_0. Their homotopy classes form the elements of *the fundamental group*, $\pi_1(X,x_0)$, with multiplication defined by (5) above. The results of the last two paragraphs show that the operation $C^{\#}$ represents $\pi_1(X,x_0)$ as a group of bundle mappings of Y_0.

Choose now an admissible map ξ: $Y \to Y_0$. Since any $\alpha \varepsilon \pi_1(X,x_0)$ is an admissible map $Y_0 \to Y_0$, the composition $\xi^{-1}\alpha\xi$ is an admissible map $Y \to Y$, i.e. it is an element of G which we denote by $\chi(\alpha)$. Obviously $\chi(\alpha\beta) = \chi(\alpha)\chi(\beta)$ so that

$$\chi\colon \quad \pi_1(X,x_0) \to G \tag{7}$$

is a homomorphism.

If ζ: $Y \to Y_0$ is also admissible, then

$$\zeta^{-1}\alpha\zeta = (\zeta^{-1}\xi)(\xi^{-1}\alpha\xi)(\xi^{-1}\zeta) = g^{-1}\chi(\alpha)g$$

where $g = \xi^{-1}\zeta$. It follows that a different choice of ξ alters χ by an inner automorphism of G. Conversely if g and ξ are given, define ζ to be the composition of g: $Y \to Y$ followed by ξ, then $\chi(\alpha)$ is altered to $g^{-1}\chi(\alpha)g$. It follows that *χ is precisely determined up to its equivalence class under inner automorphisms of G.*

The equivalence class of χ under inner automorphisms we call *the characteristic class of the bundle* $\mathfrak{B}$ and denote it by $\chi(\mathfrak{B})$.

It is worth noting that $\chi(\mathfrak{B})$ is independent of the choice of x_0 in a natural sense. Suppose x_1 is another base point, ξ_1: $Y \to Y_1$, and χ_1 is the corresponding homomorphism $\pi_1(X,x_1) \to G$. Since X is arcwise connected, there is a path D from x_0 to x_1. If C is a closed path from x_1, then DCD^{-1} is a closed path from x_0. As is well known, D induces in this way an isomorphism D_*: $\pi_1(X,x_1) \approx \pi_1(X,x_0)$. If C

represents $\alpha \in \pi_1(X,x_1)$, then

$$\begin{aligned}\chi D_*(\alpha) &= \xi^{-1}(DCD^{-1})^{\#}\xi = \xi^{-1}D^{\#}C^{\#}D^{\#-1}\xi \\ &= (\xi^{-1}D^{\#}\xi_1)(\xi_1^{-1}C^{\#}\xi_1)(\xi_1^{-1}D^{\#-1}\xi) = g^{-1}\chi_1(\alpha)g.\end{aligned}$$

Thus the homomorphisms χD_* and χ_1 differ by an inner automorphism of G.

13.6. We derive now the relations between the operations $C^{\#}$ and the coordinate transformations. Suppose first that the curve C from x_0 to x_1 lies wholly in the coordinate neighborhood V_j. Let $h_0(y) = \phi_j(x_0,y)$. Then $h(t,y) = \phi_j(C(t),y)$ is a map $I \times Y \rightarrow \mathfrak{B}$ which starts with h_0 and covers C. It follows that $C^{\#}$: $Y_1 \rightarrow Y_0$ is given by $C^{\#}(b) = \phi_j(x_0,p_j(b))$. Stated otherwise:

When C lies wholly in V_j, then $p_jC^{\#}\phi_{j,x_1}$ is the identity element of G.

Now let C_1 be a curve in V_i from x_0 to x_1, and let C_2 be a curve in V_j from x_1 to x_2. Using the above result, we obtain

$$\begin{aligned}p_i(C_1C_2)^{\#}\phi_{j,x_2} &= (p_iC_1^{\#}\phi_{i,x_1})p_i\phi_{j,x_1}(p_jC_2^{\#}\phi_{j,x_2}) \\ &= g_{ij}(x_1).\end{aligned}$$

13.7. Equivalence theorem. *Let X be arcwise connected and arcwise locally connected. Let $\mathfrak{B},\mathfrak{B}'$ be two bundles over X having the same group G which is totally disconnected. Then $\mathfrak{B},\mathfrak{B}'$ are associated bundles if and only if their characteristic classes are equal: $\chi(\mathfrak{B}) = \chi(\mathfrak{B}')$.*

Suppose $\mathfrak{B}$ and $\mathfrak{B}'$ are associated. Then their associated principal bundles $\tilde{\mathfrak{B}},\tilde{\mathfrak{B}}'$ are equivalent. Using such an equivalence it is easy to prove that $\chi(\tilde{\mathfrak{B}}) = \chi(\tilde{\mathfrak{B}}')$. It suffices therefore to prove $\chi(\mathfrak{B}) = \chi(\tilde{\mathfrak{B}})$.

Let C be a closed curve based at x_0, let h_0: $Y \rightarrow Y_0$ be admissible, let h: $I \times Y \rightarrow \mathfrak{B}$ cover C and start with h_0, and let $\tilde{h}$: $I \times G \rightarrow \tilde{\mathfrak{B}}$ be the associated map. Let h_t: $Y \rightarrow \mathfrak{B}$ be defined by $h_t(y) = h(t,y)$. Define $\tilde{h}_t$ similarly. Then, by (2), we have

$$\begin{aligned}C^{\#}(b) &= h_0h_1^{-1}(b), \qquad & b \in Y_0,\\ \tilde{C}^{\#}(\tilde{b}) &= \tilde{h}_0\tilde{h}_1^{-1}(\tilde{b}), \qquad & \tilde{b} \in G_0,\end{aligned}$$

and therefore

$$\begin{aligned}h_0^{-1}C^{\#}h_1(y) &= y, \qquad & y \in Y,\\ \tilde{h}_0^{-1}\tilde{C}^{\#}\tilde{h}_1(g) &= g, \qquad & g \in G.\end{aligned}$$

If we choose $\xi = h_1$, $\tilde{\xi} = \tilde{h}_1$ in the definition of χ for $\mathfrak{B}$ and $\tilde{\mathfrak{B}}$, we have

$$\begin{aligned}\chi(C) &= h_1^{-1}C^{\#}h_1 = h_1^{-1}h_0\colon \ Y \rightarrow Y,\\ \tilde{\chi}(C) &= \tilde{h}_1^{-1}\tilde{C}^{\#}\tilde{h}_1 = \tilde{h}_1^{-1}\tilde{h}_0\colon \ G \rightarrow G.\end{aligned}$$

Formula (3) of 13.1 can be written

$$h_t(g\cdot y) = P(\tilde{h}_t(g),y).$$

Take $t = 1$ and $g = \bar{h}_1^{-1}\bar{h}_0(e)$ where e = identity of G, then

$$h_1([\bar{h}_1^{-1}\bar{h}_0(e)]\cdot y) = P(\bar{h}_0(e),y).$$

But for $t = 0$ and $g = e$, the same formula gives

$$h_0(y) = P(\bar{h}_0(e),y).$$

Therefore

$$[\bar{h}_1^{-1}\bar{h}_0(e)]\cdot y = h_1^{-1}h_0(y),$$

or

$$\bar{h}_1^{-1}\bar{h}_0(e) = h_1^{-1}h_0 = \chi(C).$$

The map $\bar{h}_1^{-1}\bar{h}_0$ of G on itself is a left translation by some element which, of course, must be $\bar{h}_1^{-1}\bar{h}_0(e)$. Therefore $\tilde{\chi}(C) = \bar{h}_1^{-1}\bar{h}_0(e) = \chi(C)$, and the first part of the theorem is proved.

To prove the second half, suppose $\chi(\mathcal{B}) = \chi(\mathcal{B}')$. By the first half, this implies $\chi(\tilde{\mathcal{B}}) = \chi(\tilde{\mathcal{B}}')$. If we can show that this implies $\tilde{\mathcal{B}}$ equivalent to $\tilde{\mathcal{B}}'$, it follows by definition that $\mathcal{B}$ and $\mathcal{B}'$ are associated.

Suppose therefore that $\mathcal{B},\mathcal{B}'$ are principal G-bundles with $\chi(\mathcal{B}) = \chi(\mathcal{B}')$. For any curve C from x_0 to x_1, denote by $C^\#,C'^\#$ the maps $G_1 \to G_0$ and $G'_1 \to G'_0$, respectively, induced by C. Choose admissible maps ξ: $G \to G_0$ and ξ': $G \to G'_0$. Then χ,χ': $\pi_1(X,x_0) \to G$ are thereby determined. By virtue of $\chi(\mathcal{B}) = \chi(\mathcal{B}')$, we can suppose that ξ,ξ' are so chosen that $\chi = \chi'$.

For any point $x \varepsilon X$, let C be a curve from x_0 to x, and define h_x: $G_x \to G'_x$ by

$$h_x = C'^{\#-1}\xi'\xi^{-1}C^\#. \tag{8}$$

If C_1 is another curve from x_0 to x, and $h_{1,x}$ is the corresponding map, then

$$\begin{aligned} h_x^{-1}h_{1,x} &= C^{\#-1}\xi\xi'^{-1}C'^\# C_1'^{\#-1}\xi'\xi^{-1}C_1^\# \\ &= (C^{\#-1}\xi)[\xi'^{-1}(CC_1^{-1})^\#\xi'](\xi^{-1}C_1^\#) \\ &= (C^{\#-1}\xi)[\chi'(CC_1^{-1})](\xi^{-1}C_1^\#) \\ &= (C^{\#-1}\xi)[\chi(CC_1^{-1})](\xi^{-1}C_1^\#) \\ &= (C^{\#-1}\xi)[\xi^{-1}(CC_1^{-1})^\#\xi](\xi^{-1}C_1^\#) = \text{identity}. \end{aligned}$$

This proves that (8) does not depend on the choice of C.

Let C_1 be a curve from x_0 to x_1, and suppose x_1 is in V_j. Since X is locally arcwise connected, we can suppose, by passing to a refinement if necessary, that each V_j is arcwise connected. Let C be a path in V_j from x_1 to a point x. Using the path C_1C from x_0 to x, a short calculation shows that

$$h_x = C'^{\#-1}h_{x_1}C^\#.$$

Since C lies wholly in V_j, we can apply the result of §13.6 to obtain

$$h_x(b) = \phi_j'(x, p_j' h_{x_1} \phi_j(x_1, p_j(b))), \qquad b \in G_x. \tag{9}$$

If we define h: $B \to B'$ by $h(b) = h_x(b)$ where $x = p(b)$, then (9) shows that h is continuous at x_1.

Finally we must check conditions (17) and (18) of §2.5. Suppose $x_1 \in V_j \cap V_k'$ and C is a curve from x_1 to x in $V_j \cap V_k'$. As in the preceding paragraph, we have

$$\begin{aligned} \bar{g}_{kj}(x) &= p_k' h_x \phi_{j,x} = p_k' C'^{\#-1} h_{x_1} C^{\#} \phi_{j,x} \\ &= (p_k' C'^{\#-1} \phi_{k,x_1}') (p_k' h_{x_1} \phi_{j,x_1}) (p_j C^{\#} \phi_{j,x}). \end{aligned}$$

Use now the first result of §13.6 to yield

$$\bar{g}_{kj}(x) = \bar{g}_{kj}(x_1).$$

This shows that $\bar{g}_{kj}$ is continuous; for $V_j \cap V_k'$ contains an arcwise connected neighborhood of x_1. Therefore h is an equivalence, and the theorem is proved.

13.8. Existence of a bundle with a prescribed χ. A space X is said to be *semi locally 1-connected* if, for each point x in X, there is a neighborhood V of x such that each closed curve in V is homotopic to a constant in X leaving its end points fixed.

THEOREM. *Let X be arcwise connected, arcwise locally connected, and semi locally 1-connected. Let G be a totally disconnected group, and $x_0 \in X$. Corresponding to any homomorphism χ: $\pi_1(X,x_0) \to G$, there exists a principal bundle $\mathcal{B}$ over X with group G such that χ is an element of* $\chi(\mathcal{B})$.

By hypothesis there is an indexed covering $\{V_j\}$ of X by open sets such that each V_j is arcwise connected and any closed path in V_j is contractible in X. For each j, choose a representative path C_j from x_0 to a point x_j in V_j.

For any point $x \in V_i \cap V_j$, choose a curve D from x_i to x in V_i, and a curve E from x_j to x in V_j and define

$$g_{ji}(x) = \chi(C_j E D^{-1} C_i^{-1}). \tag{10}$$

If D',E' are different choices of D,E, then the closed paths $D'D^{-1}, E'E^{-1}$ are contractible in X. This implies that D',E' are homotopic to D,E, respectively, with end points fixed. It follows that (10) is independent of the choice of D and E.

To prove continuity, suppose $x \in V_i \cap V_j$. From local connectedness, there is an arcwise connected neighborhood N of x in $V_i \cap V_j$. Let D,E be as in (10), and let C be a curve in N from x to a point x'. Then DC and EC are paths in V_i and V_j respectively. It follows from the independence shown above that

$$g_{ji}(x') = \chi(C_jECC^{-1}D^{-1}C_i^{-1}).$$

Since CC^{-1} is contractible over itself into x, it follows from (10) that $g_{ji}(x') = g_{ji}(x)$. Thus g_{ji} is constant over N. It is therefore continuous.

Suppose now that $x \varepsilon V_i \cap V_j \cap V_k$. Let D,E be as in (10), and let F be a path in V_k from x_k to x. Then

$$\begin{aligned} g_{kj}(x)g_{ji}(x) &= \chi(C_kFE^{-1}C_j^{-1})\chi(C_jED^{-1}C_i^{-1}) \\ &= \chi(C_kFE^{-1}C_j^{-1}C_jED^{-1}C_i^{-1}) \\ &= \chi(C_kFD^{-1}C_i^{-1}) = g_{ki}(x). \end{aligned}$$

This proves that the $\{g_{ji}\}$ are coordinate transformations in X. By §3.2, there is a principal bundle $\mathfrak{B}$ having these coordinate transformations.

Let V_0 denote a particular coordinate neighborhood of $\mathfrak{B}$ containing x_0. We can suppose moreover that the corresponding curve C_0 is the constant path. Let ξ: $G \to G_0$ be the map ϕ_{0,x_0}. Then $p_0 = \xi^{-1}$. Define χ': $\pi_1(X,x_0) \to G$ by

$$\chi'(C) = p_0C^{\#}\xi$$

for any closed path C based at x_0. To complete the proof we shall show that $\chi(C) = \chi'(C)$ for any C.

Since I is compact, there exists a finite set

$$0 = t_0 < t_1 < \cdots < t_{n+1} = 1$$

such that C maps $[0,t_1]$ and $[t_n,1]$ into V_0 and maps $[t_i,t_{i+1}]$ ($i = 1, \cdots, n-1$) into a single coordinate neighborhood, denoted by V_i for convenience. In this way

$$C = C'_0C'_1 \cdots C'_n$$

where C'_0 and C'_n are paths in V_0, and C'_i is a path in V_i from x'_i to x'_{i+1} ($i = 1, \cdots, n-1$). Let D_i be a path in V_i joining the reference point x_i to the point x'_i. Let C_i be the reference curve from x_0 to x_i. Clearly we have the homotopy

$$C \simeq (C'_0D_1^{-1}C_1^{-1})(C_1D_1C'_1D_2^{-1}C_2^{-1}) \cdots (C_{n-2}D_{n-2}C'_{n-2}C_{n-1}^{-1}D_{n-1}^{-1})$$
$$(C_{n-1}D_{n-1}C'_{n-1}C'_n).$$

Each block in parentheses is a closed path based at x_0; and, by definition of the coordinate transformations in $\mathfrak{B}$, we have

(11) $$\chi(C) = g_{01}(x'_1)g_{12}(x'_2) \cdots g_{n-1,0}(x'_n).$$

Introduce the abbreviation ξ_i for $\phi_{i,x_i'}$: $G \to G_{x_i'}$. Then

$$(12) \qquad \begin{aligned} \chi'(C) &= p_0 C^\# \xi = p_0 C_1'^\# C_2'^\# \cdots C_n'^\# \xi \\ &= (p_0 C_1'^\# \xi_1)(p_1 C_2'^\# \xi_2) \cdots (p_{n-1}[C_{n-1}' C_n']^\# \xi). \end{aligned}$$

By the second result in §13.6, we have

$$p_{i-1} C_i'^\# \xi_i = g_{i-1,i}(x_i'), \qquad (i = 1, \cdots, n).$$

Substituting these in (12) and comparing with (11) we obtain $\chi(C) = \chi'(C)$.

Combining the results of the last two sections, we obtain the

13.9. Classification theorem. *Let X and G be as in §13.8. Then the equivalence classes of principal bundles over X with group G are in 1-1 correspondence with the equivalence classes (under inner automorphisms of G) of homomorphisms of $\pi_1(X)$ into G.*

As pointed out in §8, the classification of bundles with prescribed X,Y and G is equivalent to classifying principal bundles over X with group G. This means that the above theorem solves the classification problem for arbitrary Y if X and G are as indicated. It solves the problem at least in the sense of reducing it to the familiar problems of computing $\pi_1(X)$ and finding all equivalence classes of homomorphisms $\pi_1(X) \to G$.

COROLLARY. *If X is arcwise connected, arcwise locally connected, and simply connected, and if G is totally disconnected, then any bundle over X with group G is equivalent to the product bundle.*

It is only necessary to observe that simply connected implies semi locally 1-connected.

13.10. Definition of χ for a general G. Now let G be any topological group and $\mathcal{B}$ a bundle over the arcwise connected space X with group G. The component G_e of the identity of G is a closed invariant subgroup of G. The factor group G/G_e is totally disconnected (for Lie groups it is discrete). Let η: $G \to G/G_e$ be the natural map. Let $\mathcal{B}'$ be the weakly associated principal bundle over X with group G/G_e. We define the *characteristic class* $\chi(\mathcal{B})$ to be the characteristic class of $\mathcal{B}'$. Then $\chi(\mathcal{B})$ is an equivalence class of homomorphisms of $\pi_1(X)$ into G/G_e.

It follows directly that any two associated bundles have the same characteristic class. Thus χ may be used to distinguish between bundles. For example, the last statement of §12.4 can be checked by showing that χ is non-trivial. For the case of a general group, $\chi(\mathcal{B}) = \chi(\mathcal{B}')$ is not enough to imply that the bundles are associated. Higher dimensional invariants are involved.

Later on we will define characteristic *cohomology* classes of a bundle for various dimensions. The one considered here reduces to the 1-dimensional cohomology class whenever G/G_e is abelian.

§14. Covering Spaces

14.1. Definition of covering. We review here the standard notion of covering space and show its relation to the work of the preceding article.

Let X be an arcwise connected and arcwise locally connected space. A map p: $B \to X$ is called a *covering* if (i) $p(B) = X$, and (ii), for each $x \varepsilon X$, there is an arcwise connected neighborhood V of x such that each component of $p^{-1}(V)$ is open in B and maps topologically onto V under p. The space B is called the *covering space*.

Choose an indexed covering $\{V_j\}$ of X by neighborhoods satisfying (ii). For any $b \varepsilon p^{-1}(V_j)$, let $V_j(b)$ be the component of $p^{-1}(V_j)$ containing b. If C is a curve in V_j from x_0 to x_1, and $p(b_0) = x_0$, the homeomorphism $p|V_j(b_0)$ provides a curve C' in $V_j(b_0)$ issuing from b_0 and covering C. Since $V_j(b_0)$ is open in B, it follows that there is only one curve C' in B which issues from b_0 and covers C.

Since the interval I is compact, any curve C from a point x_0 to a point x_1 may be broken up into a finite number of small curves $C = C_1C_2 \cdots C_n$ each contained wholly in some V_j. If $p(b) = x_0$, a stepwise procedure provides a *unique* curve C' in B which covers C and issues from b. If $Y_i = p^{-1}(x_i)(i = 0,1)$, define a map $C^{\#-1}$: $Y_0 \to Y_1$ by assigning to $b \varepsilon Y_0$ the end point in Y_1 of the curve C' issuing from b and covering C.

Now Y_0 is a discrete space. For if $x_0 \varepsilon V_j$ and $p(b) = x_0$, then $V_j(b)$ is an open set of B containing no point of Y_0 other than b. It follows that $C^{\#-1}$ is continuous.

Using the inverse path $\bar{C}(t) = C(1 - t)$ from x_1 to x_0 we obtain a map $Y_1 \to Y_0$. If C' covers C, then $\bar{C}'$ covers $\bar{C}$. It follows from this that $C^{\#-1}$ is a 1-1 map and $\bar{C}^{\#-1}$ is its inverse. We now drop the exponent -1 and write $C^{\#}$: $Y_1 \to Y_0$.

If C_1 is a curve from x_0 to x_1, and C_2 is a curve from x_1 to x_2, let C_1' be a curve covering C_1 from b_0 in Y_0 to a point b_1 in Y_1, and let C_2' be a curve covering C_2 from b_1 to b_2 in Y_2. Clearly $C_1'C_2'$ covers C_1C_2. It follows from this that $(C_1C_2)^{\#} = C_1^{\#}C_2^{\#}$.

Consider now a homotopy of a curve C from x_0 to x_1 leaving its end points fixed. Let C' be a curve covering C. As in the proof of the covering homotopy theorem (§11.3), the homotopy of C may be decomposed into a finite succession of small homotopies for each of which such motion as does occur takes place in a single V_j. Using the local inverse maps $V_j \to V_j(b)$, the homotopies are lifted one at a time providing a final, complete homotopy of C' which covers the homotopy of C. Since the end points of C are fixed, the end points of C' remain in Y_0 and Y_1. But these sets are discrete, therefore the end points of C' remain fixed during the covering homotopy. It follows that $C^\#$ *depends only on the homotopy class of the path* C.

14.2. The transformation group of Y_0. An immediate consequence of the above results is that the fundamental group $\pi_1(X,x_0)$ becomes a group of transformations of the fibre Y_0 over x_0 under the operation $C^\#$.

For any point $b \in Y_0$, the map p induces a homomorphism

$$p_*: \quad \pi_1(B,b) \to \pi_1(X,x_0)$$

defined by assigning to each closed curve C' based at b, the image curve pC' based at x_0. In fact, p_* maps $\pi_1(B,b)$ isomorphically into $\pi_1(X,x_0)$; for, if pC' is contractible to x_0 leaving its ends fixed, a covering homotopy does the same for C'.

Let H be the intersection of the image groups $p_*(\pi_1(B,b))$ as b ranges over Y_0. We assert: $\alpha \in \pi_1(X,x_0)$ *induces the identity transformation of* Y_0 *if and only if* α *is in* H. For if α is in H, then, for any b in Y_0, α is represented by a curve pC' where C' is a closed curve issuing from b. By definition, $(pC')^\#$ must map b on itself. Conversely suppose C represents α, and α leaves b fixed, then the curve C' issuing from b and covering C must end at b. Hence $\alpha \in p_*(\pi_1(B,b))$. If this holds for each b, then $\alpha \in H$.

Let G be the factor group $\pi_1(X,x_0)/H$, and let χ: $\pi_1(X,x_0) \to G$ be the natural homomorphism. The above result shows that G *is an effective transformation group of* Y_0. We assign to G the *discrete* topology so that G is a topological transformation group of Y_0.

14.3. Bundle structure theorem. *The covering map* p: $B \to X$ *admits a bundle structure with fibre* Y_0, *group* G *and characteristic class* χ *as defined above.*

We use $\{V_j\}$ defined in §14.1 as coordinate neighborhoods. For each j, let C_j be a curve in X from a point x_j in V_j to x_0. For any x in V_j, and $y \in Y_0$, choose a curve D in V_j from x to x_j and define

$$\phi_j(x,y) = D^\# C_j^\#(y).$$

Now a curve D' covering D and ending at $C_j^\#(y)$ must lie in the component $V_j(C_j^\#(y))$. Since p maps this component topologically onto V_j, we obtain two results: (i) $\phi_j(x,y)$ is independent of the choice of D, and (ii) for a fixed y, ϕ_j is the inverse of the map p of $V_j(C_j^\#(y))$ onto V_j. As $V_j \times y$ is open in $V_j \times Y_0$, it follows that ϕ_j is continuous. Clearly $p\phi_j(x,y) = x$. The map p_j: $p^{-1}(V_j) \to Y_0$ is obtained by mapping each component of $p^{-1}(V_j)$ into its intersection with $p^{-1}(x_j)$ and then applying $C_j^{\#-1}$. Then $p_j\phi_j(x,y) = y$. Therefore ϕ_j is a coordinate function.

Suppose now that $x \varepsilon V_i \cap V_j$. Let D_i,D_j be curves in V_i,V_j from x to x_i,x_j respectively. By definition

$$g_{ji}(x)\cdot y = C_j^{\#-1}D_j^{\#-1}D_i^{\#}C_i^{\#}(y) = (C_j^{-1}D_j^{-1}D_iC_i)^{\#}(y)$$
$$= \chi(C_j^{-1}D_j^{-1}D_iC_i)\cdot y.$$

Hence $g_{ji}(x)$ is in G. Choose now an arcwise connected neighborhood N of x in $V_i \cap V_j$. If $x' \varepsilon N$, let E be a curve in N from x' to x. Then

$$C_j^{-1}D_j^{-1}E^{-1}ED_iC_i \simeq C_j^{-1}D_j^{-1}D_iC_i,$$

which shows that $g_{ji}(x') = g_{ji}(x)$. Thus g_{ji} is constant over N; and therefore continuous over $V_i \cap V_j$.

This establishes the existence of the bundle structure. The bundle structure provides an operation $C^{\#}$ as defined in §13.3. Due to the uniqueness of curves C' covering C, this operation must coincide with the $C^{\#}$ defined in §14.1. Therefore the χ of §14.2 coincides with that of §13.5.

14.4. The classification of coverings. Any bundle over X having a discrete fibre is easily seen to be a covering of X. As a consequence there is a complete equivalence between coverings and a special class of bundles. One who is familiar with the classical theory of covering spaces will recognize the classification theorems of §13 as an extension to bundles of the similar theorems for covering spaces. We have merely repeated the classical arguments step-by-step and made the observation at each stage that bundle structures are preserved. We shall review some additional facts about covering spaces and reinterpret them in terms of the bundle structure.

14.5. Suppose the covering B of X is arcwise connected (this is sometimes incorporated in the definition of a covering). Let b_0,b_1 be two points over x_0, and let C' be a curve in B from b_1 to b_0. If $C = pC'$, it follows from the uniqueness of covering curves that $C^{\#}$ carries b_0 into b_1. This means that G operates *transitively* on Y_0. Fixing a point b_0, the map τ: $G \rightarrow Y_0$ given by $\tau(g) = g\cdot b_0$ provides a representation of Y_0 as a left coset space of G. Then the composition $\tau\chi$: $\pi_1(X,x_0) \rightarrow Y_0$ represents Y_0 as a left coset space of $\pi_1(X,x_0)$. The subgroup of this representation is readily identified as the isomorphic image of $\pi_1(B,b_0)$ under p_*. Denote it by $H(b_0)$. For a curve C operates trivially on b_0 if and only if it is covered by a closed curve issuing from b_0.

If the base point b_0 is changed to b_1 in Y_0, the subgroup $H(b_0)$ will usually change. Let C' be a curve in B from b_1 to b_0, C the curve it covers, and α the element of the fundamental group represented by C.

Let D be a closed curve representing β in $\pi_1(B,b_1)$. Then $C'^{-1}D'C'$ is a closed curve issuing from b_0 and represents an element γ in $\pi_1(B,b_0)$. Since $C'^{-1}D'C'$ covers $C^{-1}DC$ it follows that

$$p_*(\gamma) = \alpha^{-1}p_*(\beta)\alpha.$$

This means that *$H(b_1)$ is conjugate to $H(b_0)$ under α.* Conversely, if $H_1 = \alpha H(b_0)\alpha^{-1}$, and we set $b_1 = \alpha(b_0)$, it follows quickly that $H_1 = H(b_1)$.

Summarizing, *if B is arcwise connected, then B determines a class of pairwise conjugate subgroups of $\pi_1(X,x_0)$, these are the isomorphic images of $\pi_1(B,b_0)$ under p_* for all b_0 in Y_0.*

Conversely, if we start with X,x_0 and a conjugacy class $\{H\}$ of subgroups of $\pi_1(X,x_0)$, we choose one of them, say H_1, and define Y_0 to be the left coset space of π_1 by H_1. Let H_0 be the intersection of the subgroups $\{H\}$, $G = \pi_1/H_0$, and $\chi\colon\ \pi_1 \to G$ the natural map. If X is semi locally 1-connected, then the existence theorem 13.8 provides a principal bundle with χ in its characteristic class. The associated bundle with fibre Y_0 will then be a covering of X such that $\{p_*\pi_1(B,b_0)\} = \{H\}$.

The classification theorem 13.9 yields now the classical result that *equivalence classes of coverings of X are in 1-1 correspondence with conjugacy classes of subgroups of $\pi_1(X)$.*

14.6. Regular coverings. A covering $p\colon\ B \to X$ is said to be *regular* if the group $G = \pi_1/H_0$ is simply transitive on Y_0. This means that G may be identified with Y_0 so that its operations on Y_0 correspond to left translations. It follows that the corresponding bundle structure is that of a *principal* bundle. The converse is evident. Therefore *regular coverings of X coincide with principal bundles over X which are arcwise connected and have discrete groups.*

If the covering is regular, then H_0 is the kernel of $\tau\chi\colon\ \pi_1 \to Y_0$. As shown in §14.5, this kernel is $H(b_0)$. Since this holds for each b_0 in Y_0, it follows that all the subgroups $H(b)$ coincide with the invariant subgroup H_0. Conversely, if $H(b_0) = H_0$, then the kernels of χ and $\tau\chi$ coincide; so τ is a 1-1 map. Thus *regular coverings of X correspond to invariant subgroups of $\pi_1(X)$.*

If the covering is regular, we have seen that it is a principal bundle. According to §8.11 the right translations of B by elements of G map each fibre on itself and provide an anti-representation of G as a transformation group of B. In the classical theory, these right translations are referred to as *covering transformations* (deckbewegungen) of B.

14.7. The universal covering. Let H and H' be two subgroups of $\pi_1(X)$ such that $H \supset H'$, and let B,B' be the corresponding coverings.

According to §9.6, there is a natural map ν: $B' \to B$ such that $p\nu = p'$, and ν is the projection of a bundle structure having H/H' for fibre. Since H/H' is discrete, ν: $B' \to B$ is a covering. Thus, the lattice of subgroups of $\pi_1(X)$ corresponds to a lattice of spaces and covering maps. The covering of X, which corresponds to H = the identity element of $\pi_1(X)$, is called the *universal covering.* It covers every arcwise connected covering.

Part II. The Homotopy Theory of Bundles

§15. Homotopy Groups

15.1. The role played by the fundamental group in §13 indicates that their higher dimensional analogs—the homotopy groups of Hurewicz—should be of considerable importance in the study of bundles. This section is a brief résumé of homotopy groups. Basic definitions and principal properties will be stated in detail. Proofs will only be indicated. For a fuller exposition see [55].

15.2. Definitions. The definition of the nth homotopy group of a space, $\pi_n(X,x_0)$, is strictly analogous to that of the fundamental group. We replace the interval $I = [0,1]$ by the *n-cube* I^n consisting of points $t = (t_1, \cdots, t_n)$ in euclidean n-space such that $0 \leqq t_i \leqq 1$ $(i = 1, \cdots, n)$. An $(n-1)$-*face* of I^n is obtained by setting some $t_i = 0$ or 1. The union of the $(n-1)$-faces forms the *boundary* $\dot{I}^n$ of I^n. We consider maps of I^n into X which carry $\dot{I}^n$ into x_0, then the elements of $\pi_n(X,x_0)$ are homotopy classes of such maps.

If the boundary of an n-cube is pinched to a point, we obtain a configuration topologically equivalent to an n-sphere S^n and a reference point y_0 on S^n. It follows that one might equally well define an element of π_n as a homotopy class of maps of S^n into X with y_0 mapped into x_0. Although the use of the n-sphere as "anti-image" is pictorial and suggestive, it does not lend itself well to the various constructions which must be made. Also, when $n > 1$, a new element enters. One may define a *relative* homotopy group of X modulo a subset A, analogous to the relative homology group; and this requires the use of the n-cube as anti-image. The relative groups include the non-relative as a special case; hence we define only the former.

The initial $(n-1)$-face of I^n, denoted by I^{n-1}, is defined by $t_n = 0$. The union of all the remaining $(n-1)$-faces of I^n is denoted by J^{n-1}. Then

$$\dot{I}^n = I^{n-1} \cup J^{n-1}, \qquad \dot{I}^{n-1} = I^{n-1} \cap J^{n-1}.$$

Let X be a space, A a subspace of X, and x_0 a point of A. By a map

$$(1) \qquad f\colon\ (I^n,I^{n-1},J^{n-1}) \to (X,A,x_0)$$

is meant a continuous function from I^n to X which maps I^{n-1} into A, and J^{n-1} into x_0. In particular, it carries $\dot{I}^n$ into A and $\dot{I}^{n-1}$ into

x_0. We denote by $F^n(X,A,x_0)$ (briefly: F^n) the set of all functions.

If f_1,f_2 are in F^n, their *sum* $f_1 + f_2$ is defined by

$$(2) \qquad [f_1 + f_2](t) = \begin{cases} f_1(2t_1, t_2, \cdots, t_n) & 0 \leqq t_1 \leqq 1/2, \\ f_2(2t_1 - 1, t_2 \cdots, t_n) & 1/2 \leqq t_1 \leqq 1. \end{cases}$$

If $n \geqq 2$, and $t_1 = 1/2$, then both lines reduce to x_0. Hence $f_1 + f_2$ is in F^n when $n \geqq 2$. This is also true when $n = 1$ providing $A = x_0$.

Two maps f_0,f_1 of F^n are *homotopic* in F^n, in symbols $f_0 \simeq f_1$, if there exists a map f: $I^n \times I \rightarrow X$ (where I is the interval $0 \leqq \tau \leqq 1$) such that $f(t,0) = f_0(t)$, $f(t,1) = f_1(t)$, and, for each τ, the map f_τ: $I^n \rightarrow X$, defined by $f_\tau(t) = f(t,\tau)$, is in F^n. With a suitable function-space topology in F^n, this can be expressed by saying that f_0 and f_1 are joined by a curve in F^n.

The homotopy relation is reflective, symmetric and transitive. It thereby divides F^n into mutually exclusive equivalence classes called *homotopy classes*. These classes are the elements of the set $\pi_n(X,A,x_0)$.

If $f_i \simeq f'_i$ $(i = 1,2)$ in F^n, one can combine the two homotopies to provide a homotopy $f_1 + f_2 \simeq f'_1 + f'_2$. Therefore, if α,β are elements of π_n, all sums $f_1 + f_2$ for f_1 in α and f_2 in β lie in a single homotopy class γ of π_n. We define *addition* in π_n by setting $\alpha + \beta = \gamma$.

With respect to this addition, π_n is a group. The associative law is proved by exhibiting a homotopy $(f_1 + f_2) + f_3 \simeq f_1 + (f_2 + f_3)$. This is based on a homotopy of the t_1-axis which stretches [0,1/4] into [0,1/2], translates [1/4,1/2] into [1/2,3/4], and contracts [1/2,1] into [3/4,1].

The *zero* of the group is the homotopy class of the constant map: $f_0(I^n) = x_0$. The relation $f_0 + f \simeq f$ for any f is proved by deforming the t_1-interval so that [0,1/2] shrinks to 0 and [1/2,1] expands into [0,1].

Even more, a map f in F^n which carries I^n into A represents the zero. This is seen as follows. Let h be a homotopy of I^n over itself which contracts it into the face $t_n = 1$. Such a homotopy is given by

$$h(t,\tau) = (t_1, \cdots, t_{n-1},(1 - \tau)t_n + \tau).$$

Then $h^1(t,\tau) = f(h(t,\tau))$ is a homotopy in F^n of f into f_0.

If f is in F^n, then

$$\bar{f}(t) = f(1 - t_1, t_2, \cdots, t_n)$$

is also in F^n, and $f + \bar{f}$ and $\bar{f} + f$ are both homotopic to the constant map. Since $\bar{\bar{f}} = f$, it is enough to prove this for $f + \bar{f}$. The construction of the homotopy is indicated by Fig. 3. Along the dotted line

$h(t,\tau)$ is constant and has the value $f(2t_1', t_2, \cdots, t_n)$. As t_1' ranges from 0 to 1/2, the dotted line sweeps out the (t_1,τ) square and defines h completely. It follows that $\bar{f}$ represents the negative of the element represented by f.

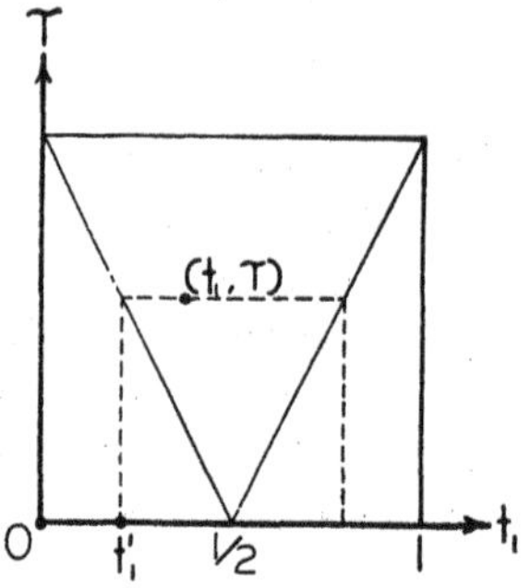

FIG. 3.

15.3. Commutativity. The group $\pi_n(X,A,x_0)$ is called *the n dimensional, relative homotopy group of X mod A with base point* x_0. It is always defined for $n \geqq 2$. In case $A = x_0$, it is defined for $n = 1$ and coincides with the fundamental group $\pi_1(X,x_0)$. In general when $A = x_0$, we write $\pi_n(X,x_0)$.

The additive notation has been used because $\pi_n(X,x_0)$ is abelian for $n > 1$, and $\pi_n(X,A,x_0)$ is abelian for $n > 2$. The proof of this for $n = 3$ is indicated by Fig. 4. We choose a homeomorphism between

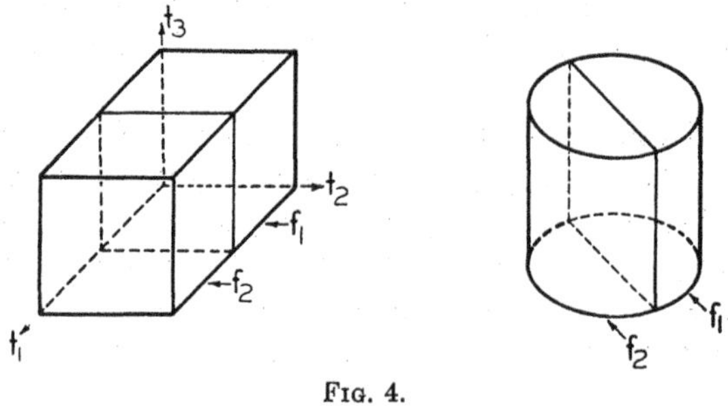

FIG. 4.

the 3-cube and a solid cylinder so that the plane $t_1 = 1/2$ corresponds to a diametral plane. A rotation of the cylinder through 180° will interchange the two halves. Under the homeomorphism this corresponds to a "rotation" of the cube. If f is any map of the cube, the composition of f and this rotation is a homotopy of f into a map f^r. Since the rotation interchanges the two halves we obtain, for f_1,f_2 in F^3,

$$f_1 + f_2 \simeq (f_1 + f_2)^r = f_2^r + f_1^r \simeq f_2 + f_1$$

which is the desired result.

When $n = 2$, the rotation homotopy does not keep the set J^1 at x_0. However it does move it in A. So if $A = x_0$, the same argument shows that $\pi_2(X,x_0)$ is abelian. The proof for $n > 3$ is the same; the rotation is in the (t_1,t_2)-plane, the remaining variables do not enter the construction.

15.4. The boundary operator. The *boundary* homomorphism

$$\partial\colon\ \pi_n(X,A,x_0) \to \pi_{n-1}(A,x_0) \tag{3}$$

is defined by choosing a map f representing α in π_n and restricting f to the initial face I^{n-1} of I^n. Since f maps J^{n-1} into x_0, it maps $\dot{I}^{n-1}$ into x_0; therefore f restricted to I^{n-1} is a map $\partial f\colon\ (I^{n-1},I^{n-2},J^{n-2}) \to (A,x_0,x_0)$.

A homotopy of f_0 into f_1 in F^n restricted to $I^{n-1} \times I$ provides a homotopy of ∂f_0 into ∂f_1 in $F^{n-1}(A,x_0)$. Therefore $f \to \partial f$ induces a mapping of homotopy classes. It is obvious that $\partial(f_1 + f_2) = \partial f_1 + \partial f_2$. Thus (3) is defined and is a homomorphism.

15.5. The induced homomorphism. Suppose h is a map of X into Y which carries A into a subset B of Y and x_0 into y_0 (this is written $h\colon\ (X,A,x_0) \to (Y,B,y_0)$). For any f in $F^n(X,A,x_0)$, the composition hf is in $F^n(Y,B,y_0)$. A homotopy f of f_0 into f_1 in $F^n(X,A,x_0)$ composes with h to provide a homotopy hf of hf_0 into hf_1 in $F^n(Y,B,y_0)$. In this way h defines a mapping

$$h_*\colon\ \pi_n(X,A,x_0) \to \pi_n(Y,B,y_0) \tag{4}$$

of homotopy classes. Since $h(f_1 + f_2) = hf_1 + hf_2$ it follows that h_* is a homomorphism. It is called *the homomorphism induced by h.*

15.6. Elementary properties. The groups π_n and the two types of homomorphisms, ∂ and h_*, have basic properties similar to those possessed by homology groups and their homomorphisms. To state these we need one definition. Let $i\colon\ (A,x_0) \to (X,x_0)$ and $j\colon\ (X,x_0,x_0) \to (X,A,x_0)$ be inclusion maps (i.e. $i(x) = j(x) = x$). The infinite sequence of groups and homomorphisms

$$\begin{aligned} &\cdots \xrightarrow{\partial} \pi_n(A,x_0) \xrightarrow{i_*} \pi_n(X,x_0) \xrightarrow{j_*} \pi_n(X,A,x_0) \xrightarrow{\partial} \pi_{n-1}(A,x_0) \to \\ &\cdots \xrightarrow{j_*} \pi_2(X,A,x_0) \xrightarrow{\partial} \pi_1(A,x_0) \xrightarrow{i_*} \pi_1(X,x_0) \end{aligned} \tag{5}$$

is called *the homotopy sequence* of (X,A,x_0). The basic properties are as follows:

1°. *If h is the identity map of (X,A,x_0), then h_* is the identity map of $\pi_n(X,A,x_0)$.*

2°. *If* h: $(X,A,x_0) \to (Y,B,y_0)$ *and* k: $(Y,B,y_0) \to (Z,C,z_0)$, *then* $(kh)_* = k_*h_*$ *for each dimension.*

3°. *If* h: $(X,A,x_0) \to (Y,B,y_0)$ *and* h': $(A,x_0) \to (B,y_0)$ *is the map* h *restricted to* A, *then* $\partial h_* = h'_*\partial$.

This means that commutativity holds in the following diagram for each $n \geqq 2$.

$$\begin{array}{ccc} \pi_n(X,A,x_0) & \xrightarrow{h_*} & \pi_n(Y,B,y_0) \\ \downarrow\partial & & \downarrow\partial \\ \pi_{n-1}(A,x_0) & \xrightarrow{h'_*} & \pi_{n-1}(B,y_0) \end{array}$$

4°. *The homotopy sequence is exact.*

This means that, at each term of the sequence (5) except the last, the image of the homomorphism on the left coincides with the kernel of the homomorphism on the right.

5°. *If the maps* h *and* k *of* (X,A,x_0) *into* (Y,B,y_0) *are connected by a homotopy which maps* $A \times I$ *into* B *and* $x_0 \times I$ *into* y_0, *then* h_* *and* k_* *coincide for each* n.

6°. *If* X *consists of a single point* x_0, *then for each* n, $\pi_n(X,x_0)$ *contains only the zero element.*

The proofs of these six properties are entirely straightforward. The most difficult is 4° which requires proving "kernel = image" in three different cases. As an example, let us prove that the image of j_* equals the kernel of ∂ in $\pi_n(X,A,x_0)$. Suppose f in $F^n(X,x_0)$ represents α in $\pi_n(X,x_0)$. Then f is an element of $F^n(X,A,x_0)$ and represents $j_*\alpha$. By definition of $F^n(X,x_0)$, f maps I^{n-1} into x_0. Therefore $\partial f = f|I^{n-1}$ maps I^{n-1} into x_0, so ∂f represents the zero of $\pi_{n-1}(A,x_0)$. This proves $\partial j_*\alpha = 0$, or image $(j_*) \subset$ kernel (∂).

Suppose now that f in $F^n(X,A,x_0)$ represents α and that $\partial\alpha = 0$. Then ∂f is homotopic to the constant map. Let h: $(I^{n-1} \times I, \dot{I}^{n-1} \times I) \to (A,x_0)$ be such a homotopy. Extend h over $I^n \times 0$ by $h(t,0) = f(t)$, and extend h over $J^{n-1} \times I$ by $h(t,\tau) = f(t) = x_0$. Then h is defined on $E = (I^n \times 0) \cup (\dot{I}^n \times I)$. Now E is just an n-cell on the boundary of the $(n+1)$-cell $I^n \times I$. Therefore there is a retraction r of $I^n \times I$ into E (see §12.1). Then rh: $(I^n \times I, I^{n-1} \times I, J^{n-1} \times I) \to (X,A,x_0)$ is a homotopy in $F^n(X,A,x_0)$ of f into a map f' which carries $\dot{I}^n$ into x_0. Then f' is in $F^n(X,x_0)$ and represents an element β of $\pi_n(X,x_0)$ such that $j_*\beta = \alpha$. Thus image $(j_*) \supset$ kernel (∂). This completes the proof of exactness at the term $\pi_n(X,A,x_0)$.

There is a very useful extension of the notion of homotopy sequence to that of a *triple* (X,A,B) where $X \supset A \supset B$ and the base point x_0 is in B. It is the sequence

$$\cdots \to \pi_n(A,B) \xrightarrow{i_*} \pi_n(X,B) \xrightarrow{j_*} \pi_n(X,A) \xrightarrow{\partial} \pi_{n-1}(A,B) \to \cdots$$

where i and j are the indicated inclusion maps. The operator ∂ is the composition

$$\pi_n(X,A) \xrightarrow{\partial} \pi_{n-1}(A) \xrightarrow{k_*} \pi_{n-1}(A,B)$$

where k is the inclusion map. The sequence ends with $\pi_2(X,A)$. It reduces to the sequence of a pair when B is a point. Just as 4° is proved, one can show:

7°. *The homotopy sequence of a triple is exact.*

This may also be derived in a purely algebraic fashion from properties 1° to 4°.

15.7. A map h: $(X,A,x_0) \to (X',A',x_0')$ induces maps h_1: $(X,x_0) \to (X',x_0')$ and h_2: $(A,x_0) \to (A',x_0')$. All three induce homomorphisms of their corresponding homotopy groups. This leads to the diagram

$$(6)\qquad \begin{array}{ccccccccc} \to & \pi_n(A,x_0) & \xrightarrow{i_*} & \pi_n(X,x_0) & \xrightarrow{j_*} & \pi_n(X,A,x_0) & \xrightarrow{\partial} & \pi_{n-1}(A,x_0) & \to \\ & \downarrow h_{2*} & & \downarrow h_{1*} & & \downarrow h_* & & \downarrow h_{2*} & \\ \to & \pi_n(A',x_0') & \xrightarrow{i'_*} & \pi_n(X',x_0') & \xrightarrow{j'_*} & \pi_n(X',A',x_0') & \xrightarrow{\partial} & \pi_{n-1}(A',x_0') & \to \end{array}$$

Properties 2° and 3° above imply that commutativity holds in each square of the diagram. This is called the *homomorphism* of the homotopy sequence of (X,A,x_0) into that of (X',A',x_0') induced by h.

If h is a homeomorphism, it follows from properties 1° and 2° that h induces an isomorphism of the homotopy sequence onto the other.

15.8. Homotopy groups of cells and spheres. A map h: $(X,A,x_0) \to (X',A',x_0')$ is called a *homotopy equivalence* if there is a map k: $(X',A',x_0') \to (X,A,x_0)$ such that kh and hk are both homotopic to the identity maps of (X,A,x_0) and (X',A',x_0') respectively. (The homotopies must move the subsets A,A' on themselves and leave x_0,x_0' fixed.) It follows quickly from 5° that *a homotopy equivalence induces an isomorphism of the one homotopy sequence onto that of the other.*

In particular, if the identity map of (X,A,x_0) is homotopic to the constant map of (X,A,x_0) into x_0, then, by 6°, all the homotopy groups of (X,A,x_0) contain only the zero.

An open or closed q-cell E is contractible to any one of its points, therefore $\pi_n(E,x_0) = 0$ for all n and any x_0.

Let (E,S) be a closed q-cell and its boundary, i.e. a homeomorph of $(I^q,\dot{I}^q)$. Let x_0 be a point of S. In the section

$$\pi_n(E) \xrightarrow{j_*} \pi_n(E,S) \xrightarrow{\partial} \pi_{n-1}(S) \xrightarrow{i_*} \pi_{n-1}(E)$$

of the homotopy sequence, the vanishing of $\pi_n(E)$ implies that the image of j_* is zero. By the exactness property, the kernel of ∂ is zero. The vanishing of $\pi_{n-1}(E)$ implies that $\pi_{n-1}(S)$ is the kernel of i_*. By the exactness property, the image of ∂ is $\pi_{n-1}(S)$. It follows that

$$\partial: \ \pi_n(E,S) \approx \pi_{n-1}(S). \tag{7}$$

The argument just given proves a more general statement: *if every third term of an exact sequence is zero, then the remaining adjacent pairs are isomorphic.*

The homotopy groups $\pi_q(S^n)$ of an n-sphere are zero for $q < n$. To prove this, one triangulates I^q and S^n. Then, for any f in F^q, the simplicial approximation theorem provides a homotopic map f' which is simplicial. Hence $f'(I^q)$ lies in the q-dimensional skeleton of S^n. Since $q < n$, $f'(I^q)$ fails to contain some point x of S^n. But $S^n - x$ is an open n-cell and is contractible. It follows that f' is homotopic to a constant.

15.9. *The first non-zero homotopy group of S^n is the nth and this group is infinite cyclic.*

This is a special case of a more general result. Let $H_n(X,A)$ denote the nth relative homology group (in the singular sense) of X mod A based on integer coefficients. The group $H_n(I^n,\dot{I}^n)$ (abbreviated by G_n) is infinite cyclic. A generator u_n of G_n is just an "orientation" of I^n. We select the generators u_n, for each n, so that the orientation of I^{n-1} is positively incident to that of I^n. If α is in $\pi_n(X,A,x_0)$, and f represents α, then f induces a homomorphism $f_*: \ G_n \to H_n(X,A)$. Define $\phi(\alpha) = f_* u_n$. Since f_* depends only on the homotopy class of f, ϕ is uniquely defined. We obtain in this way a map

$$\phi: \ \pi_n(X,A,x_0) \to H_n(X,A) \tag{8}$$

called *the natural homomorphism.*

To prove that it is homomorphic, observe that the plane section Q defined by $t_1 = 1/2$ divides I^n into two n-cells C_1,C_2. Then $H_n(I^n,\dot{I}^n \cup Q)$ decomposes into the direct sum of two subgroups G'_n,G''_n isomorphic under the inclusion maps to $H_n(C_1,\dot{C}_1)$ and $H_n(C_2,\dot{C}_2)$. These groups are infinite cyclic. One may select generators u_1,u_2 of these groups which map into u_n under the maps h_1: $C_1 \to I^n$, h_2: $C_2 \to I^n$ defined by $t_1 \to 2t_1$ and $t_1 \to 2t_1 - 1$ respectively. Let u', u'' correspond to u_1,u_2 in G'_n, G''_n. One proves immediately that the inclusion map k: $(I^n,\dot{I}^n) \to (I^n,\dot{I}^n \cup Q)$ carries u_n into $u' + u''$. If f_1,f_2 are in F^n, then $f_1 + f_2$ carries Q into x_0, defining thereby a map f: $(I^n,\dot{I}^n \cup Q) \to (X,A)$. Using standard properties of the homomorphisms of homology groups induced by mappings, we have

$$\begin{aligned}(f_1 + f_2)_* u_n &= (fk)_* u_n = f_* k_* u_n = f_*(u' + u'') \\ &= f_* u' + f_* u'' = (f_1 h_1)_* u' + (f_2 h_2)_* u'' \\ &= f_{1*} h_{1*} u' + f_{2*} h_{2*} u'' = f_{1*} u_n + f_{2*} u_n.\end{aligned}$$

The homomorphism ϕ is natural in the sense that it commutes with the operations h_* and ∂ of both homotopy and homology groups. Specifically commutativity holds in the diagrams

$$\begin{array}{ccccccc} \pi_n(X,A,x_0) & \xrightarrow{h_*} & \pi_n(Y,B,y_0), & \qquad & \pi_n(X,A,x_0) & \xrightarrow{\partial} & \pi_{n-1}(A,x_0) \\ \downarrow\phi & & \downarrow\phi & & \downarrow\phi & & \downarrow\phi \\ H_n(X,A) & \xrightarrow{h_*} & H_n(Y,B), & & H_n(X,A) & \xrightarrow{\partial} & H_{n-1}(A) \end{array}$$

In particular ϕ is a homomorphism of the homotopy sequence of (X,A,x_0) into the homology sequence of (X,A).

We may now state the

15.10. Isomorphism theorem of Hurewicz. *Let the subspace A of X be arcwise connected, and let X and A be simply-connected. Let $\pi_i(X,A,x_0) = 0$ for $2 \leqq i < n$. Then*

$$\phi\colon\ \pi_n(X,A,x_0) \approx H_n(X,A).$$

The proof is too long to give here (see [55]).

The result implies $H_p(X,A) = 0$ for $1 \leqq p < n$. For, since $\pi_i = 0$ for $i < p$, it follows that ϕ maps π_p isomorphically onto H_p. Conversely, if the first two hypotheses on X and A are satisfied and $H_i = 0$ for $2 \leqq i < n$, the result may be iterated to prove that $\pi_i = 0$ for $i = 2$, then $i = 3$ and so forth up to n. It follows that the third hypothesis can be replaced by $H_i(X,A) = 0$ for $2 \leqq i < n$, and the conclusion still holds.

The Hurewicz theorem can be paraphrased by saying that *the first non-zero homology group and the first non-zero homotopy group have the same dimension and are isomorphic under ϕ.*

The result applies immediately to the n-sphere S^n to show that

$\pi_n(S^n) \approx H_n(S^n)$ is an infinite cyclic group. In the same way, if (E^n,S^{n-1}) is an n-cell and its boundary, $\pi_n(E^n,S^{n-1},x_0) \approx H_n(E^n,S^{n-1})$ is infinite cyclic. In Part II, we shall use only these two cases of the Hurewicz theorem.

In general, the problem has not been solved of computing the homotopy groups of even simple spaces such as complexes and spheres. Only by special devices in special cases have answers been obtained. As will be shown later, certain bundles play an important role in computing several homotopy groups. (See App. sect. 4.)

15.11. The use of cells and spheres as anti-images. There are a number of standard homotopy "tricks" used in connection with homotopy groups. We give these here.

An *n-cell and its boundary*, denoted by (E^n,S^{n-1}), is a space and subspace homeomorphic with $(I^n,\dot{I}^n)$. In particular the *euclidean* n-cell, defined by $\Sigma_1^n t_i^2 \leqq 1$, and its boundary, $\Sigma_1^n t_i^2 = 1$, is such a pair; and it is regarded as the prototype. One therefore speaks of S^{n-1} as an *$(n - 1)$-sphere;* also of a point x_0 interior to E^n as an *origin;* and of *radial lines* from x_0 to points of S^{n-1}. This language is based on a definite homeomorphism with the euclidean n-cell.

Let (E,S) be an oriented n-cell, i.e. an orientation of (E,S) is a selection of a generator v_n of $H_n(E,S)$. Let y_0 be a reference point of S; and let f be a map $(E,S,y_0) \to (X,A,x_0)$. We choose a map

$$k\colon\ (I^n,I^{n-1},J^{n-1}) \to (E,S,y_0)$$

such that $k_*u_n = v_n$. We may even suppose that k maps $I^{n-1} - \dot{I}^{n-1}$ topologically onto $S - y_0$ and $I^n - \dot{I}^n$ topologically onto $E - S$; for if J^{n-1} is pinched to a point, the resulting image space of (I^n,I^{n-1},J^{n-1}) is homeomorphic to (E,S,y_0). Compose f and k to obtain fk in F^n. If k' is a second map with the properties of k, then $k_*u_n = k'_*u_n$ implies, by 15.10, that both k and k' represent the generator of $\pi_n(E,S,y_0)$. Therefore $k \simeq k'$ in $F^n(E,S,y_0)$. Thus, the homotopy class of fk depends only on f and determines a unique element $c(f)$ in $\pi_n(X,A,x_0)$ which is called the element of $\pi_n(X,A,x_0)$ *represented* by f. A homotopy of f keeping S in A and y_0 at x_0 provides a homotopy of fk. Thus $c(f)$ depends only on the homotopy class of f.

If a map f' in $F^n(X,A,x_0)$ is given, let $f = f'k^{-1}$. Although k^{-1} is not single-valued, f is single-valued and continuous. It follows that any f' is of the form fk. Therefore any element of $\pi_n(X,A,x_0)$ is a $c(f)$.

All of this means that we might have used homotopy classes of maps of (E,S,y_0) into (X,A,x_0) in defining π_n. The virtue of the fixed choice of the n-cube lies in the ease of defining the addition of functions and the boundary operator.

In the original definition that Hurewicz gave of the absolute group $\pi_n(X,x_0)$, the elements were homotopy classes of maps $(S^n,y_0) \to (X,x_0)$ where S^n is a fixed n-sphere. The correspondence is set up in a similar way. Let S^n be oriented by a choice of a generator v_n of $H_n(S^n)$. Choose a map k: $(I^n,\dot{I}^n) \to (S^n,y_0)$ such that $k_* u_n = v_n$, and k maps $I^n - \dot{I}^n$ topologically onto $S^n - y_0$. Any f: $(S^n,y_0) \to (X,x_0)$ composes to give fk in $F^n(X,x_0)$. As above the homotopy class of fk depends only on f, so that f *represents* a unique element $c(f)$ in $\pi_n(X,x_0)$. It depends only on the homotopy class of f. If f' in F^n is given, then $f = f'k^{-1}$ is a single-valued continuous map; therefore any element of $\pi_n(X,x_0)$ is a $c(f)$. *Thus $c(f)$ sets up a 1-1 correspondence between $\pi_n(X,x_0)$ and homotopy classes of maps of (S^n,y_0) into (X,x_0).* It is to be emphasized that $c(f)$ depends on the orientation of S^n. A reversal of orientation replaces $c(f)$ by its negative.

Let E be an $(n+1)$-cell whose boundary is S^n, and f: $(S^n,y_0) \to (X,x_0)$. If f is extendable to a map of E into X, then $c(f) = 0$; for S^n is contractible over E into y_0, and the image of this homotopy contracts f into the constant map. Conversely a homotopy of f to the constant map yields an extension of f over E. One maps the center of E into x_0 and each radial line into the path followed by its end point under the homotopy. *Thus, $c(f) = 0$ if and only if f is extendable over E.*

15.12. Direct sum theorems. We derive now three useful consequences of the exactness of homotopy sequences.

Let A be a retract of X, and f: $X \to A$ a retraction. Let i: $A \to X$ and j: $X \to (X,A)$ be inclusion maps, and let $x_0 \in A$. Then

$$\pi_n(X) \approx \pi_n(A) + \pi_n(X,A), \qquad n \geqq 2.$$

Precisely, $\pi_n(X)$ decomposes into the direct sum of two subgroups M and N such that i_* maps $\pi_n(A)$ isomorphically onto M, and j_* maps N isomorphically onto $\pi_n(X,A)$.

Define M to be the image of i_* and N to be the kernel of f_*: $\pi_n(X) \to \pi_n(A)$. Since fi is the identity map, so is $(fi)_* = f_* i_*$. This proves that $\pi_n(X)$ decomposes into the direct sum $M + N$ and that i_* maps $\pi_n(A)$ isomorphically onto M. Since the kernel of i_*: $\pi_{n-1}(A) \to \pi_{n-1}(X)$ is zero, by exactness, so also is the image of ∂: $\pi_n(X,A) \to \pi_{n-1}(A)$. Therefore the kernel of ∂ = image of j_* is the whole of $\pi_n(X,A)$. Since M is the kernel of j_*, it follows that j_* maps N isomorphically onto $\pi_n(X,A)$.

15.13. *Let the identity map of X be homotopic, leaving $x_0 \varepsilon A$ fixed, to a map f: $X \to A$, then*

$$\pi_n(A) \approx \pi_n(X) + \pi_{n+1}(X,A), \qquad n \geqq 2.$$

Precisely, $\pi_n(A)$ decomposes into the direct sum of two subgroups M and N such that i_* maps M isomorphically onto $\pi_n(X)$, and ∂ maps $\pi_{n+1}(X,A)$ isomorphically onto N.

Define M to be the image of f_*: $\pi_n(X) \to \pi_n(A)$. Since if is homotopic to the identity map of X, $(if)_* = i_*f_*$ is the identity. Therefore i_* maps M isomorphically onto $\pi_n(X)$. If N denotes the kernel of i_*, it follows that $\pi_n(A) = M + N$. Since i_* maps $\pi_{n+1}(A)$ onto $\pi_{n+1}(X)$, exactness implies that $\pi_{n+1}(X)$ is the kernel of j_*, and, therefore, zero is the image of j_*. Exactness again implies that ∂ maps $\pi_{n+1}(X,A)$ isomorphically into $\pi_n(A)$. Since N = kernel (i_*) = image (∂), the result is proved.

In the case $n = 1$, the same argument carries through except that M need not be an invariant subgroup. However N is invariant and we have the isomorphisms of M and N with $\pi_1(X)$ and $\pi_2(X,A)$ respectively. Furthermore each element of $\pi_1(A)$ is uniquely expressible as a product of an element of M with an element of N.

15.14. *Let the inclusion map i: $A \to X$ be homotopic in X, leaving x_0 fixed, to the constant map $f(A) = x_0$. Then*

$$\pi_n(X,A) \approx \pi_n(X) + \pi_{n-1}(A), \qquad n \geqq 2.$$

Precisely, $\pi_n(X,A)$ decomposes into the direct sum of two subgroups M and N such that ∂ maps M isomorphically onto $\pi_{n-1}(A)$ and j_* maps $\pi_n(X)$ isomorphically onto N.

Define N to be the image of j_*. The constant map f induces the zero homomorphism of $\pi_p(A)$ into $\pi_p(X)$. Since $i \simeq f$ the same is true of i. Hence kernel (i_*) = image (∂) is $\pi_p(A)$. Thus ∂ is a homomorphism onto for each p. Since image $(i_*) = 0$ so also is the kernel of j_*. Taking $p = n$, we have that j_* maps $\pi_n(X)$ isomorphically onto N. Thus by the use of exactness, we have shown that $\pi_n(X,A)$ is a group extension of $\pi_{n-1}(A)$ by $\pi_n(X)$. To establish a direct sum we must use the definition of the homotopy group.

Let h: $A \times I \to X$ be a homotopy of i into f so that $h(x_0,\tau) = x_0$ for all τ. If g is in $F^{n-1}(A,x_0)$, define

$$(hg)(t_1, \cdots, t_n) = h(g(t_1, \cdots, t_{n-1}),t_n).$$

It follows that hg is in $F^n(X,A,x_0)$. It is easily checked that the operation $g \to hg$ preserves addition and the relation of homotopy. Thus, h induces a homomorphism h_*: $\pi_{n-1}(A) \to \pi_n(X,A)$. Since $\partial hg = g$ for every g, it follows that ∂h_* is the identity map. Define M to be the image of h_*. If $n \geqq 3$, all groups considered are abelian, and the proposition has been proved. When $n = 2$, we must show that M is invariant. If $f \in F^2(X,A,x_0)$, define $kf \in F^2(X,x_0)$ by

$$kf(t_1,t_2) = \begin{cases} h(f(t_1,0),1 - 2t_2), & 0 \leqq t_2 \leqq 1/2, \\ f(t_1,2t_2 - 1), & 1/2 \leqq t_2 \leqq 1. \end{cases}$$

Then k induces a homomorphism $k_*\colon\ \pi_n(X,A) \to \pi_n(X)$, and M is the kernel of k_*.

15.15. Comparison of homotopy and homology groups. At this point it is worthwhile to compare homotopy groups with homology groups. Just as in the case of homotopy groups, there are relative homology groups $H_n(X,A)$. However they do not involve the base point. In addition they are defined for the dimensions 0 and 1, and are abelian in all dimensions. A boundary operator ∂ and induced homomorphisms h_* are also defined for homology groups. Furthermore the system H_n,∂,h_* satisfies the analogs of properties 1° to 6°.

Up to this point homology theory and homotopy theory bear a strong formal resemblance. Homology satisfies a seventh property called *invariance under excision*. If $X = A \cup B$, and f is the inclusion map of $(A,A \cap B)$ in (X,B), then $f_*\colon\ H_n(A,A \cap B) \approx H_n(X,B)$ for every dimension. In our axiomatic approach to homology theory, Eilenberg and the author have shown that all *seven* properties characterize homology theory completely for triangulable spaces. The homotopy groups are not invariant under excisions. A counterexample is provided by the 2-sphere S^2 where A and B are upper and lower hemispheres. We shall see in article 21 that $\pi_3(S^2,B) \approx \pi_3(S^2)$ is an infinite cyclic group, and that $\pi_3(A,A \cap B) \approx \pi_2(S^1)$ is zero.

Their different behaviors under an excision is the chief distinguishing feature of homology and homotopy. The fact that homology groups of triangulable spaces are readily computable while the homotopy groups are not is just a reflection of this difference. An unsolved problem is to determine properties of homotopy groups (additional to the six listed) which are characteristic of homotopy theory.

§16. THE OPERATIONS OF π_1 ON π_n

16.1. The isomorphism of π_n induced by a curve. If A is arcwise connected, we will show that $\pi_n(X,A,x_0)$ is independent of the choice of the base point x_0 in A. Precisely, if $C\colon\ I \to A$ is a curve from x_0 to x_1 in A (A need not be connected), we can assign to C an isomorphism

(1) $$C^\#\colon\ \pi_n(X,A,x_1) \approx \pi_n(X,A,x_0)$$

with the following two properties: If C_1 is a curve from x_0 to x_1, and C_2 is a curve from x_1 to x_2, then

$$(2) \qquad (C_1C_2)^\# = C_1^\# C_2^\#.$$

If C and C' are two curves from x_0 to x_1 and C is homotopic to C' leaving its end points fixed, then $C^\# = C'^\#$.

The idea of the construction is simple. If f represents an element of $\pi_n(X,A,x_1)$, we construct a homotopy of f which (i) moves J^{n-1} along C^{-1} into x_0 keeping the image of J^{n-1} a point at each stage, and (ii) deforms I^{n-1} over A. The final map represents therefore an element of $\pi_n(X,A,x_0)$.

To construct the homotopy h, we set

$$(3) \qquad \begin{aligned} h(t,0) &= f(t), && t \in I^n, \\ h(t,\tau) &= C(1-\tau), && t \in J^{n-1},\ 0 \leqq \tau \leqq 1. \end{aligned}$$

Then h is defined on the subset

$$(4) \qquad K = (I^n \times 0) \cup (J^{n-1} \times I)$$

of $I^n \times I$. The extension of h over $I^n \times I$ is based on a useful lemma which we prove first.

16.2. LEMMA. *If (E,S) is a cell and its boundary, then $(E \times 0) \cup (S \times I)$ is a retract of $E \times I$.*

We let E be the cell of radius 1 with center at the origin in a euclidean n-space contained in an $(n+1)$-space, and let I be the unit interval on the axis orthogonal to E. Let P be the point on this axis at the distance 2 from E (see Fig. 5). The retraction is simply the projection from P of $E \times I$ onto $E \times 0 \cup S \times I$.

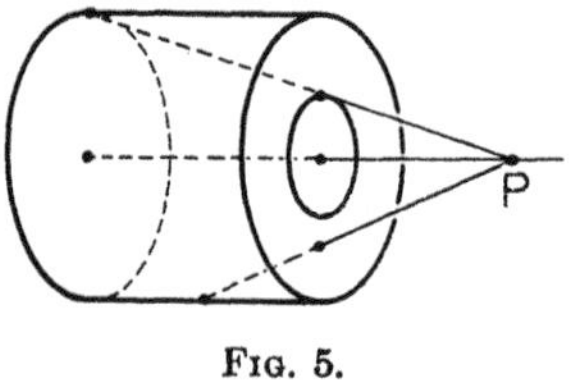

FIG. 5.

Precisely, if Q is in $E \times I$, the ray from P through Q meets $E \times 0 \cup S \times I$ in just one point $r(Q)$. If (t,τ) are the coordinates of Q (where t is a vector in E, and τ a real coordinate in I), then the coordinates (t',τ) of $r(Q)$ are given by

$$\begin{aligned} t' &= t/|t|, & \tau' &= 2 - (2-\tau)/|t| && \text{when } |t| \geqq 1 - \tau/2, \\ t' &= 2t/(2-\tau), & \tau' &= 0 && \text{when } |t| \leqq 1 - \tau/2. \end{aligned}$$

Having proved the lemma, we return to the extension of the homotopy h, defined on K by (3), to all of $I^n \times I$. Let r_1 be a retraction of $I^n \times I$ into $I^n \times 0 \cup \dot{I}^n \times I$. Let r_2 be a retraction of $I^{n-1} \times I$ into $I^{n-1} \times 0 \cup \dot{I}^{n-1} \times I$. Define $r_2(t,\tau) = (t,\tau)$ for $t \varepsilon J^{n-1}$. The so extended r_2 is a retraction of $I^n \times 0 \cup \dot{I}^n \times I$ into K. Let r be the composition r_2r_1 which retracts $I^n \times I$ into K. Then the composition of r followed by $h|K$ is an extension h of $h|K$ to all of $I^n \times I$. Since r retracts $I^{n-1} \times I$ into $I^{n-1} \times 0 \cup \dot{I}^{n-1} \times I$, and $h|K$ maps this set into A, we have that h deforms I^{n-1} over A. Thus h satisfies condition (ii). Condition (i) follows from (3).

Now let C,C' be two curves from x_0 to x_1 which are homotopic in A leaving the end points fixed; and let h,h' be homotopies of f along C,C' respectively satisfying conditions (i) and (ii). Let $g(t) = h(t,1)$ and $g'(t) = h'(t,1)$. We must show that g and g' represent the same element of $\pi_n(X,A,x_0)$. Define

$$h''(t,\tau) = \begin{cases} h(t,1-2\tau) & 0 \leqq \tau \leqq 1/2, \\ h'(t,2\tau-1) & 1/2 \leqq \tau \leqq 1. \end{cases}$$

Then h'' is a homotopy $g \simeq g'$ under which the image of J^{n-1} is a point describing $C^{-1}C'$. Let I' be another unit interval $0 \leqq \tau' \leqq 1$. By hypothesis, there is a homotopy η: $I \times I' \to A$ which shrinks $C^{-1}C'$ to x_0 leaving x_0 fixed. Set

$$k(t,\tau,\tau') = \begin{cases} h''(t,\tau) & \text{when } \tau' = 0, \\ g(t) & \text{when } \tau = 0, \\ g'(t) & \text{when } \tau = 1, \\ \eta(h(t,\tau),\tau') & \text{when } t \varepsilon J^{n-1}. \end{cases}$$

Then k is defined on the subset

$$L = (I^n \times I \times 0) \cup [(I^n \times 0 \cup I^n \times 1 \cup J^{n-1} \times I) \times I']$$

of $I^n \times I \times I'$. Lemma 16.2 gives a retraction

$$r_1\colon\ I^n \times I \times I' \to (I^n \times I \times 0) \cup [(I^n \times I)^{\cdot} \times I'].$$

It also gives a retraction

$$r_2\colon\ I^{n-1} \times I \times I' \to (I^{n-1} \times I \times 0) \cup [(I^{n-1} \times I)^{\cdot} \times I'].$$

Extend the latter over $L \cup (I^{n-1} \times I \times I')$ by setting $r_2 =$ the identity on L. Then r_2r_1 is a retraction of $I^n \times I \times I'$ into L. We extend $k|L$ to a map k of $I^n \times I \times I'$ by composing r_2r_1 with $k|L$. If we now set $\tau' = 1$ in k, we obtain a homotopy of g into g' lying in $F^n(X,A,x_0)$.

If we specialize by setting $C = C'$, it follows that the homotopy class in $\pi_n(X,A,x_0)$ obtained by deforming f along C^{-1} does not depend on the choice of the deformation. In particular, if we first subject f to a homotopy in $F^n(X,A,x_1)$, and then deform along C^{-1}, the composed homotopy is a deformation of f along C^{-1}. It follows that homotopic maps in $F^n(X,A,x_1)$ deform along C^{-1} into homotopic maps in $F^n(X,A,x_0)$. Thus, deformation along C^{-1} is a class operation $C^\#$ of $\pi_n(X,A,x_1)$ into $\pi_n(X,A,x_0)$. The argument of the preceding paragraph shows that $C^\#$ depends only on the homotopy class of C.

If $f_1,f_2 \varepsilon F^n(X,A,x_1)$ are deformed along C^{-1} into f_1',f_2' so that the point image of J^{n-1} moves at the same rate for both deformations, then we may add the functions obtained at each stage of the homotopy. This clearly provides a deformation of $f_1 + f_2$ along C^{-1} into $f_1' + f_2'$. It follows that $C^\#$ is a homomorphism.

Let C be a curve from x_0 to x_1 in A, and C' a curve from x_1 to x_2 in A. If $f \varepsilon F^n(X,A,x_2)$, and we deform f along C'^{-1} into f' and then deform f' along C^{-1} into f'', the composition of the deformations is a deformation of f along $C'^{-1}C^{-1}$ into f''. This proves (2) above.

If, in (2), we set $C_2 = C_1^{-1}$, then C_1C_2 is homotopic to the constant path C_0. Hence $C_1^\# C_2^\# = C_0^\# =$ identity. It follows that $C^\#$ is always an isomorphism.

We have thus proved the initial statements of §16.1.

16.3. Special case of the absolute homotopy groups. In the special case $A = x_0$, the operation $C^\#$ becomes trivial. However there is a similar operation which when applied to a curve C from x_0 to x_1 in X yields an isomorphism

$$C^\#: \quad \pi_n(X,x_1) \approx \pi_n(X,x_0) \tag{5}$$

which depends only on the homotopy class of C, and satisfies (2) for curves C,C' in X.

It is not necessary to repeat the entire construction for this case. We need only observe that the preceding construction restricted to the principal face I^{n-1} of I^n yields an isomorphism $C^\#$: $\pi_{n-1}(A,x_1) \approx \pi_{n-1}(A,x_0)$ with the required two properties. If we replace A by X, and $n - 1$ by n, the desired results follow.

If the path C lies in A, then it induces isomorphisms of the homotopy groups of X, A and (X,A) based at x_1 into the same at x_0. This leads to the diagram

$$\begin{array}{ccccccc}
\pi_n(A,x_1) & \xrightarrow{i_*} & \pi_n(X,x_1) & \xrightarrow{j_*} & \pi_n(X,A,x_1) & \xrightarrow{\partial} & \pi_{n-1}(A,x_1) \\
\downarrow C^\# & & \downarrow C^\# & & \downarrow C^\# & & \downarrow C^\# \\
\pi_n(A,x_0) & \xrightarrow{i_*} & \pi_n(X,x_0) & \xrightarrow{j_*} & \pi_n(X,A,x_0) & \xrightarrow{\partial} & \pi_{n-1}(A,x_0)
\end{array}$$

It is easy to show that commutativity holds in each square of the diagram. This means that $C^\#$ is an isomorphism of the homotopy sequence of (X,A,x_1) onto that of (X,A,x_0).

16.4. Automorphisms induced by closed curves. A path C from x_0 to x_0 in A induces an automorphism $C^\#$ of the homotopy sequence of (X,A,x_0). Since it depends only on the homotopy class of C, and since (2) holds, it follows that *the operations* $C^\#$ *represent* $\pi_1(A,x_0)$ *as a group of automorphisms of the homotopy sequence of* (X,A,x_0). In the same way $\pi_1(X,x_0)$ *is a group of automorphisms of* $\pi_n(X,x_0)$.

It is customary to use the multiplicative notation for the fundamental group. The effect of the operation of γ in $\pi_1(A,x_0)$ on α in $\pi_n(X,A,x_0)$, $\pi_n(X,x_0)$ or $\pi_n(A,x_0)$ is written as $\gamma(\alpha)$. Then

$$(\gamma_1\gamma_2)(\alpha) = \gamma_1(\gamma_2(\alpha)), \qquad \gamma(\alpha + \beta) = \gamma(\alpha) + \gamma(\beta).$$

The results of the preceding section imply that the operation γ commutes with all the homomorphisms of the homotopy sequence:

$$\begin{aligned} \partial\gamma(\alpha) &= \gamma(\partial\alpha), & \alpha &\in \pi_n(X,A,x_0), \\ j_*\gamma(\alpha) &= \gamma(j_*\alpha), & \alpha &\in \pi_n(X,x_0), \\ i_*\gamma(\alpha) &= \gamma(i_*\alpha), & \alpha &\in \pi_n(A,x_0). \end{aligned} \tag{6}$$

In particular, γ operates on $\pi_1(A,x_0)$. Reference to the definition of $C^\#$ shows that

$$\begin{aligned} \gamma(\alpha) &= \gamma\alpha\gamma^{-1}, & \alpha &\in \pi_1(A,x_0), \\ \gamma(\alpha) &= (i_*\gamma)\alpha(i_*\gamma)^{-1}, & \alpha &\in \pi_1(X,x_0). \end{aligned} \tag{7}$$

16.5. n-simplicity. The space X is said to be *n-simple* if it is arcwise connected, and, for any two points x_1,x_2 and curves C_1,C_2 from x_1 to x_2, the isomorphisms $C_1^\#,C_2^\#$ of $\pi_n(X,x_2)$ onto $\pi_n(X,x_1)$ coincide, i.e. the isomorphism is independent of the path.

Clearly, if X is n-simple, then, for each x_0, $\pi_1(X,x_0)$ operates trivially on $\pi_n(X,x_0)$. Conversely, suppose, for some x_0, that $\pi_1(X,x_0)$ operates trivially on $\pi_n(X,x_0)$, and suppose X is arcwise connected. Let x_1,x_2,C_1,C_2 be points and paths as above. Let C be a path from x_0 to x_1. Then the closed path $CC_1C_2^{-1}C^{-1}$ operates trivially on $\pi_n(X,x_0)$. Hence $C_1C_2^{-1}$ operates trivially on $\pi_n(X,x_1)$. Therefore $C_1^\# = C_2^\#$, and X is n-simple.

It follows from the preceding result that an arcwise connected space X is 1-simple if and only if $\pi_1(X)$ is abelian.

Another corollary is that, if $\pi_1(X) = 0$, then X is n-simple for every n.

A useful feature of an n-simple space X is that *a map f of an oriented n-sphere S^n in X determines a unique element of $\pi_n(X,x_0)$ for any x_0.*

Choose a reference point y_1 in S^n and let $x_1 = f(y_1)$. Then f determines an element of $\pi_n(X,x_1)$, and this in turn an element of $\pi_n(X,x_0)$. We must show that the resulting element is independent of the choice of y_1. Let y_2 be a second choice. There is a rotation of S^n carrying y_1 into y_2. The image of this rotation under f is a homotopy of f which moves y_1 along a curve C in X from x_1 to x_2. Then the elements of $\pi_1(X,x_1)$ and $\pi_1(X,x_2)$ determined by the choices y_1 and y_2 are equivalent under $C^{\#}$.

16.6. The homotopy groups of a topological group. The homotopy groups of a topological group G have special properties. Briefly stated, the fundamental group is abelian and operates trivially on the higher homotopy groups. If e is the identity element, and G_e the arcwise connected component of e, then G/G_e operates on the homotopy groups of G_e. We define $\pi_0(G) = G/G_e$. In this way we again have a lowest dimensional homotopy group which has non-trivial operations on the higher homotopy groups. The details follow.

16.7. LEMMA. *If f_1 and f_2 are in $F^n(G,e)$, then $f_1 + f_2$ is homotopic (in F^n) to $f_1 \cdot f_2$ where $(f_1 \cdot f_2)(t) = f_1(t) \cdot f_2(t)$.*

Let f_0 in F^n be the constant map. Then

$$f_1 + f_0 \simeq f_1, \qquad f_0 + f_2 \simeq f_2 \qquad \text{(see §15.2).}$$

By multiplying two such homotopies, one obtains a homotopy

$$(f_1 + f_0) \cdot (f_0 + f_2) \simeq f_1 \cdot f_2.$$

Since $f_0(t) = e$ for every t, reference to the definition of addition, §15.2 (2), shows that

$$f_1 + f_2 = (f_1 + f_0) \cdot (f_0 + f_2),$$

and the lemma is proved.

16.8. LEMMA. *Let C be a curve in G from g_0 to e. The isomorphism*

$$C^{\#}: \quad \pi_n(G,e) \approx \pi_n(G,g_0)$$

coincides with the isomorphism induced by either the left or the right translation of G by g_0.

If $f \varepsilon F^n(G,e)$, it is clear that $h(t,\tau) = C(1 - \tau) \cdot f(t)$ is a homotopy of f in G which moves the point image of $\dot{I}^n$ along C^{-1}. Putting $\tau = 1$ gives $g_0 \cdot f(t)$ in $F^n(G,g_0)$ as the result of deforming f along C^{-1}. Therefore $C^{\#}$ is equivalent to left translation by g_0. Right translation is handled similarly.

16.9. THEOREM. *For any base point g_0 in G, $\pi_1(G,g_0)$ is abelian and operates trivially on $\pi_n(G,g_0)$, i.e. G is n-simple for every n.*

Consider first the case $g_0 = e$. If C is a closed path based at e, then, by §16.8, $C^\#$ must be the identity automorphism of $\pi_n(G,e)$. Therefore π_1 operates trivially on π_n. Since this holds also for $n = 1$, and π_1 operates on itself by inner automorphisms (see §16.4, (7)), it follows that π_1 is abelian. For any other base point g_0, right translation by g_0 maps (G,e) homeomorphically onto (G,g_0) and thereby induces isomorphisms of the homotopy groups and in such a way as to preserve the operations of π_1.

16.10. Automorphisms induced by inner automorphisms. Let G_e denote the set of all elements of G which can be joined to e by a curve in G. It is easily proved that G_e is a subgroup of G and it is invariant. Define

$$\pi_0(G,e) = G/G_e.$$

Any element of G operates on G as an inner automorphism, and e remains fixed. It thereby induces an automorphism of $\pi_n(G,e)$, and G is represented as a group of automorphisms of π_n. If g_0,g_1 in G are joined by a curve $g_\tau (0 \leqq \tau \leqq 1)$ in G, then $h(g,\tau) = g_\tau g g_\tau^{-1}$ is a homotopy of the inner automorphism corresponding to g_0 into that corresponding to g_1, and e remains fixed. Therefore g_0 and g_1 induce the same automorphism of π_n. In particular each element of G_e operates trivially on π_n. Therefore:

G/G_e is a group of operators on $\pi_n(G,e)$.

These operations are generally non-trivial. As an example, let G be the group of all rotations and reflections of the circle. Then G has two components and G_e consists of all rotations. If g is a reflection, it is easy to see that conjugation of G_e by g is a reflection of the circle G_e. But $\pi_1(G_e)$ is infinite cyclic, and reflection carries each element of the group into its inverse. Therefore g operates in a non-trivial fashion.

16.11. THEOREM. *If B is a Lie group, and G is a closed connected subgroup, then B/G is n-simple for every n.*

Let p: $B \to B/G$ be the natural map, and let $x_0 = p(G)$. It suffices to prove that $\pi_1(B/G,x_0)$ operates trivially on $\pi_n(B/G,x_0)$; for B operates transitively on B/G. Let $f \in F^n(B/G,x_0)$, and let $C(\tau)$ $(0 \leqq \tau \leqq 1)$ be a closed curve in B/G based at x_0. If we regard $C(\tau)$ as a homotopy of x_0, a covering homotopy yields a curve $C'(\tau)$ such that $C'(0) = e$ and $pC'(\tau) = C(\tau)$. Then $C'(1)$ is in G. If we adjoin a curve in G from $C'(1)$ to e, we obtain a closed curve D' such that $pD' = D$ is homotopic to C. Now $h(t,\tau) = D'(1 - \tau)\cdot f(t)$ is a homotopy of f around D^{-1} back into f. Hence $D^\#$ operates trivially on the element of π_n represented by f. As f is arbitrary, the theorem is proved.

§17. THE HOMOTOPY SEQUENCE OF A BUNDLE

17.1. Fundamental theorem. *Let* $\mathfrak{B}$ *be a bundle over* X, $A \subset X$, $B_0 = p^{-1}(A)$, $y_0 \varepsilon B_0$, *and* $x_0 = p(y_0)$. *Then*

$$p_*: \quad \pi_n(B,B_0,y_0) \approx \pi_n(X,A,x_0), \qquad n \geqq 2.$$

Suppose $p_*(\alpha) = 0$, and f in $F^n(B,B_0,y_0)$ represents α. Then there is a homotopy h in $F^n(X,A,x_0)$ of pf into the constant map. By 11.7, there is a covering homotopy h' of f which is stationary with h. Since $h(J^{n-1} \times I) = x_0$, it follows that $h'(J^{n-1} \times I) = y_0$. Since $h(I^{n-1} \times I)$ is in A, $h'(I^{n-1} \times I)$ is in B_0. Thus h' is a homotopy in $F^n(B,B_0,x_0)$ of f into a map f': $(I^n,I^{n-1},J^{n-1}) \to (Y_0,Y_0,y_0)$ where Y_0 is the fibre over x_0. Define

(1) $$k(t_1, \cdots, t_n,\tau) = (t_1, \cdots, t_{n-1},(1-\tau)t_n + \tau).$$

Then k is a homotopy of I^n over itself into the face $t_n = 1$. This face lies in J^{n-1} and J^{n-1} is also deformed over itself. Let $k'(t,\tau) = f'(k(t,\tau))$. Then k' is a homotopy in $F^n(B,B_0,y_0)$ of f' into the constant map. It follows that $\alpha = 0$. Therefore the kernel of p_* is zero.

Now let $\beta \varepsilon \pi_n(X,A,x_0)$ and f a representative of β. Set $h(t,\tau) = f(k(t,\tau))$ where k is defined by (1). Then h is a homotopy of f into the constant map (it is not generally a homotopy in $F^n(X,A,x_0)$). Let f' be the map of I^n into y_0. Then $pf'(t) = h(t,1)$. There exists then a homotopy h' covering h which is stationary with h, and $h'(t,1) = f'(t)$ (naturally, one applies §11.7 to the reverse homotopy as t varies from 1 to 0). Let $f''(t) = h'(t,0)$. Then $pf'' = f$. Since h leaves J^{n-1} at x_0, h' leaves J^{n-1} at y_0. Therefore f'' is in $F^n(B,B_0,y_0)$, and represents an element α in $\pi_n(B,B_0,y_0)$ such that $p_*\alpha = \beta$. Thus p_* is an isomorphism.

17.2. COROLLARY. $p_*: \quad \pi_n(B,Y_0,y_0) \approx \pi_n(X,x_0)$, $n \geqq 2$.

REMARK. The property §17.1 of homotopy groups is not enjoyed by homology groups. If one is seeking an axiomatic characterization of homotopy groups, then §17.1 is a candidate to replace the excision property (see §15.15).

17.3. Definition of homotopy sequence of a bundle. Let $\mathfrak{B} = \{B,p,X,Y,G\}$ be a bundle, Y_0 the fibre over x_0 in X, and $y_0 \varepsilon Y_0$. Let i: $Y_0 \to B$ and j: $B \to (B,Y_0)$ be inclusion maps. Then *the homotopy sequence of* (B,Y_0,y_0) is

$$\cdots \to \pi_n(Y_0) \xrightarrow{i_*} \pi_n(B) \xrightarrow{j_*} \pi_n(B,Y_0) \xrightarrow{\partial} \pi_{n-1}(Y_0) \to \cdots$$

Let p_1 denote p regarded as a map $(B,Y_0,y_0) \to (X,x_0,x_0)$. Then p_1j is just the map p: $(B,y_0) \to (X,x_0)$. By §17.2, we can define

(2) $$\Delta = \partial(p_{1*})^{-1}: \quad \pi_n(X,x_0) \to \pi_{n-1}(Y_0,y_0).$$

The sequence of groups and homomorphisms

$$
(3)\qquad
\begin{array}{l}
\cdots \to \pi_n(Y_0) \xrightarrow{i_*} \pi_n(B) \xrightarrow{p_*} \pi_n(X) \xrightarrow{\Delta} \pi_{n-1}(Y_0) \to \cdots \\
\cdots \to \pi_2(X) \xrightarrow{\Delta} \pi_1(Y_0) \xrightarrow{i_*} \pi_1(B) \xrightarrow{p_*} \pi_1'(X)
\end{array}
$$

is called *the homotopy sequence of the bundle* $\mathfrak{B}$ *based at* y_0.

17.4. Basic properties of the homotopy sequence.

THEOREM. *The homotopy sequence of a bundle is exact.*

The homotopy sequence of (B,Y_0,y_0) is exact (§15.6). In forming the homotopy sequence of $\mathfrak{B}$ we have replaced the terms $\pi_n(B,Y_0)$ in the homotopy sequence of (B,Y_0,y_0) by the isomorphic groups $\pi_n(X)$, and adjoined the new term $\pi_1'(X)$. The replacements do not affect exactness. It remains to prove exactness at $\pi_1(B)$. Clearly p_*i_* is trivial since p maps a path in Y_0 into the point x_0. Suppose C is a closed path in B based at y_0 and pC is contractible to x_0 leaving its end points fixed. A covering homotopy will contract C into a curve lying in Y_0 and its end points will remain fixed. This proves exactness at $\pi_1(B)$.

17.5. Let h be a map of $\mathfrak{B}$ into $\mathfrak{B}'$. Let $\bar{h}$: $X \to X'$ be the induced map of the base space. Let $x_0 \in X$, $x_0' = \bar{h}(x_0)$, let Y_0, Y_0' be the fibres over x_0, x_0' respectively, let $y_0 \in Y_0$ and $y_0' = h(y_0)$. Finally let h_0: $Y_0 \to Y_0'$ be $h|Y_0$. We obtain then a *homomorphism* of the homotopy sequence of $\mathfrak{B}$ at y_0 into that of $\mathfrak{B}'$ at y_0':

$$
(4)\qquad
\begin{array}{ccccccc}
\pi_n(Y_0) & \xrightarrow{i_*} & \pi_n(B) & \xrightarrow{p_*} & \pi_n(X) & \xrightarrow{\Delta} & \pi_{n-1}(Y_0) \\
\downarrow h_{0*} & & \downarrow h_* & & \downarrow \bar{h}_* & & \downarrow h_{0*} \\
\pi_n(Y') & \xrightarrow{i_*'} & \pi_n(B') & \xrightarrow{p_*'} & \pi_n(X') & \xrightarrow{\Delta'} & \pi_{n-1}(Y_0')
\end{array}
$$

Commutativity in the middle and left squares follows from the commutativity of the maps: $hi = i'h_0$ and $\bar{h}p = p'h$. Expand the right square according to the definition of Δ:

$$
(5)\qquad
\begin{array}{ccccc}
\pi_n(X) & \xleftarrow{p_{1*}} & \pi_n(B,Y_0) & \xrightarrow{\partial} & \pi_{n-1}(Y_0) \\
\downarrow \bar{h}_* & & \downarrow h_{1*} & & \downarrow h_{0*} \\
\pi_n(X') & \xleftarrow{p_{1*}'} & \pi_n(B',Y_0') & \xrightarrow{\partial'} & \pi_{n-1}(Y_0')
\end{array}
$$

Commutativity on the left follows from $\bar{h}p_1 = p_1'h_1$, and, on the right, from §15.6 property 3°. Since p_{1*}, p_{1*}' are isomorphisms, commutativity follows for the right-hand square of (4).

17.6. Covering space theorem. *If p: $B \to X$ is a covering (see 14.1), $b_0 \varepsilon B$ and $x_0 = p(b_0)$, then*

$$p_*: \quad \pi_n(B,b_0) \approx \pi_n(X,x_0), \qquad n \geqq 2,$$

and p_* maps $\pi_1(B,b_0)$ isomorphically into $\pi_1(X,x_0)$.

According to §14.3, the covering admits a bundle structure $\mathfrak{B}$ with a *discrete* fibre Y_0. Then $\pi_n(Y_0) = 0$ for all n. Thus every third term of the homotopy sequence of $\mathfrak{B}$ is zero. As remarked in §15.8, exactness implies that the remaining adjacent pairs must be isomorphic. This is the desired result if $n > 1$. The case $n = 1$ was proved in §14.2. It is also a trivial consequence of $\pi_1(Y_0) = 0$ and exactness at $\pi_1(B)$.

17.7. Direct sum theorems.

THEOREM. *If the bundle $\mathfrak{B}$ admits a cross-section, then we have the direct sum relation*

$$\pi_n(B) \approx \pi_n(X) + \pi_n(Y), \qquad n \geqq 2,$$

and $\pi_1(B)$ contains two subgroups M and N such that M is invariant and isomorphic to $\pi_1(Y)$, p_* maps N isomorphically onto $\pi_1(X)$ and each element of $\pi_1(B)$ is uniquely representable as the product of an element of M with an element of N.

This theorem should be compared with §15.12. Their proofs are similar. Referring to (3), let f be a cross-section, and

$$M = \text{image } i_* = \text{kernel } p_*$$
$$N = \text{image } f_*: \quad \pi_n(X) \to \pi_n(B).$$

Since pf = identity, p_* maps N isomorphically onto $\pi_n(X)$. Since M is the kernel of p_*, it follows that $\pi_n(B) = M + N$ (except in the case $n = 1$ when N may not be invariant). Since p_* is onto, exactness requires that the image of Δ is zero. Therefore the kernel of i_* is zero. Hence i_* maps $\pi_n(Y_0)$ isomorphically onto M; and the proof is complete.

An example where the exceptional behavior for $n = 1$ actually occurs is provided by the Klein bottle as a bundle (§1.4). In this case Y and X are circles and $\mathfrak{B}$ admits a cross-section. If $\pi_1(B)$ were a direct product of $\pi_1(X)$ and $\pi_1(Y)$ which are infinite cyclic, then it would be an abelian group. But this is not the case, it is a group on two generators a,b with the sole relation $ab = b^{-1}a$.

The importance of §17.7 is that it provides a strong necessary condition for the existence of a cross-section. If the direct sum relation fails to hold in some dimension, no cross-section exists.

17.8. COROLLARY. $\pi_n(X \times Y) \approx \pi_n(X) + \pi_n(Y)$, $n \geqq 1$.

This follows since the product space is a bundle and admits a cross-section. In the case $n = 1$, we have a direct product representation since N is the kernel of the projection $X \times Y \to Y$ and is therefore invariant.

17.9. THEOREM. *If, in the bundle* $\mathcal{B}$, *the fibre* Y_0 *is a retract of* B, *then the conclusions of* §17.7 *hold.*

Application of §15.12 gives

$$\pi_n(B) \approx \pi_n(Y_0) + \pi_n(B, Y_0),$$

and the result follows from §17.2.

17.10. THEOREM. *If* $\mathcal{B}$ *is a bundle, and the fibre* Y_0 *over* x_0 *is contractible in* B *to the point* y_0 *in* Y_0 *leaving* y_0 *fixed, then*

$$\pi_n(X) \approx \pi_{n-1}(Y) + \pi_n(B), \qquad n \geqq 2.$$

We may apply §15.14 to obtain

$$\pi_n(B, Y_0) \approx \pi_{n-1}(Y_0) + \pi_n(B).$$

The result follows from §17.2.

The analog of §15.13 for bundles is left to the reader.

17.11. The homotopy sequence of a principal bundle. In the case of a principal bundle $\mathcal{B}$, we can extend the homotopy sequence by an extra term in a significant way. Let G_0 be the fibre over x_0 and y_0 the base point in G_0. There is a unique admissible map ξ: $G \to G_0$ such that $\xi(e) = y_0$ (e = identity). Using ξ we define a multiplication in G_0 so that it is a group having y_0 for the identity element, and ξ is an isomorphism. Define $\pi_0(G_0)$ to be the factor group of G_0 by the invariant subgroup of elements which can be joined to y_0 by curves in G_0 (see §16.10 for the definition of $\pi_0(G)$). Each element of $\pi_0(G_0)$ is an *arc-component* of G_0 (i.e. two points belong to the same arc-component if they can be joined by a curve). The map ξ carries arc-components into such and thereby induces an isomorphism

$$\xi_*: \quad \pi_0(G) \approx \pi_0(G_0). \tag{6}$$

Corresponding to ξ we have the homomorphism

$$\chi: \quad \pi_1(X, x_0) \to G/G_e = \pi_0(G) \tag{7}$$

(see §13.10). Define

$$\Delta = \xi_*\chi: \quad \pi_1(X, x_0) \to \pi_0(G_0, y_0)$$

We extend the homotopy sequence of the principal bundle so that it terminates in

$$(8) \qquad \cdots \xrightarrow{i_*} \pi_1(B,y_0) \xrightarrow{p_*} \pi_1(X,x_0) \xrightarrow{\Delta} \pi_0(G_0,y_0).$$

For convenient use of Δ, we derive an alternative definition. Let η: $G \to G/G_e$ be the natural homomorphism. Let $\mathcal{B}'$ be the bundle associated with $\mathcal{B}$ having G/G_e as fibre. By §9.6, we have an associated map

$$\bar{\eta}: \quad B \to B'$$

with $p'\bar{\eta} = p$. It is easily checked that $\xi' = \bar{\eta}\xi\eta^{-1}$ is a single-valued admissible map of G/G_e onto the fibre G_0' over x_0 in B'. We may assume that ξ' was used in defining χ within its automorphism class (see §13.5).

Now let C be a closed curve representing α in $\pi_1(X,x_0)$. Let h: $I \times G \to \mathcal{B}$ be a translation of G_0 around the curve C (see §13.1). We can suppose that $h_1 = \xi$. Then $h' = \bar{\eta}h\eta^{-1}$ is a translation of G_0' around C, and $h_1' = \xi'$. As shown in the proof of §13.7, $\chi(C) = h_1'^{-1}h_0'(e')$. Then

$$\Delta\alpha = \xi_*\chi(C) = \xi'h_1'^{-1}h_0'(e') = h_0'(e') = \bar{\eta}h_0(e).$$

The curve $D(t) = h(t,e)$ covers C, ends at y_0, and begins at $h_0(e)$. If D_1,D_2 are two curves in B which cover C and end at y_0, then $\bar{\eta}D_1$ and $\bar{\eta}D_2$ must coincide due to the uniqueness in B' of covering curves. Therefore the initial points of D_1,D_2 lie in the same arc-component of G_0. We have therefore the desired alternative definition of Δ:

If C represents α in $\pi_1(X,x_0)$, and the curve D in B covers C and ends at y_0, then $\Delta\alpha$ is the arc-component of G_0 which contains the initial point of D.

This description is strictly analogous to the definition of Δ for the higher dimensional cases.

We can now prove the exactness of the augmented homotopy sequence. Let D be a closed curve in B representing α in $\pi_1(B,y_0)$. Then pD represents $p_*\alpha$ in $\pi_1(X,x_0)$. By the above result, the initial point y_0 of D belongs to $\Delta p_*\alpha$. Hence $\Delta p_*\alpha = e'$. Conversely, suppose C represents α in $\pi_1(X)$ and $\Delta\alpha = e'$. Then C is covered by a curve D which ends at y_0 and begins in the arc-component of y_0. Let E be a curve in G_0 from y_0 to the initial point of D. Then ED represents some β in $\pi_1(B)$. Since pED is homotopic to C, it follows that $p_*\beta = \alpha$. This proves exactness of the augmented sequence at $\pi_1(X)$.

17.12. Characteristic homomorphisms. The preceding section exhibits a relation between the characteristic class χ and the 1-dimensional operator Δ. The extension of χ to all dimensions is now obvious. Let $\mathcal{B}$ be a principal bundle, G_0 the fibre over x_0, and ξ: $G \to G_0$ an admissible map. Let $y_0 = \xi(e)$. Let Δ be the boundary operator of the homotopy sequence of $\mathcal{B}$ based at y_0. Define *the characteristic homomorphism*

$$\chi\colon\ \pi_n(X,x_0) \to \pi_{n-1}(G,e), \qquad n \geqq 1 \tag{9}$$

to be the composition

$$\chi = \xi_*^{-1}\Delta. \tag{10}$$

We have seen (§16.10) that $G/G_e = \pi_0(G)$ operates on $\pi_{n-1}(G,e)$ through inner automorphisms of G.

LEMMA. *An alteration of the choice of ξ alters χ simultaneously in all dimensions by an operation of an element of $\pi_0(G)$ on $\{\pi_{n-1}(G,e)\}$. Conversely any such alteration of χ can be achieved by a change in the choice of ξ.*

Suppose $\zeta\colon\ G \to G_0$ is also admissible and $y_1 = \zeta(e)$. Let $g_0 = \xi^{-1}(y_1)$. Let T denote the right translation of B corresponding to g_0 (see §8.12). It follows quickly that T maps y_0 into y_1. It induces an isomorphism of the homotopy sequence based at y_0 onto that based at y_1. Let T' be the inner automorphism of G determined by g_0. We have then the diagram:

$$\begin{array}{ccccc}
 & & & \xi_*^{-1} & \\
 & & \pi_{n-1}(G_0,y_0) & \longrightarrow & \pi_{n-1}(G,e) \\
 & \Delta\nearrow & & & \\
\pi_n(X) & & \downarrow T_* & & \downarrow T'_* \\
 & \Delta\searrow & & \zeta_*^{-1} & \\
 & & \pi_{n-1}(G_0,y_1) & \longrightarrow & \pi_{n-1}(G,e)
\end{array}$$

Commutativity holds in the triangle since T maps each fibre on itself. Commutativity in the square follows from $T\xi = \zeta T'$. Therefore

$$T'_*\xi_*^{-1}\Delta = \zeta_*^{-1}\Delta.$$

Since T'_* is the operation determined by $\eta(g_0)$, the first part of the lemma is proved. The second half follows quickly. If g_0 and ξ are given, define $\zeta(g) = \xi(g_0g)$. Then χ will be altered by the operation of $\eta(g_0)$.

The sequence of homomorphisms (9) we call *characteristic homomorphisms*. The equivalence class of this sequence under the operations of $\pi_0(G)$ is called *the characteristic class* and is denoted by $\chi(\mathcal{B})$.

If $\mathcal{B},\mathcal{B}'$ are equivalent principal bundles, then $\chi(\mathcal{B}) = \chi(\mathcal{B}')$.

Let $h\colon\ \mathcal{B} \to \mathcal{B}'$ be an equivalence. Then h induces an isomorphism of the homotopy sequence of $\mathcal{B}$ at y_0 into that of $\mathcal{B}'$ at $y'_0 = h(y_0)$. As in (4) of §17.5,

$$h_{0*}\Delta = \Delta'\bar{h}_* = \Delta'$$

since $\bar{h}$ is the identity. Then

$$\xi_*'^{-1}\Delta' = \xi_*'^{-1}h_{0*}\Delta = \xi_*^{-1}\Delta,$$

and the result is proved.

We define the *characteristic class* of any bundle $\mathfrak{B}$ by $\chi(\mathfrak{B}) = \chi(\tilde{\mathfrak{B}})$ where $\tilde{\mathfrak{B}}$ is the associated principal bundle. It follows immediately that, *if $\mathfrak{B}$ and $\mathfrak{B}'$ are associated bundles, then they have the same characteristic class.*

17.13. LEMMA. *Let $\tilde{\mathfrak{B}}$ be the principal bundle of $\mathfrak{B}$, and let h: $(\tilde{B},G_0,g_0) \to (B,Y_0,y_0)$ be the principal map defined by a point of Y (see §8.12). Then the kernel of i_*: $\pi_n(Y_0) \to \pi_n(B)$ is contained in the image of h'_*: $\pi_n(G_0) \to \pi_n(Y_0)$ where $h' = h|G_0$.*

Consider the diagram

$$\begin{array}{ccccccc}
 & & & \tilde{\partial} & & & \\
 & & \pi_{n+1}(\tilde{B},G_0) & \longrightarrow & \pi_n(G_0) & & \\
 & \tilde{p}_* \swarrow & & & & & \\
\pi_{n+1}(X) & & \downarrow h_* & & \downarrow h'_* & & \\
 & p_* \nwarrow & & \partial & & i_* & \\
 & & \pi_{n+1}(B,Y_0) & \longrightarrow & \pi_n(Y_0) & \longrightarrow & \pi_n(B)
\end{array}$$

Since $ph = \tilde{p}$, and $p_*,\tilde{p}_*$ are isomorphisms, so is h_*. Therefore:

$$\begin{aligned}
\text{kernel } i_* &= \text{image } \partial = \text{image } \partial h_* \\
&= \text{image } h'_*\tilde{\partial} \subset \text{image } h'_*.
\end{aligned}$$

§18. THE CLASSIFICATION OF BUNDLES OVER THE n-SPHERE

18.1. Normal form of a bundle over S^n. In case the base space is an n-sphere S^n, we will show that the characteristic homomorphism χ: $\pi_n(S^n) \to \pi_{n-1}(G)$ of a bundle is indeed characteristic of the bundle. In essence, this reduces the classification problem for the pair S^n,G to the computation of $\pi_{n-1}(G)$. These results are due to Feldbau [32].

Let S^{n-1} be a great $(n-1)$-sphere on S^n and let E_1,E_2 be the closed hemispheres of S^n determined by S^{n-1}. For $i = 1,2$, let V_i be an open n-cell on S^n containing E_i and bounded by an $(n-1)$-sphere parallel to S^{n-1}. Then V_1,V_2 cover S^n and $V_1 \cap V_2$ is an equatorial band containing S^{n-1}. Let x_0 be a reference point on S^{n-1}. A coordinate bundle $\mathfrak{B}$ over S^n is said to be in *normal form* if its coordinate neighborhoods are V_1,V_2, and $g_{12}(x_0) = e$.

Any bundle $\mathfrak{B}$ over S^n is strictly equivalent to a bundle in normal form. This is proved as follows. Since V_i is a cell, anv bundle over V_i is equivalent to a product bundle (§11.6). This is true of the portion $\mathfrak{B}_i$ of $\mathfrak{B}$ over V_i. Hence there exist bundle maps

$$\phi'_i\colon \quad V_i \times Y \to \mathfrak{B}_i, \qquad i = 1,2.$$

Then ϕ'_1,ϕ'_2 are coordinate functions of a bundle $\mathcal{B}'$ strictly equivalent to $\mathcal{B}$. If $g'_{12}(x_0) = a$, we alter $\mathcal{B}'$ to a strictly equivalent bundle by setting $\lambda_1(x) = e$ for $x \in V_1$, and $\lambda_2(x) = a$ for $x \in V_2$, and applying §2.10. The resulting bundle is in normal form.

If $\mathcal{B}$ is a bundle in normal form, the map

$$T = g_{12}|S^{n-1}$$

which maps S^{n-1} into G is called *the characteristic map of* $\mathcal{B}$. It is to be noted that $T(x_0) = e$, and we regard T as a map $(S^{n-1},x_0) \to (G,e)$, and shall allow only such homotopies of T which keep the image of x_0 at e.

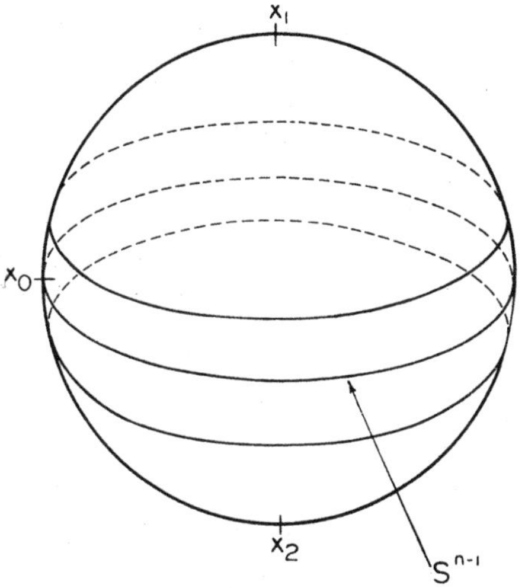

FIG. 6.

18.2. LEMMA. *Any map T: $(S^{n-1},x_0) \to (G,e)$ is the characteristic map of some bundle over S^n in normal form.*

Let r: $V_1 \cap V_2 \to S^{n-1}$ be the retraction which maps x into the intersection with S^{n-1} of the great circle through x orthogonal to S^{n-1}. Define $g_{12}(x) = T(r(x))$. Setting $g_{11} = g_{22} = e$, and $g_{21} = g_{12}^{-1}$, the desired bundle is provided by §3.2.

18.3. Equivalence theorem. *Let $\mathcal{B},\mathcal{B}'$ be bundles over S^n in normal form and having the same fibre and group. Let T,T' be their characteristic maps. Then $\mathcal{B},\mathcal{B}'$ are equivalent if and only if there exists an element $a \in G$ and a homotopy $T' \simeq aTa^{-1}$. If G is arcwise connected, then $\mathcal{B},\mathcal{B}'$ are equivalent if and only if $T' \simeq T$.*

If $\mathcal{B},\mathcal{B}'$ are equivalent, we have maps λ_1,λ_2 as in §2.10. Let $\mu_i = \lambda_i|S^{n-1}$. It follows that

$$T'(x) = \mu_1(x)T(x)\mu_2(x)^{-1}.$$

Since $T(x_0) = T'(x_0) = e$, we have that $\mu_1(x_0) = \mu_2(x_0) = a$. Now S^{n-1} is contractible over E_i into x_0 leaving x_0 fixed $(i = 1,2)$. The image of this homotopy under λ_i is a homotopy of μ_i into the constant map $S^{n-1} \to a$ keeping x_0 at a. Let h_i: $(S^{n-1} \times I, x_0 \times I) \to (G,a)$ be this homotopy. Then $h'(x,t) = h_1(x,t)T(x)h_2(x,t)^{-1}$ is a homotopy of T' into aTa^{-1} keeping x_0 at e.

Conversely, suppose $a \varepsilon G$ and the homotopy $T' \simeq aTa^{-1}$ are given. If we define $\lambda_i(x) = a$ for x in V_i $(i = 1,2)$, and apply §2.10, we obtain a bundle equivalent to $\mathfrak{B}$ whose characteristic map is aTa^{-1}. We may therefore suppose that $a = e$, and $T' \simeq T$.

Now $T' \simeq T$ implies that $T'T^{-1}$ is homotopic to a constant map; and therefore $T'T^{-1}$ is extendable to a map ν: $E_1 \to G$. Define

$$\lambda_1(x) = \begin{cases} g'_{12}(x)g_{12}(x)^{-1} & \text{for } x \varepsilon E_2 \cap V_1, \\ \nu(x) & \text{for } x \varepsilon E_1. \end{cases}$$

Since the two lines agree when x is in S^{n-1}, λ_1 is continuous. Let V'_2 be the interior of E_2. If, in $\mathfrak{B}$, we replace V_2 by V'_2 and ϕ_2 by $\phi_2|V'_2 \times Y$ we obtain a strictly equivalent bundle $\mathfrak{B}_1$. Likewise $\mathfrak{B}'$ is strictly equivalent to $\mathfrak{B}'_1$ obtained by the analogous substitution. Now let $\lambda_2(x) = e$ for $x \varepsilon V'_2$. Then

$$g'_{12}(x) = \lambda_1(x)g_{12}(x)\lambda_2(x)^{-1} \qquad \text{for } x \varepsilon V_1 \cap V'_2.$$

By §2.10, this implies $\mathfrak{B}_1 \sim \mathfrak{B}'_1$, and therefore $\mathfrak{B} \sim \mathfrak{B}'$.

If G is arcwise connected, join a to e by a curve f: $I \to G$. Then $h(x,t) = f(t)T(x)f(t)^{-1}$ is a homotopy $aTa^{-1} \simeq T$.

18.4. THEOREM. *Let $\mathfrak{B}$ be a principal bundle over S^n in normal form, and T its characteristic map. Let x_1 in E_1 be the pole of S^{n-1}, $G_1 = p^{-1}(x_1)$, $\xi = \phi_{1,x_1}$ and $y_1 = \xi(e)$. Let α be the generator of $\pi_n(S^n,x_1)$ corresponding to an orientation of S^n. Orient E_2 concordantly with S^n, and orient S^{n-1} so as to be positively incident with E_2. Then ξT: $(S^{n-1},x_0) \to (G_1,y_1)$ represents the element $\Delta(\alpha)$ of $\pi_{n-1}(G_1,y_1)$ (see §17.3); and therefore T represents $\chi(\alpha) = \xi_*^{-1}\Delta(\alpha)$.*

As a first step, we define a map

$$h: \quad (E_2,S^{n-1}) \to (S^n,x_1).$$

Let x_2 be the antipode of x_1. For each $x \varepsilon E_2$, $h(x)$ lies on the great circle arc $C(x) = x_2xx_1$, and its arc length from x_2 is twice that of x. Clearly h maps the quarter circle $C(x) \cap E_2$ topologically onto the semicircle $C(x)$. Hence h maps $E_2 - S^{n-1}$ topologically onto $S^n - x_1$. Furthermore h has degree $+1$. This follows since h is the end result of a homotopy of S^n in which E_1 contracts over itself into x_1 and each point x of E_2 moves along $C(x)$ from x to $h(x)$ linearly with time.

Next, define a map

$$h' \colon \quad (E_2, S^{n-1}, x_0) \to (B, G_1, y_1)$$

which covers h as follows. For each x in E_2 other than x_2, let $k(x)$ be the point $C(x) \cap S^{n-1}$. Let

$$h'(x) = \begin{cases} \phi_1(h(x), Tk(x)) & \text{when } h(x) \in E_1, \\ \phi_2(h(x), e) & \text{when } h(x) \in E_2. \end{cases}$$

The two parts of the definition overlap when $h(x)$ is in S^{n-1}. In this case $h(x) = k(x) = x'$ say. Then

$$\phi_1(x', Tx') = \phi_1(x', g_{12}(x')) = \phi_2(x', e),$$

and the two definitions agree. It follows that h' is continuous over E_2. Obvious relations are

$$ph' = h, \qquad h'(S^{n-1}) \subset G_1, \qquad h'(x_0) = y_1.$$

When x is in S^{n-1}, then $h(x) = x_1$, and $k(x) = x$. Therefore

$$h'(x) = \phi_1(x_1, T(x)) = \xi T(x).$$

Since h has degree $+1$, it represents α in $\pi_n(S^n, x_1)$. Since $ph' = h$, h' represents $p_*^{-1}(\alpha)$. It follows that $h'|S^{n-1} = \xi T$ represents $\partial p_*^{-1}(\alpha) = \Delta(\alpha)$. This completes the proof.

In view of §16.10, the result of §18.3 can be restated: $\mathcal{B} \sim \mathcal{B}'$ if and only if the elements of $\pi_{n-1}(G)$ represented by T and T' are equivalent under the operations of $\pi_0(G)$. Combining this with the above result (that T represents $\chi(\alpha)$) yields

18.5. Classification theorem. *The equivalence classes of bundles over S^n with group G are in 1-1 correspondence with equivalence classes of elements of $\pi_{n-1}(G)$ under the operations of $\pi_0(G)$. Such a correspondence is provided by $\mathcal{B} \to \chi(\alpha)$ where α is a generator of $\pi_n(S^n)$ and $\chi\colon\ \pi_n(S^n) \to \pi_{n-1}(G)$ is a characteristic homomorphism of $\mathcal{B}$.*

18.6. COROLLARY. *If G is arcwise connected, then the set of equivalence classes of bundles over S^n with group G is in 1-1 correspondence with $\pi_{n-1}(G)$.*

These results reduce the bundle classification problem for spheres to a familiar problem of algebraic topology. This does not solve the problem; for homotopy groups are not generally calculable. In the sequel (§§22–25), we shall compute some of the homotopy groups of various groups and obtain applications of these results.

18.7. Weak equivalence of bundles. Two bundles $\mathcal{B}, \mathcal{B}'$ over X are called *weakly equivalent* if there exists a map $h\colon\ \mathcal{B} \to \mathcal{B}'$ which induces a homeomorphism of X on itself.

THEOREM. *Let $\mathcal{B},\mathcal{B}'$ be bundles over S^n and let α,α' be their conjugacy classes in $\pi_{n-1}(G)$ under a fixed correspondence. Then $\mathcal{B}$ and $\mathcal{B}'$ are weakly equivalent if and only if $\alpha = \pm\alpha'$.*

First suppose a weak equivalence h: $\mathcal{B} \to \mathcal{B}'$ is given. Then $\bar{h}$: $S^n \to S^n$ has degree $\epsilon = \pm 1$. If $\epsilon = 1$, then $\bar{h}$ is homotopic to the identity map. A covering homotopy leads to a map h': $\mathcal{B} \to \mathcal{B}'$ which induces the identity map of S^n. Then $\mathcal{B},\mathcal{B}'$ are equivalent and $\alpha = \alpha'$. Suppose $\epsilon = -1$, then $\bar{h}$ is homotopic to an orthogonal map $\bar{h}'$ which leaves x_1 fixed and maps S^{n-1} on itself with degree -1. A covering homotopy leads to h': $\mathcal{B} \to \mathcal{B}'$ covering $\bar{h}'$. We may identify $\mathcal{B}$ with the bundle induced by $\mathcal{B}'$ and $\bar{h}'$ (see §10.3). With $\mathcal{B}'$ in normal form, the induced (coordinate) bundle $\mathcal{B}$ is also in normal form. By definition of the induced bundle, $g_{12}(x) = g'_{12}(\bar{h}'(x))$. Therefore $T = T'\bar{h}'$. Since $\bar{h}'$ has degree -1, T and T' represent elements of $\pi_{n-1}(G)$ of opposite signs; hence $\alpha = -\alpha'$.

For the converse, $\alpha = \alpha'$ implies that $\mathcal{B},\mathcal{B}'$ are equivalent. Suppose $\alpha = -\alpha'$. Let h be an orthogonal map of S^n leaving x_1 fixed and of degree -1 on S^{n-1}. Let $\mathcal{B}_1$ be the bundle induced by $\mathcal{B}$ and h. Then $\mathcal{B}$ and $\mathcal{B}_1$ are weakly equivalent. By the argument above, $\mathcal{B}_1$ corresponds to $-\alpha$. Hence $\mathcal{B}_1$ and $\mathcal{B}'$ are equivalent, and the theorem is proved.

§19. UNIVERSAL BUNDLES AND THE CLASSIFICATION THEOREM

19.1. Complexes. In this article we restrict attention to bundles for which the base space is a finite cell complex. A *q-dimensional cell* σ^q (briefly: q-cell) is any homeomorph of the set $\Sigma_1^q x_i^2 \leqq 1$ in cartesian q-space. Its *boundary* $\dot{\sigma}^q$ is the part which corresponds to the $(q - 1)$-sphere $\Sigma_1^q x_i^2 = 1$. Its *interior* is the complement $\sigma^q - \dot{\sigma}^q$ of the boundary.

A *finite n-dimensional complex K* (briefly: n-complex) is a topological space $|K|$ and a collection $\{\sigma_i^q\}$ $(i = 1, \cdots, \alpha_q; q = 0, 1, \cdots, n)$ of closed subsets such that (i) each σ_i^q is a q-cell, (ii) if $|K^p|$ denotes the union of all q-cells for $q \leqq p$, then $|K| = |K^n|$, (iii) $\sigma_i^q \cap |K^{q-1}|$ is the boundary $\dot{\sigma}_i^q$ of σ_i^q, and it is an exact union of cells called the *faces* of σ_i^q, and (iv) if $i \neq j$ then the interiors of σ_i^q and σ_j^q have no point in common.

The set $|K^p|$ and the collection $\{\sigma_i^q, q \leqq p\}$ is itself a complex, denoted by K^p, called the *p-dimensional skeleton of K*.

A *subcomplex L* of K consists of a subspace $|L|$ of $|K|$ and a subcollection of the cells of K satisfying the conditions for a complex. Any collection of cells of K determines a subcomplex if each face of a cell of the collection is also in the collection.

When no confusion can arise we shall abbreviate $|K|$ by K. Any q-complex will be regarded as an n-complex with $\alpha_p = 0$ for $p > q$.

Since each cell is a compact space, and K is a finite union of cells, it follows that K is compact.

If K_1 and K_2 are cell complexes, their product $K_1 \times K_2$ consists of the product space $|K_1| \times |K_2|$ and the subsets of products of cells of K_1 and K_2. In particular if the interval $I = [0,1]$ is regarded as a cell complex consisting of one 1-cell I and the 0-cells 0 and 1, then $K \times I$ is a cell complex for any K. The dimension of $K_1 \times K_2$ is the sum of the dimensions of the factors.

19.2. Definition of universal bundle. Let $\mathfrak{B}$ be a principal bundle over a space X (not necessarily a complex) with group G. We say that $\mathfrak{B}$ is *universal for the dimension n* (briefly: n-universal) if, for any n-complex K, subcomplex L of K, principal bundle $\mathfrak{B}'$ over K with group G, and any map h of $\mathfrak{B}'|L$ into $\mathfrak{B}$, there exists an extension of h to a map of $\mathfrak{B}'$ into $\mathfrak{B}$.

The requirement can be paraphrased by saying that any partial map of a bundle into $\mathfrak{B}$ is extendable to the whole bundle.

If $\mathfrak{B}$ is n-universal, and $\mathfrak{B}'$ is any bundle over the n-complex K, then there is a map $\mathfrak{B}' \to \mathfrak{B}$. This follows if we take L to be vacuous.

19.3. Classification theorem. *Let $\mathfrak{B}$ be an $(n+1)$-universal bundle with group G, let X be its base space, and let K be an n-complex. The operation of assigning to each map f: $K \to X$ its induced bundle (see §10.1) sets up a 1-1 correspondence between homotopy classes of maps of K into X and equivalence classes of principal bundles over K with group G.*

By §11.5, two homotopic maps of K into X induce equivalent bundles. Therefore, to each homotopy class is assigned a unique equivalence class of bundles.

As observed in §19.2, any principal bundle $\mathfrak{B}'$ over K admits a map h: $\mathfrak{B}' \to \mathfrak{B}$. If f: $K \to X$ is the induced map, then $\mathfrak{B}'$ is equivalent to the bundle induced by f and $\mathfrak{B}$ (see §10.3).

To complete the proof we must show that, if two maps f_0,f_1: $K \to X$ induce bundles $\mathfrak{B}_0,\mathfrak{B}_1$ which are equivalent, then $f_0 \simeq f_1$. Let

$$h_i\colon\ \mathfrak{B}_i \to \mathfrak{B}, \qquad i = 0,1,$$

be the induced maps (see §10.1), and let

$$h\colon\ \mathfrak{B}_0 \to \mathfrak{B}_1$$

be an equivalence. Form the bundle $\mathfrak{B}' = \mathfrak{B}_0 \times I$ (see §11.1) and let r: $\mathfrak{B}_0 \times I \to \mathfrak{B}_0$ be the natural map $r(b,t) = b$. Let $\mathfrak{B}'_0,\mathfrak{B}'_1$ be the parts of $\mathfrak{B}'$ over $K \times 0$ and $K \times 1$ respectively, and let $r_i = r|\mathfrak{B}'_i$ ($i =$

0,1). Define

$$h': \mathfrak{B}'_0 \cup \mathfrak{B}'_1 \to \mathfrak{B}$$

by

$$h'|\mathfrak{B}'_0 = h_0 r_0, \qquad h'|\mathfrak{B}'_1 = h_1 h r_1.$$

It is clear that h' is a bundle mapping. Now $K \times I$ is an $(n + 1)$-complex and $\mathfrak{B}$ is $(n + 1)$-universal. Therefore h' is extendable to a map $\mathfrak{B}' \to \mathfrak{B}$. The induced map $K \times I \to X$ is the required homotopy.

19.4. Characterization of a universal bundle.

THEOREM. *A principal bundle $\mathfrak{B}$ is n-universal if and only if B is arcwise connected and $\pi_i(B) = 0$ for $1 \leqq i < n$.*

Suppose $\mathfrak{B}$ is n-universal. Let E be an $(i + 1)$-cell and S its boundary i-sphere. Let f: $S \to B$. Define f': $S \times G \to \mathfrak{B} \times G$ by $f'(y,g) = (f(y),g)$. Let P: $\mathfrak{B} \times G \to \mathfrak{B}$ be the principal map (see §8.7). By §8.8 the composition Pf' is a bundle map of $S \times G$ into $\mathfrak{B}$. Let $i < n$. Since $\mathfrak{B}$ is n-universal, Pf' extends to a bundle map h: $E \times G \to \mathfrak{B}$. By (3) of §8.12, $P(b,e) = b$ (e = identity of G). Therefore $Pf'(y,e) = f(y)$. Hence $h(y,e)$ is an extension of f to a map of E into B. Thus any map of an i-sphere into B ($i = 0, 1, \cdots, n - 1$) is contractible to a point. This implies that B is arcwise connected and $\pi_i(B) = 0$ for $1 \leqq i < n$ (see §15.11).

Conversely suppose B satisfies the stated conditions. Let (E,S) be an $(i + 1)$-cell and its boundary ($i < n$). Let h: $S \times G \to \mathfrak{B}$ be a bundle map. Let $f(y) = h(y,e)$. Then f: $S \to B$. Since $\pi_i(B) = 0$, f is extendable to a map f': $E \to B$. Define $h'(y,g) = P(f'(y),g)$ for y in E. Then h' is a bundle map $E \times G \to \mathfrak{B}$. We assert that h' is an extension of h. Note first that any two admissible maps of G into G_x differ by a left translation of G; hence any two which agree on e must coincide. But $h'(y,e) = f'(y) = f(y) = h(y,e)$ for y in S. Since both h and h' are bundle maps, they must coincide on $S \times G$. Thus we have proved that any bundle map h: $S \times G \to \mathfrak{B}$ is extendable to h': $E \times G \to \mathfrak{B}$.

Let $\mathfrak{B}'$ be a bundle over K and $\mathfrak{B}''$ the part over a subcomplex L, and suppose h: $\mathfrak{B}'' \to \mathfrak{B}$. For any 0-cell v of K not in L, choose any two admissible maps ξ: $G \to G_v$ and ζ: $G \to G_x$ for some x in X and extend h to be $\zeta\xi^{-1}$ on G_v. This extends h over $p'^{-1}(K^0 \cup L)$. Suppose, inductively, that h is defined on $p'^{-1}(K^i \cup L)$ and $i < n$. Let E be an $(i + 1)$-cell of K not in L. Since E is contractible, the portion of $\mathfrak{B}'$ over E is equivalent to the product bundle $E \times G$. Hence the part of $\mathfrak{B}'$ over the boundary S of E is likewise a product $S \times G$. But h is defined over the latter part. By the result of the preceding paragraph h is extendable over the portion of the bundle over E. This stepwise extension leads to an extension of h over $p'^{-1}(K^n \cup L)$. It follows that $\mathfrak{B}$ is n-universal.

19.5. Existence of universal bundles. We adopt now the notations of 7.6 where O_m is the real orthogonal group on m variables, and O_n, $1 \leqq n \leqq m$, is the subgroup which operates trivially on the last $m - n$ variables.

LEMMA. *If $1 \leqq n \leqq m$, then the left coset space O_m/O_n is arcwise connected and $\pi_i(O_m/O_n) = 0$ for $1 \leqq i < n$.*

Since $n \geqq 1$, O_n contains an element of determinant -1. Since O_m has just two components (both arcwise connected) and O_n contains points from both, it follows that O_m/O_n is arcwise connected.

Let e be the identity of O_m, and let f be an element of $F^i(O_k,O_n,e)$, $n < k \leqq m$ (see §15.2). Let p be the natural projection of O_k onto $O_k/O_{k-1} = S^{k-1}$ (see §7.6). Since $i < k - 1$, we have, by §15.8, that $\pi_i(S^{k-1}) = 0$. Therefore there is a homotopy of pf into the constant map and $pf(\dot{I}^i)$ remains at $p(e)$ during the homotopy. A covering homotopy deforms f into a map f' in $F^i(O_{k-1},O_n,e)$ and $f|\dot{I}^i$ remains fixed.

If we begin with f in $F^i(O_m,O_n,e)$ and apply the above argument successively for $k = m, m - 1, \cdots, n + 1$, we obtain a succession of homotopies which combine to give a homotopy of f into a map of I^i into O_n and $f|\dot{I}^i$ remains fixed during the homotopy. This implies that $\pi_i(O_m,O_n) = 0$ for $2 \leqq i < n$. Since O_m is a bundle over O_m/O_n with fibre O_n (see §7.5), it follows from §17.2 that $\pi_i(O_m/O_n) = 0$ for $2 \leqq i < n$.

The case $i = 1$ remains. Since $i < n$, we must have $n \geqq 2$. Let C be a closed curve in O_m/O_n based at the point corresponding to O_n. Cover C by an open curve C' in O_m which starts at the identity and ends at a point of O_n. The argument of the preceding paragraph shows that C' may be contracted into O_n leaving its end points fixed in O_n (note that $n \geqq 2$ is necessary for this to be true). The image of this homotopy is a contraction of C to a point, and the lemma is proved.

§19.6. THEOREM. *If G is a compact Lie group, then, for each integer n, there exists an n-universal bundle $\mathfrak{B}$ with group G.*

By the classical result [12, p. 211], G is isomorphic to a subgroup of an orthogonal group O'_k for k sufficiently large. We can suppose that $G \subset O'_k$. Adopting the notation of the preceding section, let $m = n + k$, and let O'_k be the subgroup of O_m which operates trivially on the first n coordinates. Then the subgroups O_n and O'_k of O_m commute; and one may identify their direct product $O_n \times O'_k$ with a subgroup of O_m. Since $G \subset O'_k$, the same is true of $O_n \times G$. Let

$$B = O_m/O_n, \qquad X = O_m/(O_n \times G) \tag{1}$$

be left coset spaces, and let p: $B \to X$ be the natural projection. According to §7.4 and §7.5, $\mathcal{B} = \{B,p,X\}$ admits a bundle structure with fibre $O_n \times G/O_n' \approx G$. The largest subgroup of O_n invariant in $O_n \times G$ is O_n. Hence the group of the bundle is also isomorphic to G. Thus $\mathcal{B}$ is a principal bundle with group G. By §19.5 and §19.4 it follows that $\mathcal{B}$ is n-universal.

19.7. The above argument proves more than is stated in the theorem. It is to be noted that O_{n+k}/O_n *is the space of the n-universal bundle for any closed subgroup G of* O_k'. In §7.7, we have called O_{n+k}/O_n the Stiefel manifold $V_{n+k,k}$ of orthogonal k-frames in a cartesian $(n + k)$-space. The base space of the n-universal bundle is a coset space of O_{n+k}.

For k-sphere bundles (see §7.8), we may take $G = O_{k+1}'$. Then *the base space of the n-universal k-sphere bundle is the Grassmann manifold* $M_{n+k+1,k+1}$ *of* $(k + 1)$*-planes through the origin in* $(n + k + 1)$*-space* (see §7.9).

19.8. For a Lie group G which is not compact but is connected, we have the result stated in §12.14 that G is a product space $H \times E$ of a compact subgroup H and a euclidean space E, and equivalence classes of bundles with group G are in a natural 1-1 correspondence with equivalence classes of bundles with group H.

Let $\mathcal{B}$ be an n-universal bundle for H. Since H operates on G by left translations there is an associated bundle $\mathcal{B}'$ of $\mathcal{B}$ having fibre G and group H. Let $\mathcal{B}''$ be the bundle obtained from $\mathcal{B}'$ by enlarging its group to G. Then $\mathcal{B}''$ is a principal bundle with group G and its base space is that of $\mathcal{B}$. It is a reasonable conjecture that $\mathcal{B}''$ is an n-universal bundle for G. But it is not necessary to prove this to obtain the conclusion of the classification theorem §19.3. The combination of the 1-1 correspondence between classes of G-bundles and classes of H-bundles followed by the 1-1 correspondence between the latter and the homotopy classes of maps of a complex into the base space of $\mathcal{B}$(= base space of $\mathcal{B}''$) is a 1-1 correspondence which reduces the bundle classification problem to a homotopy classification problem.

19.9. THEOREM. *Let $\mathcal{B}$ be an n-universal bundle with base space X and group G. Then, in the notation of §17.12,*

$$\chi: \quad \pi_i(X) \approx \pi_{i-1}(G) \quad (i = 1, \cdots, n - 1).$$

Since $\chi = \xi_*^{-1}\Delta$, and ξ is a homeomorphism, it suffices to prove that Δ: $\pi_i(X) \approx \pi_{i-1}(G_0)$. By §19.4, every third term $\pi_i(B)$ of the homotopy sequence of $\mathcal{B}$ is zero up to $i = n$. By the exactness of the homotopy sequence, it follows the Δ is an isomorphism onto for $i = 2, \cdots, n - 1$, and is an isomorphism into for $i = 1$ (see the argument in §15.8). To prove the *onto* part for $i = 1$, let u be an element of $\pi_0(G_0)$. Then u is a component of G_0. Since B is arcwise connected, there is a path D in B starting at $\xi(e)$ and ending at a point of u. Then pD is a closed path based at x_0. As such it represents an element α of $\pi_1(X)$. Using the alternative description of Δ in §17.11, it follows that $\Delta\alpha = u$.

HISTORICAL NOTE. The result that any k-sphere bundle over a complex can be generated by a mapping into a suitable Grassmann manifold was first proved by Whitney [104]. The classification theorem in full for k-sphere bundles was proved independently by the author [88] and by Pontrjagin [75]. The generalization to Lie groups was found independently by the author, G. W. Whitehead and jointly by S. Chern and Y. Sun [10].

§20. THE FIBERING OF SPHERES BY SPHERES

20.1. The Hopf map $S^3 \to S^2$. We shall describe now the various fiberings of spheres by spheres discovered by Hopf [49], and exhibit the precise sense in which they are bundles.

The simplest of these is the map p: $S^3 \to S^2$ of a 3-sphere on a 2-sphere. Let S^3 be represented in the space C_2 of two complex variables (z_1,z_2), as the locus $z_1\bar{z}_1 + z_2\bar{z}_2 = 1$. Let S^2 be represented as the complex projective line (i.e. as pairs $[z_1,z_2]$ of complex numbers, not both zero, with the equivalence relation $[z_1,z_2] \sim [\lambda z_1,\lambda z_2]$ where $\lambda \neq 0$). The projection p is defined by $p((z_1,z_2)) = [z_1,z_2]$.

Any pair $[z_1,z_2]$ can be normalized by dividing by $\lambda = (z_1\bar{z}_1 + z_2\bar{z}_2)^{1/2}$. Therefore S^2 is the complete image of S^3. If $|\lambda| = 1$, and (z_1,z_2) is in S^3, so also is $(\lambda z_1,\lambda z_2)$, and they have the same image point in S^2. Conversely, if $p(z_1,z_2) = p(z_1',z_2')$ then $(z_1',z_2') = (\lambda z_1,\lambda z_2)$ for some λ of absolute value 1. Therefore the inverse image of a point of S^2 is obtained from any point of the inverse image by multiplying it by $e^{i\theta}$ $(0 \leqq \theta \leqq 2\pi)$. Hence the inverse image is just a great circle of S^3. In this way the 3-sphere is decomposed into a family of great circles with the 2-sphere as a decomposition space.

In the next sections it will be shown, as a consequence of a more general result, that p: $S^3 \to S^2$ is a 1-sphere bundle.

20.2. Transitive groups on spheres. Let Q denote one of the three fields of real numbers, complex numbers, or quaternions. Let Q_n be the right vector space whose elements are ordered sets of n elements of Q. Specifically $x = (x_1, \cdots, x_n)$ is in Q_n if each $x_i \in Q$; and, if q is in Q, then $xq = (x_1q, \cdots, x_nq)$. Define the *inner product* of x and y in Q_n by

$$x \cdot y = \Sigma_1^n \bar{x}_i \cdot y_i \qquad (\bar{x}_i = \text{conjugate of } x_i).$$

It follows that

$$y\cdot x = \overline{x\cdot y}, \qquad (xq)\cdot y = \bar{q}(x\cdot y), \qquad x\cdot(yq) = (x\cdot y)q.$$

In particular $x\cdot y = 0$ if and only if $y\cdot x = 0$, thus the relation of orthogonality is symmetric. Let S be the unit sphere in Q_n, i.e. the locus $x\cdot x = 1$.

Let G_n be the group of linear transformations in Q_n preserving the inner product. Precisely

$$y_i = \sum_{j=1}^{n} \sigma_{ij}x_j, \qquad \sigma_{ij} \in Q,$$

is a linear transformation $x \to \sigma(x)$. The condition $\sigma(x)\cdot\sigma(y) = x\cdot y$ leads immediately to the relations

$$\sum_{i=1}^{n} \bar{\sigma}_{ij}\sigma_{ik} = \delta_{jk}$$

which means that the inverse matrix of σ is its conjugate transpose. G_n is called the *orthogonal, unitary* or *symplectic group* according as the scalars are real, complex or quaternionic. It is a compact Lie group.

The group G_n is transitive on the unit sphere S. The standard proof in the real case generalizes as follows. If x^1 is a vector of norm 1, and $n > 1$, there is a second vector y linearly independent of x^1. Then $y - x^1(x^1\cdot y)$ is orthogonal to x^1. Multiply this vector on the right by the reciprocal of its norm, and obtain a unit vector x^2 orthogonal to x^1. If $n > 2$, there is a vector y independent of x^1,x^2. Then $y - x^1(x^1\cdot y) - x^2(x^2\cdot y)$ is orthogonal to both x^1 and x^2. Normalize and obtain x^3. In this way one obtains n mutually orthogonal unit vectors $x^1, \cdots, x^n$. Let $\sigma_{ij} = x^j_i$. Then σ belongs to G_n and σ maps $(1, 0, \cdots, 0)$ into x^1. It follows that G_n is transitive.

Let G_{n-1} be the subgroup of G_n leaving fixed the vector $x^0 = (1, 0, \cdots, 0)$. As shown in §7.3 and §7.4, the map

$$p_1\colon\ G_n \to S, \text{ defined by } p_1(\sigma) = \sigma(x^0), \tag{1}$$

represents S as the left coset space

$$S = G_n/G_{n-1}, \tag{2}$$

and G_n is a principal bundle over S with group G_{n-1}.

20.3. The sphere as a bundle over a projective space. Let M_n be the projective space associated with Q_n. Specifically, if x and y are non-zero elements of Q_n, we say that $x \sim y$ if there exists a q in Q such that $y = xq$. This relation is symmetric, reflexive and transitive. The non-zero elements of Q_n are thereby divided into equivalence classes, and these are the elements of M_n. Let

$$p\colon\ S \to M_n \tag{3}$$

be the map which assigns to each element of S its equivalence class. Since any non-zero vector can be normalized, it follows that p maps S onto M_n.

If σ is in G_n we have $\sigma(xq) = \sigma(x)q$. It follows that G_n is a transformation group of M_n in such a way that

$$\sigma(p(x)) = p(\sigma(x)), \qquad x \in S. \tag{4}$$

Since G_n is transitive on S and $p(S) = M_n$ it follows that G_n is transitive on M_n. Let $z^0 = p(x^0)$, and define

$$p_2\colon\ G_n \to M_n \qquad \text{by} \qquad p_2(\sigma) = \sigma(z^0). \tag{5}$$

If H is the subgroup of G_n which leaves z^0 fixed, then, as shown in §7.3, p_2 provides an identification

$$M_n = G_n/H. \tag{6}$$

An element σ of G_n belongs to H if and only if $\sigma(x^0) = x^0q$ for some q in Q with $|q| = 1$. This means that $\sigma_{11} = q$ and $\sigma_{1i} = \sigma_{i1} = 0$ for $i \neq 1$. Therefore the matrix σ decomposes into

$$\begin{vmatrix} q & 0 \\ 0 & \sigma' \end{vmatrix}, \qquad |q| = 1, \qquad \sigma' \in G_{n-1}$$

It follows that H may be identified with the direct product

$$H = Q' \times G_{n-1} \tag{7}$$

where Q' is the subgroup of elements of Q of absolute value 1.

One consequence of (7) is that H is a closed subgroup. Then G_n/H is an analytic manifold. We assign to M_n the topology of G_n/H under the identification (6). By (4) we have

$$p_2(\sigma) = \sigma(z^0) = \sigma(p(x^0)) = p\sigma(x^0) = pp_1(\sigma),$$

so that

$$p_2 = pp_1. \tag{8}$$

In view of (8) and the identifications (2) and (6), we may apply the bundle structure theorem 7.4 to obtain

The unit sphere S is a bundle over the projective space M_n with fibre

$$Q' \times G_{n-1}/G_{n-1} = Q'$$

and group Q' (since G_{n-1} is invariant in $Q' \times G_{n-1}$).

20.4. Special cases. The above result provides three special cases. In the case Q = real numbers, Q' is just the 0-sphere consisting of $+1$ and -1. The unit sphere in Q_n has dimension $n - 1$. And the result states that S^{n-1} is a double covering of the real projective space of $n - 1$ real dimensions.

In the case Q = complex numbers, Q' is the 1-sphere $e^{i\theta}$ ($0 \leq \theta < 2\pi$). The unit sphere S in Q_n has $2n - 1$ real dimensions. The complex projective space M_n has $2n - 2$ real dimensions. The result states that *S^{2n-1} is a 1-sphere bundle over the projective space of n homogeneous complex variables.*

In particular, when $n = 2$, M_n may be identified with the 2-sphere, and we have shown that p: $S^3 \to S^2$, given in §20.1, has a 1-sphere bundle structure.

In the case Q = quaternions, Q' is the 3-sphere S^3: $|q| = 1$. The unit sphere S in Q_n has $4n - 1$ real dimensions. The quaternionic projective space M_n has $4n - 4$ real dimensions. The result states that *S^{4n-1} is a 3-sphere bundle over the projective space of n homogeneous quaternionic variables.* Actually the result is stronger. *The group of the bundle is the group $Q' = S^3$ of unit quaternions* which is decidedly smaller than the group of all orthogonal maps of S^3.

In particular when $n = 2$, the projective space may be identified with the 4-sphere S^4 obtained by adjoining a point at infinity to Q. In this way, *S^7 is a 3-sphere bundle over S^4 with group $Q' = S^3$.*

Since, in all cases, Q' is both the group and the fibre, all the foregoing are principal bundles.

20.5. Cayley numbers. A *Cayley number* $c = (q_1,q_2)$ is an ordered pair of quaternions. They are added by adding coordinates, and multiplication is defined by

$$(q_1,q_2)(q'_1,q'_2) = (q_1q'_1 - \bar{q}'_2q_2,\ q'_2q_1 + q_2\bar{q}'_1).$$

Define the conjugate of $c = (q_1,q_2)$ to be $\bar{c} = (\bar{q}_1, -q_2)$. Then $c\bar{c} = |c|^2$ is real and non-negative, and is zero if and only if $c = 0 = (0,0)$. If $c \neq 0$, then $c^{-1} = \bar{c}/|c|^2$ is a right and left inverse of c. The pair of quaternions (1,0) is a 2-sided unit. Furthermore

$$|cd| = |c||d| \tag{9}$$

can be verified by direct calculation. Therefore $cd = 0$ implies $c = 0$ or $d = 0$. Multiplication is distributive with respect to addition. Therefore the set C of all Cayley numbers forms a division algebra.

The associative law does not hold in general. However it can be shown [13] that any two elements of C generate an associative algebra isomorphic to a subalgebra of the quaternions.

If a quaternion is represented, in the usual way, as a set of four real numbers, then C may be identified with a real 8-dimensional vector space. Its topology is that which it obtains under this identification.

Furthermore addition in C corresponds to vector addition. It is easily seen that the operation $y \to xy$ in C is a linear transformation $g(x)$. The operation $x \to g(x)$ maps the non-zero elements of C into the linear group L_8, and is called the left representation of C. It is a continuous correspondence, but is not homomorphic due to the non-associativity of C.

20.6. The bundle $S^{15} \to S^8$. The construction of sphere bundles in the preceding sections was based on the associative law for the algebras considered. For example, in defining the projective space the proof of transitivity of the equivalence relation used associativity. It appears therefore that there is no notion of a projective space based on Cayley numbers. However by proceeding in a different manner we can construct a fibering of S^{15} over S^8 with S^7 for fibre.

Let Z be the 16-dimensional real space consisting of pairs (c,d) of Cayley numbers. Let B denote the complement in Z of the pair $(0,0)$. Let X denote the 8-sphere obtained from C by adjoining a point ∞ (a neighborhood of ∞ in X is the complement in X of any compact set in C). Define p: $B \to X$ by

$$p(c,d) = \begin{cases} cd^{-1} & \text{if } d \neq 0, \\ \infty & \text{if } d = 0. \end{cases}$$

The continuity of p at a point (c,d) with $d \neq 0$ is readily verified. If a sequence d_n tends to 0, it follows from (9) that d_n^{-1} tends to ∞. If also c_n tends to $c \neq 0$, (9) implies that $c_n d_n^{-1}$ tends to ∞. Therefore p is continuous.

Let Y denote the set of non-zero Cayley numbers. Let $V_1 = C$, and let V_2 be the complement of zero in X. Define

$$\phi_1\colon\ V_1 \times Y \to B \quad \text{by} \quad \phi_1(x,y) = (xy,y)$$

$$\phi_2\colon\ V_2 \times Y \to B \quad \text{by} \quad \phi_2(x,y) = \begin{cases} (y,x^{-1}y), & x \neq \infty, \\ (y,\,0), & x = \infty, \end{cases}$$

$$p_1(c,d) = d, \qquad p_2(c,d) = c.$$

Using the associativity of the subalgebra of C generated by two elements, we have

$$\begin{aligned} p\phi_1(x,y) &= (xy)y^{-1} = x(yy^{-1}) = x \\ p\phi_2(x,y) &= y(x^{-1}y)^{-1} = y(y^{-1}x) = x, \qquad x \neq \infty, \\ p\phi_2(\infty,y) &= p(y,0) = \infty, \\ p_1\phi_1(x,y) &= y, \qquad p_2\phi_2(x,y) = y. \end{aligned}$$

Therefore ϕ_i maps $V_i \times Y$ homeomorphically onto $p^{-1}(V_i)$ $(i = 1,2)$. If x is in $V_1 \cap V_2$, then

$$\phi_{2,x}^{-1}\phi_{1,x}(y) = p_2\phi_1(x,y) = p_2(xy,y) = xy = g(x)\cdot y$$

where $g(x)$ is the left linear transformation corresponding to x. It follows that $\{B,p,X,Y,L_8,(\phi_1,\phi_2)\}$ is a coordinate bundle.

According to §12.9, any linear bundle is equivalent to an orthogonal bundle. In particular the bundle $\mathfrak{B}$ just constructed is equivalent in L_8 to a bundle $\mathfrak{B}'$ with group O_8. Let S^7 be the unit 7-sphere in Y. Since S^7 is invariant under O_8, it determines a unique subbundle $\mathfrak{B}''$ of $\mathfrak{B}'$ consisting of a 7-sphere in each fibre of $\mathfrak{B}'$. The space B'' is compact and lies in B. Since the equivalence of $\mathfrak{B}$ and $\mathfrak{B}'$ is linear, each ray through the origin of Z meets B'' in exactly one point. Therefore B'' is homeomorphic to a 15-sphere. We have proved

The 15-sphere is the bundle space of a 7-sphere bundle over the 8-sphere.

20.7. The problem of finding real division algebras. It is a classical theorem of Hurwitz that any real division algebra having a norm which satisfies (9) must be the real, complex, quaternionic or Cayley algebra. The preceding construction used (9) in two ways: to show that C is a division algebra, and to prove the continuity of p. The fact that C is a division algebra was used in an essential way. The continuity of p could be proved from the bilinearity alone. One is therefore led to seek real division algebras in order to construct more fiberings of spheres by spheres. Whether such exist other than in the dimensions 1, 2, 4 and 8 is not known. It has been shown [51] that the dimension of a division algebra must be a power of 2. (See App. sect. 4.)

20.8. Universal bundles. The bundle $S^3 \to S^2$ is principal and has S^1 as its group. Since $\pi_i(S^3) = 0$ for $i < 3$, we obtain from §19.4:

The bundle $S^3 \to S^2$ is 3-universal for the group S^1.

In a similar way:

The bundle $S^7 \to S^4$ is 7-universal for the group S^3.

By virtue of the classification theorem §19.3, we are led to the problem of enumerating the homotopy classes of maps of a complex K into a sphere. In the first case, dim $K \leqq 2$, and, in the second, dim $K \leqq 6$. The classification of maps of an n-complex into an n-sphere is given by the Hopf theorem (see §37.12). We obtain thus an effective enumeration of bundles over a 2-complex with group S^1 and bundles over a 4-complex with group S^3.

§21. THE HOMOTOPY GROUPS OF SPHERES

21.1. The results of the last two articles indicate the importance to bundle theory of solutions of various homotopy classification problems, and, in particular, the importance of computing the homotopy groups of the classical Lie groups. Only partial results have been obtained in this direction. These depend on the knowledge of homotopy groups of spheres where again the facts are only partly known.

In this article, we state the known results and give indications of their proofs.

We adopt the convention that, if G is a group, $G = \infty$ means that G is infinite cyclic, and $G = 2$ means that G is cyclic of order 2.

21.2. Relations derived from bundles. In §§15.8–15.10, we have seen that

$$\pi_i(S^n) = \begin{cases} 0 & i < n, \\ \infty & i = n. \end{cases} \tag{1}$$

The circle S^1: $|z| = 1$ is covered by the real number system under the map $z = \exp(ix)$. Since a line is contractible to a point, all its homotopy groups vanish. It follows from §17.6 that

$$\pi_i(S^1) = 0 \qquad \text{for } i > 1 \tag{2}$$

As shown in §20, S^3 is a 1-sphere bundle over S^2. Since $\pi_1(S^3) = 0$, any fibre is contractible to a point in S^3. By §17.10, we have

$$\pi_i(S^2) \approx \pi_{i-1}(S^1) + \pi_i(S^3), \qquad i \geqq 2.$$

When $i = 2$, this provides no new information. Using (2), we have

$$\pi_i(S^2) \approx \pi_i(S^3), \qquad i \geqq 3. \tag{3}$$

In particular

$$\pi_3(S^2) = \infty. \tag{4}$$

It was also shown in article 20 that S^7 is a 3-sphere bundle over S^4, and S^{15} is a 7-sphere bundle over S^8. In each case the fibres are contractible to a point; hence §17.10 implies

$$\pi_i(S^4) \approx \pi_{i-1}(S^3) + \pi_i(S^7), \qquad i \geqq 2. \tag{5}$$
$$\pi_i(S^8) \approx \pi_{i-1}(S^7) + \pi_i(S^{15}), \qquad i \geqq 2. \tag{6}$$

Using (1), we obtain

$$\pi_i(S^4) \approx \pi_{i-1}(S^3), \qquad 2 \leqq i < 7. \tag{7}$$
$$\pi_i(S^8) \approx \pi_{i-1}(S^7), \qquad 2 \leqq i < 15. \tag{8}$$
$$\pi_7(S^4) \approx \pi_6(S^3) + \infty. \tag{9}$$
$$\pi_{15}(S^8) \approx \pi_{14}(S^7) + \infty. \tag{10}$$

These are the results obtained directly from bundle relations. Except for (2) and (4), they do not give complete answers. However they indicate the complexity of the problem.

21.3. The suspension homomorphism of Freudenthal. Let f be a map $S^i \to S^n$. Let S^i and S^n be equators of S^{i+1} and S^{n+1} respectively. The *suspension* of f, denoted by Ef (called Einhängung by Freudenthal).

is the map

$$Ef\colon \quad S^{i+1} \to S^{n+1}.$$

which reduces to f on the equator and maps the upper (lower) hemisphere $E_+^{i+1}(E_-^{i+1})$ of S^{i+1} on the upper (lower) hemisphere $E_+^{n+1}(E_-^{n+1})$ of S^{n+1}. This is done by mapping the center of each hemisphere into the center of the corresponding hemisphere, and extending radially. A homotopy of f into f' may be suspended to provide a homotopy of Ef into Ef'. Therefore E carries a homotopy class into a homotopy class. In §15.11, we have shown how homotopy classes of sphere maps correspond to elements of the homotopy group. Thus E determines a map

$$E\colon \quad \pi_i(S^n) \to \pi_{i+1}(S^{n+1}). \tag{11}$$

It is a homomorphism. To show this, we present an alternative definition of E. In the diagram

$$\begin{array}{ccc} \pi_{i+1}(S^{n+1},E_-^{n+1}) & \xleftarrow{l_*} & \pi_{i+1}(E_+^{n+1},S^n) \\ \uparrow k_* & & \downarrow \partial \\ \pi_{i+1}(S^{n+1}) & \xleftarrow{E} & \pi_i(S^n) \end{array} \tag{12}$$

k and l are inclusion maps. Since the hemispheres are cells, their homotopy groups vanish. This and exactness of the homotopy sequences of (S^{n+1},E_-^{n+1}) and (E_+^{n+1},S^n) imply that k_* and ∂ are isomorphisms. It is not hard to show that

$$E = k_*^{-1} l_* \partial^{-1}. \tag{13}$$

When $i < n$, E is trivially an isomorphism since both groups are zero. When $i = n$, all groups of (12) are infinite cyclic and l_* is an isomorphism (see §15.10). It is easy to check directly that the isomorphisms (7) and (8) above are given by E. These are all special cases of the following general result of Freudenthal [36].

21.4. *If $i < 2n - 1$, then E in (11) is an isomorphism onto. If $i = 2n - 1$, then E is a homomorphism onto.*

The proof is omitted since it is difficult and not closely related to bundle theory.

In view of (5) and (6), we have the additional result: *If n is 3 or 7, then E in (11) is an isomorphism into a direct summand for all $i \geqq 1$.*

21.5. *If $f\colon \quad (E^n,S^{n-1}) \to (S^n,x_0)$ represents a generator of $\pi_n(S^n)$, then*

$$f_*\colon \quad \pi_i(E^n,S^{n-1}) \approx \pi_i(S^n) \qquad \text{for } i < 2n - 2$$

and f_* is a homomorphism onto when $i = 2n - 2$.

In order to use diagram (12) replace i,n by $i + 1$, $n + 1$. Since k_* is an isomorphism for $i + 1 = n + 1$, kf represents a generator α of $\pi_{n+1}(S^{n+1},E^{n+1})$. Since l_* is an isomorphism when $i + 1 = n + 1$, there is a topological map g: $(E^{n+1},S^n) \to (E_+^{n+1},S^n)$ such that lg also represents α. It follows that $kf \simeq lg$; therefore

$$k_*f_* = l_*g_*$$

for all dimensions of the homotopy groups. Using (13) we have

$$f_* = E\partial g_*.$$

Since g is topological, g_* is isomorphic. As observed above ∂ is also isomorphic. Hence f_* is equivalent to E. The result follows now from §21.4. (Under the substitution of $i - 1$, $n - 1$ for i,n, the inequality $i < 2n - 1$ becomes $i < 2n - 2$).

21.6. The Hopf invariant. The study of the group $\pi_{2n-1}(S^n)$, and of the kernel of E, when $i = 2n - 1$. is based on the *Hopf invariant.* Let an element α in $\pi_{2n-1}(S^n)$ be represented by a map f: $S^{2n-1} \to S^n$. We may suppose that f is a simplicial map relative to some triangulations. If x_1,x_2 are interior points of n-simplexes of S^n, then $f^{-1}(x_i) = \gamma_i$ ($i = 1,2$) is an $(n - 1)$-manifold in S^{2n-1}. Using orientations of S^n and S^{2n-1} we assign a natural orientation to γ_i. In this way γ_1,γ_2 are $(n - 1)$-cycles in S^{2n-1}. Their linking number $H(\alpha)$ is proved to depend only on α, and is called the Hopf invariant. In this way H is a homomorphism

$$H: \quad \pi_{2n-1}(S^n) \to \text{the group of integers.}$$

When n is odd, H is always zero due to the anti-commutativity of linking numbers. When n is even, Hopf has shown [49] that there always exist elements having an even invariant. For $n = 2,4$ and 8, the bundle maps given in article 20 have Hopf invariant 1. This is intuitively clear if it is observed that $p^{-1}(x)$ is a great $(n - 1)$-sphere of S^{2n-1} for each x, and any two such necessarily link one another once.

According to Freudenthal's result (§21.4), E maps $\pi_3(S^2)$ onto $\pi_4(S^3)$. He also proved that the kernel of E consists of all elements of even Hopf invariant. Since H maps $\pi_3(S^2)$ isomorphically onto the integers, it follows that $\pi_4(S^3)$ is a cyclic group of order 2. This and §21.4 imply

For $n \geqq 3$, $\pi_{n+1}(S^n) = 2$. Its non-zero element is represented by an $(n - 2)$-fold suspension of the Hopf map $S^3 \to S^2$.

From this and (3) above, we obtain

21.7. *The group $\pi_4(S^2) = 2$. Its non-zero element is represented by the composition $p\cdot Ep$ where p: $S^3 \to S^2$ is the Hopf map.*

Numerous additional results concerning homotopy groups of spheres have been obtained. They are mainly to the effect that certain groups are not zero. Up to the present writing, the only groups determined completely are $\pi_i(S^1)$ for all i, $\pi_i(S^2)$ for $i \leqq 4$, and $\pi_i(S^n)$ for $i \leqq n + 1$ and all n. (See App. sect. 4.)

Pontrjagin [72] has announced that $\pi_{n+2}(S^n)$ is zero for $n \geqq 3$, but a complete proof has not appeared. Freudenthal has shown [36] that it is either zero or a cyclic group of order 2 isomorphic to $\pi_4(S^2)$ under $(n - 2)$-fold suspension.

ADDED IN PROOF. It has been shown that $\pi_{n+2}(S^n)$ is cyclic of order 2 by Pontrjagin [*C. R. Acad. Sci. URSS*, 70 (1950), 957–959] and G. Whitehead [*Annals of Math. 52* (1950), 245–247].

§22. HOMOTOPY GROUPS OF THE ORTHOGONAL GROUPS

22.1. Conventions. Much of the material of this and the next two articles are covered in detail in papers of G. W. Whitehead [99] and B. Eckmann [15]. (See App. sect. 5.)

We adopt notations similar to those of §§7.6–7.10. We assume an infinite set of real variables $(t_0, t_1, t_2, \cdots)$. C^{n+1} denotes the cartesian space of the variables $(t_0, \cdots, t_n)$. S^n is the unit n-sphere in C^{n+1}. The point x_n is the unit point on the t_n-axis. For any point x of S^n, its antipode is denoted by $-x$. O_{n+1} is the orthogonal group in C^{n+1}. It is assumed to operate trivially on the variables $t_{n+1}, \cdots$ so that $O_{n+1} \subset O_{n+2}$ for all n. The identity element of O_{n+1} is denoted by e.

As observed in §7.8, O_{n+1} has just two components, and the component of e is R_{n+1} (the rotation group of S^n). Therefore $\pi_0(O_{n+1}) = O_{n+1}/R_{n+1}$ is a cyclic group of order 2. The non-zero element of this group is denoted by α_0. It is represented by any orthogonal map of determinant -1.

We shall use e as the base point for all homotopy groups. Since R_{n+1} is the component of e in O_{n+1}, we have $\pi_i(R_{n+1}) = \pi_i(O_{n+1})$ for $i \geqq 1$. We shall therefore restrict attention to the rotation groups with the exception that we must consider the operations of α_0 on $\pi_i(R_{n+1})$ (see §16.10).

Since $R_n \subset R_{n+1}$ for each n, we have numerous inclusion maps, e.g. $(R_k,R_n) \subset (R_m,R_n)$ for $n \leqq k \leqq m$. Any such induces homomorphisms of the homotopy groups. It is to be understood in the sequel that any homorphism $\pi_i(R_k,R_n) \to \pi_i(R_m,R_n)$ is induced by the appropriate inclusion map.

We adopt the convention (§21.1) that equating a group to ∞ means it is infinite cyclic, and to 2 means it is cyclic of order 2.

22.2. The group R_2. The group R_2 is, topologically, a 1-sphere. *Therefore $\pi_1(R_2)$ is infinite cyclic and $\pi_i(R_2) = 0$ for $i > 1$* (see §21.2). As observed in §16.10, the element α_0 of $\pi_0(O_2)$ operates in $\pi_1(R_2)$ by sending each element into its inverse.

22.3. The group R_3. *The group R_3 is topologically equivalent to real projective 3-space P^3.*

There are various ways of showing this of which the following one is quite useful. Let S^3 denote the group of quaternions of absolute value 1. The subset of S^3 of those quaternions whose real (scalar) part is zero is a 2-sphere S^2. It is the intersection with S^3 of the 3-plane spanned by the units i, j, and k.

We define a continuous homomorphism

$$\rho\colon \quad S^3 \to R_3 \tag{1}$$

by the rule

$$\rho(q)\cdot q' = qq'q^{-1} \qquad \text{for } q' \in S^2.$$

Note that $q' \to qq'q^{-1}$ is linear in terms of the four real components of q'. Since $|q| = 1$, and the norm of the product is the product of the norms, the transformation is orthogonal. Since $q1q^{-1} = 1$, and S^2 is orthogonal to 1, it follows that $\rho(q)$ is in R_3.

A quaternion commutes with the pure imaginary units i, j, and k if and only if it is real. It follows that the kernel of ρ is the group of two elements 1 and -1. The cosets of this subgroup are just the pairs q, $-q$. It follows that $\rho(S^3)$ is a projective 3-space, and ρ is the standard double covering.

It remains to show that $\rho(S^3) = R_3$. Let D_i, D_j, and D_k be the 1-parameter subgroups of R_3 leaving fixed the quaternions i, j, and k respectively. It is well known that any element of R_3 is a product of elements from these subgroups. We show that each is contained in $\rho(S^3)$. By symmetry, it is enough to prove this of one of them, D_i say. Let $q = e^{i\theta}$ $(0 \leqq \theta \leqq \pi)$. Then

$$\begin{aligned} qjq^{-1} &= e^{i\theta}je^{-i\theta} = e^{i\theta}e^{i\theta}j = e^{i2\theta}j \\ &= j\cos 2\theta + k \sin 2\theta. \end{aligned}$$

Thus $\rho(q)$ is a rotation in the (j,k)-plane through the angle 2θ. This completes the proof.

Let $p\colon\ R_3 \to S^2$ assign to each r in R_3 the element $r(i)$. Then $p\rho\colon\ S^3 \to S^2$ is the factoring of S^3 into the left cosets of D_i. It is not hard to show directly that $p\rho$ is topologically equivalent to the Hopf map (§20.1).

Having shown that R_3 is covered twice by S^3, it follows that

$$\pi_1(R_3) \text{ is cyclic of order 2.} \tag{2}$$

By §17.6, we have

$$(3) \qquad \pi_i(R_3) \approx \pi_i(S^3), \qquad i \geqq 2.$$

Applying the known facts for S^3 (§21), we have

$$(4) \qquad \pi_2(R_3) = 0,$$
$$(5) \qquad \pi_3(R_3) = \infty,$$
$$(6) \qquad \pi_4(R_3) = 2.$$

A generator of $\pi_3(R_3)$ is represented by the map ρ of (1) above, and the non-zero element of $\pi_4(R_3)$ is represented by the composition of the essential map $S^4 \rightarrow S^3$ followed by ρ.

The operations of α_0 on $\pi_i(R_3)$ are trivial. This is a special case of

22.4. The operation α_0. *If n is odd, α_0 in $\pi_0(O_n)$ operates trivially on $\pi_i(R_n)$ for each i.*

Since n is odd, the scalar matrix with -1's down the diagonal is an element of O_n of determinant -1. Conjugation of R_n by this element is the identity transformation. It therefore induces the identity transformation of $\pi_i(R_n)$, and the assertion is proved.

22.5. *If n is odd, and $k = 1, 2, \cdots$, then α_0 in $\pi_0(O_{n+k})$ operates trivially on the image of $\pi_i(R_n)$ in $\pi_i(R_{n+k})$ for each i.*

Let r be the diagnonal matrix having the first n diagonal elements equal to -1 and the rest to 1. Then its determinant is -1. Conjugation of R_{n+k} by r is non-trivial, however the subgroup R_n remains pointwise fixed since r commutes with R_n. This implies the result.

22.6. The group R_4. Define the homomorphism

$$(7) \qquad \sigma: \quad S^3 \rightarrow R_4$$

by

$$\sigma(q)\cdot q' = qq'.$$

If the bundle projection p: $R_4 \rightarrow S^3$ is given by $p(r) = r\cdot 1$, it follows that $p\sigma(q) = q$. Thus, as in §8.6, *R_4 is equivalent to the product bundle $S^3 \times R_3$.*

By §17.8, this implies

$$(8) \qquad \pi_i(R_4) \approx \pi_i(S^3) + \pi_i(R_3), \qquad i \geqq 1.$$

Applying the results already obtained for S^3 and R_3:

$$(9) \qquad \pi_1(R_4) = 2,$$
$$(10) \qquad \pi_2(R_4) = 0,$$
$$(11) \qquad \pi_3(R_4) = \infty + \infty,$$
$$(12) \qquad \pi_4(R_4) = 2 + 2.$$

The direct sum relation (8) is natural in that the inclusion $R_3 \subset R_4$ induces an isomorphism of $\pi_i(R_3)$ onto a subgroup of $\pi_i(R_4)$, and the map σ induces an isomorphism of $\pi_i(S^3)$ onto a subgroup of $\pi_i(R_4)$. Finally, $\pi_i(R_4)$ is the direct sum of these two subgroups. Because of this we can give explicit generators for the groups of (11) and (12) as follows:

22.7. Generators for π_3 and π_4 of R_4. *The maps ρ (see (1) above) and σ (see (7)) of the oriented sphere S^3 into R_4 determine elements α_3 and β_3, respectively, which generate $\pi_3(R_4)$. If h: $S^4 \to S^3$ is an essential map, then the compositions ρh and σh represent generators α_4 and β_4, respectively, of $\pi_4(R_4)$.*

By §22.5, we have, for α_0 in $\pi_0(O_4)$,

$$(13) \qquad \alpha_0(\alpha_3) = \alpha_3, \qquad \alpha_0(\alpha_4) = \alpha_4.$$

Let $r \in O_4$ be the transformation $q \to \bar{q}$ (conjugate). Then r has determinant -1. To obtain the effect of α_0 operating on β_3, we conjugate σ by r:

$$(r\sigma(q)r^{-1})\cdot q' = r\sigma(q)\cdot\bar{q}' = r\cdot(q\bar{q}') = q'\bar{q}.$$

On the other hand

$$\rho(q)\sigma(q)^{-1}\cdot q' = \rho(q)\cdot(q^{-1}q') = q(q^{-1}q')q^{-1} = q'\bar{q}.$$

Since these two equations hold for all q', we have

$$(14) \qquad r\sigma(q)r^{-1} = \rho(q)\sigma(q)^{-1}.$$

By §16.7, this implies

$$(15) \qquad \alpha_0(\beta_3) = \alpha_3 - \beta_3.$$

Formula (14) implies the relation obtained by replacing ρ,σ by the compositions ρh and σh respectively. Therefore

$$(16) \qquad \alpha_0(\beta_4) = \alpha_4 - \beta_4 = \alpha_4 + \beta_4$$

since β_4 is of order 2.

22.8. The constancy of $\pi_n(R_m)$ for large m. *Inclusion maps induce a homomorphism of $\pi_n(R_{n+1})$ onto $\pi_n(R_{n+2})$, and isomorphisms $\pi_n(R_{n+2}) \approx \pi_n(R_m)$ for $m > n + 2$.*

As shown in §19.5

$$\pi_i(R_{m+1},R_{k+1}) \approx \pi_i(O_{m+1}/O_{k+1}) = 0, \qquad i \leqq k \leqq m.$$

In particular $\pi_n(R_{n+2},R_{n+1}) = 0$. This and exactness of the homotopy sequence of (R_{n+2},R_{n+1}) imply the first statement. Similarly, in the section

$$\pi_{n+1}(R_m,R_{n+2}) \to \pi_n(R_{n+2}) \to \pi_n(R_m) \to \pi_n(R_m,R_{n+2})$$

of the homotopy sequence of (R_m,R_{n+2}), the end terms are zero. This and exactness imply that the middle homomorphism is an isomorphism.

Using this result with §22.2 and formula (2) above, we have

22.9. *$\pi_1(R_n)$ is a cyclic group of order 2 for $n > 2$. Its generator is represented by the inclusion map of the 1-sphere R_2 in R_n.*

In the same way we obtain from §22.2 and formula (4):

22.10. *$\pi_2(R_n) = 0$ for all n.*

To determine $\pi_3(R_n)$, $n > 4$, we must calculate the kernel of $\pi_3(R_4) \to \pi_3(R_5)$. This is accomplished in the next article.

§23. A Characteristic Map for the Bundle R_{n+1} over S^n

23.1. The kernel of $\pi_i(R_{n+1}) \to \pi_i(R_{n+2})$. We have seen, in §18, that any bundle $\mathcal{B}$ over S^n may be given a normal form which leads to a characteristic map T: $S^{n-1} \to G$. We shall exhibit a T for the bundle R_{n+1} over S^n. The importance of T in computing the homotopy groups of R_{n+1} is clear from the following general result.

23.2. Theorem. *Let $\mathcal{B}$ be a principal bundle over S^n in normal form, T its characteristic map, G_1 the fibre over the midpoint x_1 of E_1, $\xi = \phi_{1,x_1}$, and $y_1 = \xi(e)$. If $i \leqq 2n - 3$, then the kernel of the homomorphism $\pi_i(G_1) \to \pi_i(B)$ is the image group $\xi_* T_* \pi_i(S^{n-1})$.*

The exactness of the homotopy sequence of the bundle implies that the kernel of the homomorphism is the image $\Delta\pi_{i+1}(S^n)$. When $i = n - 1$ the result follows directly from §18.4. For larger values we must appeal to the Frendenthal suspension theorem 21.4.

Let h,h' be the maps constructed in the proof of §18.4. We recall that h maps $E_2 - S^{n-1}$ topologically onto $S^n - x_1$. It follows that h represents a generator of $\pi_n(S^n)$. Therefore, by §21.5, h_* maps $\pi_i(E_2,S^{n-1})$ onto $\pi_i(S^n)$ for $i \leqq 2n - 2$. Since $ph' = h$, and p_* is isomorphic for all i, we have that h'_* maps $\pi_i(E_2,S^{n-1})$ onto $\pi_i(B,G_1)$ for $i \leqq 2n - 2$. As shown in the proof of §18.4, $h'|S^{n-1} = \xi T$. By the standard property of the boundary operator, commutativity holds in the diagram

$$\begin{array}{ccc} \pi_i(E_2,S^{n-1}) & \xrightarrow{h'_*} & \pi_i(B,G_1) \\ \downarrow \partial' & & \downarrow \partial \\ \pi_{i-1}(S^{n-1}) & \xrightarrow{\xi_* T_*} & \pi_{i-1}(G_1) \end{array}$$

But ∂' is an isomorphism for each i (see §15.8, (7)). Therefore the image of $\xi_* T_*$ coincides with that of $\xi_* T_* \partial' = \partial h'_*$. Since h'_* is onto for $i \leqq 2n - 2$, the image of $\partial h'_*$ is the same as the image of ∂. We conclude that $\xi_* T_*$ and ∂ have the same image for $i \leqq 2n - 2$. The exactness property insures that the image of ∂ is the kernel of $\pi_{i-1}(G_1) \to \pi_{i-1}(B)$. If we replace $i - 1$ by i, the restriction on i becomes $i \leqq 2n - 3$, and the theorem is proved.

23.3. The normal form. We give now the normal form for the bundle p: $R_{n+1} \to S^n$ where $p(r) = r(x_n)$ (the notations of §22.1 are still in force).

Define ϕ: $S^n - (-x_n) \to R_{n+1}$ to be the map which assigns to $x \neq -x_n$ the rotation which is the identity on all points orthogonal to both x and x_n, and rotates the great circle through x and x_n so as to carry x_n into x. Clearly

$$p\phi(x) = x.$$

In terms of coordinates $x = (t_0, \cdots, t_n)$, ϕ is given by the matrix

$$(1) \qquad \phi(x) = \left\| \begin{array}{ccc|c} & & & t_0 \\ & \delta^\alpha_\beta - \dfrac{t_\alpha t_\beta}{1 + t_n} & & \vdots \\ & & & t_{n-1} \\ \hline -t_0 & \cdots & -t_{n-1} & t_n \end{array} \right\|$$

where $\delta^\alpha_\beta = 1$ or 0 is the Kronecker δ, and α,β range from 0 to $n - 1$. The proof that the matrix is orthogonal, and has the stated properties of ϕ is mechanical and is omitted. The continuity of ϕ is obvious from the matrix form.

Recall that $x_{n-1} = (0, \cdots, 0, 1, 0)$. Hence

$$\lambda = \phi(x_{n-1})^2$$

is a 180° rotation in the plane of the last two coordinates. Define

$$(2) \qquad \begin{aligned} \phi_1(x,r) &= \phi(x)r, && x \in S^n - (-x_n),\ r \in R_n, \\ \phi_2(x,r) &= \lambda\phi(\lambda(x))r, && x \in S^n - x_n,\ r \in R_n. \end{aligned}$$

The coordinate neighborhoods are $V_1 = S^n - (-x_n)$ and $V_2 = S^n - x_n$. Define

$$(3) \qquad \begin{aligned} p_1(r) &= [\phi p(r)]^{-1}r, && p(r) \in V_1, \\ p_2(r) &= [\phi\lambda p(r)]^{-1}\lambda r, && p(r) \in V_2. \end{aligned}$$

It follows quickly that

$$p\phi_i(x,r) = x, \qquad p_i\phi_i(x,r) = r \qquad (i = 1,2)$$

and

(4) $$g_{12}(x) = [\phi(x)]^{-1}\lambda\phi(\lambda(x)), \qquad x \in V_1 \cap V_2.$$

In particular, $g_{12}(x_{n-1}) = e$. Therefore, the bundle is in normal form relative to the base point x_{n-1} of S^{n-1}. By definition

(5) $$T_{n+1} = g_{12}|S^{n-1}\colon \quad (S^{n-1},x_{n-1}) \to (R_n,e)$$

is the characteristic map of this normal form. Assuming $x \in S^{n-1}$, and using the matrix form (1) of ϕ, we compute the right side of (4) and obtain the matrix form

(6) $$T_{n+1}(x) = \|\delta^\alpha_\beta - 2t_\alpha t_\beta\|\cdot\begin{Vmatrix}I_{n-1} & 0\\ 0 & -1\end{Vmatrix}, \qquad x \in S^{n-1},$$

where $\alpha,\beta = 0, 1, \cdots, n-1$, and I_{n-1} is the identity matrix of $n-1$ rows. (The subscript on T_{n+1} will be omitted frequently.)

23.4. Properties of the characteristic map.

THEOREM. *The characteristic map T_{n+1}: $S^{n-1} \to R_n$ of a normal form of the bundle R_{n+1} over S^n is given by (6) above. As a rotation of S^{n-1}, $T_{n+1}(x)$ leaves fixed all points orthogonal to x and x_{n-1}, and rotates the great circle of x,x_{n-1} through an angle twice that from x_{n-1} to x. It follows that $T(x) = T(-x)$, and $T(x) \neq T(x')$ if $x \neq \pm x'$. Hence $T_{n+1}(S^{n-1})$ is a real projective $(n-1)$-space contained in R_n, and T_{n+1} is the usual double covering. If p': $R_n \to S^{n-1}$ is defined by $p'(r) = r(x_{n-1})$, then $p'T_{n+1}$ maps S^{n-1} on itself with degree 0 if n is odd, and degree 2 if n is even.*

If a point of S^{n-1} is orthogonal to x and x_{n-1}, it is also orthogonal to the planes of x,x_n and x,x_{n-1}. It therefore remains fixed under each factor of (4), and, hence, under $T(x)$.

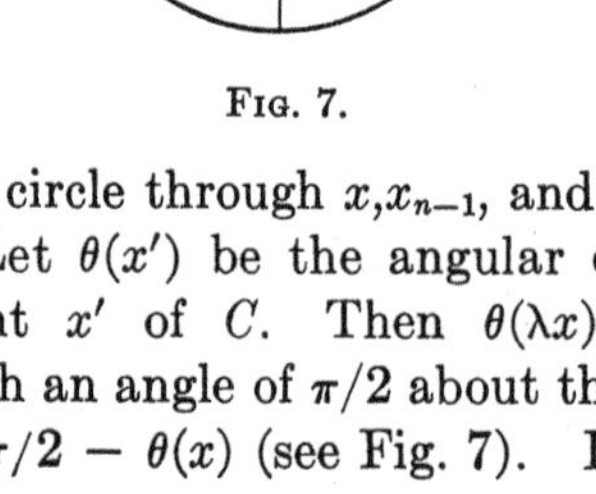

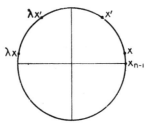

FIG. 7.

Let C be the great circle through x,x_{n-1}, and let S be the 2-sphere through C and x_n. Let $\theta(x')$ be the angular coordinate, measured from x_{n-1}, of a point x' of C. Then $\theta(\lambda x) = \pi - \theta(x)$. Hence $\phi(\lambda x)$ rotates S through an angle of $\pi/2$ about the point x' of C whose angular coordinate is $\pi/2 - \theta(x)$ (see Fig. 7). It carries x_{n-1} into the point of the arc from x_n to x' at an angle $\theta(x)$ from x_n. Then $\lambda\phi(\lambda x)\cdot x_n$ is on the arc from $-x_n$ to $\lambda x'$ at the angle $\theta(x)$ from $-x_n$. The rotation $\phi(x)^{-1}$ has $\lambda x'$ as fixed point, and carries $-x_n$ into x. Therefore $T(x)\cdot x_{n-1}$ lies on the arc from x to $\lambda x'$ at an angle of θ from x. Hence $\theta(T(x)\cdot x_{n-1}) = 2\theta(x)$. This proves the second statement. The next two sentences in §23.4 are immediate consequences.

To prove the last statement, let E,E' be the hemispheres $t_{n-1} \geqq 0$ and $t_{n-1} \leqq 0$ of S^{n-1}. By what has been shown, $p'T$ maps $E - S^{n-2}$ topologically onto $S^{n-1} - (-x_{n-1})$ and certainly with degree $+1$ if E is oriented concordantly with S^{n-1}; for $p'T|E$ is the result of a homotopy in which each point x moves along the great circle C through x_{n-1},x to the point whose angle from x_{n-1} is twice that of x; thus S^{n-1} is deformed over E' into $-x_{n-1}$. Since $T(x) = T(-x)$, $p'T$ also maps $E' - S^{n-1}$ topologically onto $S^{n-1} - (-x_{n-1})$; and its degree is that of the antipodal transformation $x \to -x$. The latter is $(-1)^n$. The total degree of $p'T$: $S^{n-1} \to S^{n-1}$ is the sum of the degrees $1 + (-1)^n$ on E,E'; and the result follows.

Remark. The map T_{n+1} was used by Hopf [49] to construct a map $S^{2n-1} \to S^n$ which, for even n, has Hopf invariant 2.

23.5. Corollary. *If n is odd, the kernel of $\pi_n(R_{n+1}) \to \pi_n(R_{n+2})$ is an infinite cyclic group and T_{n+2}: $S^n \to R_{n+1}$ represents a generator of this kernel.*

That T represents a generator of the kernel follows from §23.2. Since n is odd, $p'T$ has degree 2, and therefore represents an element of infinite order in $\pi_n(S^n)$. Hence T represents an element of infinite order.

23.6. The group $\pi_3(R_m)$.

Theorem. *If the maps ρ: $S^3 \to R_3$ and σ: $S^3 \to R_4$ are defined as in §22.3 and §22.6, then T_5: $S^3 \to R_4$ satisfies*

$$T_5(q) = \rho(q)^{-1}\sigma(q)^2.$$

Hence T_5 represents the element $-\alpha_3 + 2\beta_3$ (see §22.7). It follows that $\pi_3(R_5) \approx \pi_3(R_m)$ $(m \geqq 5)$ is an infinite cyclic group generated by the image of β_3.

As in §22.6, S^3 is regarded as the group of unit quaternions. Let the reference point x_3 be the quaternion 1. Referring to the definitions of ρ and σ, we have

$$\rho(q)^{-1}\sigma(q)^2\cdot q' = q^{-1}q^2q'q = qq'q.$$

If q' is orthogonal to 1 and q, it is easily shown that $q'q = \bar{q}q'$. Hence $qq'q = q'$. On the other hand, if q' lies on the circle through 1 and q (which is a 1-parameter subgroup), then $q'q = qq'$. Hence the transformation $q' \to qq'q = q^2q'$ is a rotation of this circle through an angle which is twice the angle from 1 to q. It follows now, from §23.4, that $T_5 = \rho^{-1}\sigma^2$. Since ρ represents α_3, and σ represents β_3, §16.7 implies that T_5 represents $-\alpha_3 + 2\beta_3$. This and §23.5 yield the conclusion of the theorem.

The next problem is to compute $\pi_4(R_5)$. As a first step we have

23.7. LEMMA. *When n is even, $\pi_n(R_n)$ maps homomorphically onto $\pi_n(R_{n+1})$. The kernel is either zero or a cyclic group of order 2. In the latter case the non-zero element of the kernel is represented by the composition of an essential map $S^n \to S^{n-1}$ followed by T_{n+1}: $S^{n-1} \to R_n$.*

Consider the following section of the homotopy sequence of (R_{n+1},R_n):

$$\pi_n(R_n) \xrightarrow{i_*} \pi_n(R_{n+1}) \xrightarrow{j_*} \pi_n(R_{n+1},R_n) \xrightarrow{\partial} \pi_{n-1}(R_n) \xrightarrow{i_*} \pi_{n-1}(R_{n+1}).$$

Now $\pi_n(R_{n+1},R_n) \approx \pi_n(S^n) = \infty$, and, by §23.2, its image under ∂ is generated by T_{n+1}. Since $n - 1$ is odd, §23.5 states that the kernel of i_* (on the right), which is the image of ∂, is infinite cyclic; hence the kernel of ∂ is zero. Therefore, by exactness, the image of j_* is zero. Again, by exactness, the image of i_* (equal to the kernel of j_*) is the entire group $\pi_n(R_{n+1})$. By §23.2, the kernel of i_* is the image group $T_{n+1*}\pi_n(S^{n-1})$. The case $n = 2$ was settled in §22.10. If $n \geqq 4$, then, by §21.6, $\pi_n(S^{n-1})$ is cyclic of order 2, and the lemma is proved.

23.8. The left distributive law. The second step in computing $\pi_4(R_5)$ is to establish a "left distributive law" for homotopy groups. Suppose, first, that S^n,S^m are oriented spheres and f_1,f_2: $S^n \to S^m$ represent elements α_1,α_2 of $\pi_n(S^m)$. Let h be a map $S^m \to X$. Then hf_1,hf_2 represent $h_*\alpha_1,h_*\alpha_2$ in $\pi_n(X)$. Since h_* is a homomorphism, we have $h_*(\alpha_1 + \alpha_2) = h_*\alpha_1 + h_*\alpha_2$. This is called the *right distributive law.*

In general, there is no left distributive law. Suppose f: $S^n \to S^m$ and h_1,h_2: $S^m \to X$. Let $h_1 + h_2$: $S^m \to X$ denote a map representing the sum of the elements of $\pi_m(X)$ represented by h_1 and h_2. Let α,α_1,α_2 be the elements of $\pi_n(X)$ represented by $(h_1 + h_2)f$, h_1f, h_2f respectively. In general, $\alpha \neq \alpha_1 + \alpha_2$. An example of this is provided by the Hopf map f: $S^3 \to S^2$ and $h_1 = h_2 =$ the identify map of S^2; in this case $\alpha = 2(\alpha_1 + \alpha_2)$ (see [49]). However, we have

LEMMA. *If f: $S^n \to S^m$ is homotopic to the suspension of a map $S^{n-1} \to S^{m-1}$ (see §21.3), then the left distributive law holds: $(h_1 + h_2)f \simeq h_1f + h_2f$.*

In §15 we have defined the addition in $\pi_m(X)$ assuming the elements are represented by maps of a cell. If they are represented by maps

h_1,h_2: $S^m \to X$, it is not hard to show that the sum is represented by a map $h_1 + h_2$: $S^m \to X$ defined as follows. Let S^{m-1} be an equator of S^m (containing the reference point). Pinch S^{m-1} to a point, obtaining two spheres S_1^m, S_2^m having a single (reference) point in common. Map S_i^m into X according to h_i ($i = 1,2$). The composition of these maps is $h_1 + h_2$: $S^m \to X$.

Suppose now that f is in the form of a suspension $f = Eg$ where g: $S^{n-1} \to S^{m-1}$. The composition of f and the pinching map k: $S^m \to S_1^m \cup S_2^m$ can be factored into the composition of the pinching map l: $S^n \to S_1^n \cup S_2^n$ followed by a map f': $S_1^n \cup S_2^n \to S_1^m \cup S_2^m$ which reduces to maps f_i: $S_i^n \to S_i^m$ ($i = 1,2$). This follows since Eg maps the upper and lower hemispheres of S^n on the corresponding hemispheres of S^m. Since these maps of hemispheres are constructed by mapping pole into pole and extending so as to carry great circles through the poles into great circles, it follows that f_i ($i = 1,2$) is a replica of Eg. Hence h_1f_1, h_2f_2 are replicas of h_1f, h_2f respectively. However l composed with replicas of h_1f, h_2f is the sum $h_1f + h_2f$. This completes the proof.

23.9. The group $\pi_4(R_5)$.

THEOREM. *The group $\pi_4(R_5)$ is cyclic of order 2. Its non-zero element is represented by the image of the generator β_4 of $\pi_4(R_4)$ (see §22.7).*

By §23.7, $\pi_4(R_4)$ maps onto $\pi_4(R_5)$; and, by §23.2, the kernel is $T_{5*}\pi_4(S^3)$. Let h: $S^4 \to S^3$ be an essential map. Then T_5h generates the kernel. But T_5 represents $-\alpha_3 + 2\beta_3$. Applying the left distributive law (§23.8), it follows that T_5h represents $-\alpha_4 + 2\beta_4$ (see §22.7). But α_4 and β_4 are of order 2. Hence α_4 generates the kernel of $\pi_4(R_4) \to \pi_4(R_5)$. This completes the proof.

23.10. The operation α_0.

LEMMA. *If α_0 is the generator of $\pi_0(O_n)$ and $\alpha \in \pi_i(R_n)$, then $\alpha_0(\alpha) - \alpha$ lies in the kernel of $\pi_i(R_n) \to \pi_i(R_{n+1})$.*

The operation α_0 is obtained by choosing a matrix μ of determinant -1, and conjugating R_n by μ. Let μ' be the matrix of order $n + 1$ obtained by bordering μ' with zeros except for a -1 on the diagonal. Then μ' has determinant $+1$; so μ' is in R_{n+1}. Since R_{n+1} is connected, μ' can be connected to e by a curve. It follows that conjugation of R_{n+1} by μ' is homotopic to the identity map. But μ' acts on $R_n \subset R_{n+1}$ exactly as does μ. This implies the assertion of the lemma.

23.11. LEMMA. *If α in $\pi_{n-1}(R_n)$ is represented by T_{n+1}: $S^{n-1} \to R_n$, and α_0 is the generator of $\pi_0(O_n)$, then $\alpha_0(\alpha) = -\alpha$.*

Let μ be the diagonal matrix with 1's down the diagonal save for a -1 in the last place. Then $\alpha_0(\alpha)$ is represented by $\mu T \mu^{-1}$. Referring to the matrix form of T (formula (6) of §23.3), we obtain the matrix identity $\mu T(x)\mu^{-1} = T(\mu(x))$. But μ maps S^{n-1} on itself with degree -1. Hence the composition of μ followed by T represents $-\alpha$.

§24. A CHARACTERISTIC MAP FOR THE BUNDLE U_n OVER S^{2n-1}

24.1. Motivation. To compute $\pi_4(R_n)$ for $n \geqq 6$, it is only necessary to compute the kernel of $\pi_4(R_5) \to \pi_4(R_6)$ (see §22.8). Since $\pi_4(R_5)$ is cyclic of order 2 generated by the image of β_4 (see §23.9), and T_6: $S^4 \to R_5$ represents a generator of the kernel, we have only to decide whether T_6 represents the zero or the image of β_4. Since T_6 composed with the projection p: $R_5 \to S^4$ gives a self map of S^4 of degree 0 (§23.4), pT_6 is homotopic to a constant. A covering homotopy will deform T_6 into a map T': $S^4 \to R_4$. We shall exhibit an explicit T', and show that it represents β_4; thus proving that $\pi_4(R_n) = 0$ for $n \geqq 6$.

The T' to be constructed is just the characteristic map associated with the unitary group U_3 considered as a bundle over S^5. We proceed therefore to generalize the construction of T_n from the orthogonal to the unitary case.

24.2. The normal form. Let U_{m+1} be the unitary group operating in the space $(z_0, z_1, \cdots, z_m)$ of $m + 1$ complex variables. If we pass to real coordinates by the rule $z_j = t_{2j+1} + it_{2j}$, any unitary transformation becomes an orthogonal transformation of determinant $+1$. In this way, we have

$$U_{m+1} \subset R_{2m+2}. \tag{1}$$

Both groups operate transitively on the unit sphere S^{2m+1}. Let x_{2m+1} be the reference point $z_j = 0$ for $j < m$, $z_m = 1$. Then the bundle projection

$$p\colon \quad U_{m+1} \to S^{2m+1}$$

given by $p(u) = u(x_{2m+1})$ agrees with p: $R_{2m+2} \to S^{2m+1}$. The fibre is U_m.

Let S^{2m} be the equator of S^{2m+1} orthogonal to x_{2m+1} (defined by $\Re z_m = 0$), and let S^{2m-1} be the equator of S^{2m} defined by $z_m = 0$. Define reference points by

$$x_{2m} = (0, \cdots, 0, i), \qquad x_{2m-1} = (0, \cdots, 0, 1, 0).$$

We parallel the normal form construction of §23.3. The map ϕ: $S^{2m+1} - (-x_{2m+1}) \to U_{m+1}$ is defined by generalizing the matrix formula (1)

$$(2) \qquad \phi(x) = \left\| \begin{array}{c|c} \delta^\alpha_\beta - \dfrac{z_\alpha \bar{z}_\beta}{1 + \bar{z}_m} & \begin{array}{c} z_0 \\ \cdot \\ \cdot \\ \cdot \\ z_{m-1} \end{array} \\ \hline -b\bar{z}_0 \quad \cdots \quad -b\bar{z}_{m-1} & z_m \end{array} \right\|$$

where

$$x = (z_0, z_1, \cdots, z_m), \qquad b = \frac{1 + z_m}{1 + \bar{z}_m}.$$

Under the assumption $x \in S^{2m+1}$, $x \neq -x_{2m+1}$ (i.e. $\Sigma z_i \bar{z}_i = 1$, $z_m \neq -1$), it is easily shown that $\phi(x)$ is a unitary transformation, it is continuous, and $p\phi(x) = x$.

Let λ be the unitary transformation

$$\lambda(z_0, \cdots, z_m) = (z_0, \cdots, z_{m-1}, -z_m).$$

Using these values of ϕ and λ we define the coordinate functions ϕ_1, ϕ_2 just as in (2) of §23.3, and arrive at the characteristic map

$$(3) \qquad T'_{m+1}\colon \quad S^{2m} \to U_m$$

defined by

$$(4) \qquad T'_{m+1}(x) = \phi(x)^{-1}\lambda\phi(\lambda(x)), \qquad x \in S^{2m}.$$

Straightforward computation gives the matrix form

$$(5) \qquad T'_{m+1}(x) = \left\| \delta^\alpha_\beta - \frac{2z_\alpha \bar{z}_\beta}{(1 + z_m)^2} \right\|, \quad (\alpha, \beta = 0, 1, \cdots, m-1).$$

One checks immediately that $T'_{m+1}(x_{2m})$ is the unit matrix.

24.3. Properties of the characteristic map T'.

THEOREM. *The characteristic map $T'_{m+1}\colon\ S^{2m} \to U_m$ for a normal form of the unitary bundle $p\colon\ U_{m+1} \to S^{2m+1}$ is given by the matrix (5) above. If $p'\colon\ U_m \to S^{2m-1}$ is the projection $p'(u) = u(x_{2m-1})$, then the map $p'T'_{m+1}\colon\ S^{2m} \to S^{2m-1}$ is essential when m is even and inessential when m is odd.*

To prove the second part of the theorem, let $(w_0, \cdots, w_{m-1})$ denote the coordinates of $p'T'_{m+1}(x)$, so that $p'T'_{m+1}$ is given by the equations

$$(6) \qquad \begin{aligned} w_\alpha &= -2z_\alpha \bar{z}_{m-1}(1 + z_m)^{-2}, \qquad \alpha = 0, \cdots, m-2, \\ w_{m-1} &= 1 - 2|z_{m-1}|^2(1 + z_m)^{-2}, \end{aligned}$$

where $\Sigma_0^m z_i \bar{z}_i = 1$ and $\mathcal{R} z_m = 0$.

Let S^{2m-2} be the equator of S^{2m-1} defined by $\mathscr{I}w_{m-1} = 0$. If we put $z_m = 0$ in (6) we obtain

$$
\begin{aligned}
w_\alpha &= -2z_\alpha\bar{z}_{m-1}, \qquad \alpha = 0, \cdots, m-2,\\
w_{m-1} &= 1 - 2|z_{m-1}|^2.
\end{aligned}
\tag{7}
$$

If we define $h = p'T'_{m+1}|S^{2m-1}$, it follows that h is a map

$$
h\colon\quad S^{2m-1} \to S^{2m-2}.
\tag{8}
$$

We assert that $p'T'_{m+1}$ *is homotopic to the suspension of* h (see §21.3). From (6) one shows that $\mathscr{I}w_{m-1}$ and $\mathscr{I}z_m$ have the same sign. Hence $p'T'$ maps the hemisphere $\mathscr{I}z_m > 0$ (< 0) of S^{2m} into the hemisphere $\mathscr{I}w_{m-1} > 0$ (< 0) of S^{2m-1}. The suspension of h does likewise. It follows that $Eh(x)$ and $p'T'(x)$ are not antipodes for any x. Then a homotopy of Eh into $p'T'$ is given by letting $F(x,t)$ be the point which divides the shortest great circle arc from $Eh(x)$ to $p'T'(x)$ in the ratio $t:(1-t)$.

Consider now the case $m = 2$. Then $h\colon\ S^3 \to S^2$ is given by

$$
w_0 = -2z_0\bar{z}_1, \qquad w_1 = 1 - 2z_1\bar{z}_1.
\tag{9}
$$

We assert that h *is the Hopf map* (see §20.1). Recall that the latter map assigns to the point (z_0,z_1) of S^3 the point $[z_0,z_1]$ (homogeneous coordinates) of S^2. First pass to an inhomogeneous coordinate in S^2 by $z = -z_0/z_1$. Then, by the stereographic projection

$$
w_0 = \frac{2z}{1 + z\bar{z}}, \qquad w_1 = 1 - \frac{2}{1 + z\bar{z}},
$$

pass to coordinates (w_0,w_1) for S^2 where w_0 is complex, w_1 is real and $|w_0|^2 + w_1^2 = 1$. One computes quickly that the final assignment $(z_0,z_1) \to (w_0,w_1)$ of the Hopf map is given by (9); and thus the assertion is proved. Since $p'T'$ is homotopic to Eh, it follows from §21.5 that $p'T'$ is essential.

When m is even and > 2, we will show that h *is homotopic to the* *(2m − 4)-fold suspension of the Hopf map.* Since the Hopf map is given by (9), it is easy to see that the $(2m-4)$-fold suspension of the Hopf map is given by

$$
\begin{aligned}
w_\alpha &= z_\alpha, && \alpha = 0, 1, \cdots, m-3,\\
w_{m-2} &= -\frac{2}{c} z_{m-2}\bar{z}_{m-1} && \text{where } c = (|z_{m-2}|^2 + |z_{m-1}|^2)^{1/2}\\
w_{m-1} &= c - \frac{2}{c}|z_{m-1}|^2 && \text{and } c \neq 0\\
w_{m-2} &= w_{m-1} = 0 && \text{where } c = 0.
\end{aligned}
\tag{10}
$$

The homotopy connecting h and the suspension is demonstrated by writing equations for it. (These equations can be derived by considering a similar problem, in the real case, for the map $p'T_n$, defining a homotopy for it by a geometric construction, deriving equations for the homotopy, and by generalizing these equations to the complex case. Regardless of how the homotopy is found, it suffices to exhibit it.) For each $0 \leqq t \leqq 1$, let

$$t' = 1 - t, \qquad \tau = [t(1-t)]^{1/2}, \qquad \sigma = [1 - t(1-c^2)]^{1/2}.$$

Then the homotopy is given by

$$\begin{aligned} w_{2j} &= -tz_{2j} + \tau \bar{z}_{2j+1} - \frac{2}{\sigma}[t' z_{2j}\bar{z}_{m-1} + \tau z_{m-1}\bar{z}_{2j+1}], \\ w_{2j+1} &= -tz_{2j+1} - \tau \bar{z}_{2j} - \frac{2}{\sigma}[t' z_{2j+1}\bar{z}_{m-1} - \tau z_{m-1}\bar{z}_{2j}], \\ &\qquad \text{for } j = 0, 1, \cdots, (m-4)/2, \\ w_{m-2} &= -\frac{2}{\sigma} z_{m-2}\bar{z}_{m-1} \\ w_{m-1} &= \sigma - \frac{2}{\sigma}|z_{m-1}|^2. \end{aligned} \tag{11}$$

Notice that the hypothesis that m is even has been used. When $t = 0$, these equations reduce to (7). It is to be noted that the equations (11) are not defined everywhere for $t = 1$. However it is easily proved that they converge, uniformly in z, to the equations (10) as $t \to 1$. That (11) gives the desired homotopy will follow if it is shown that the image of S^{2m-1} remains on S^{2m-2} during the homotopy. This requires proving that $\Sigma_0^{m-1} z_\alpha \bar{z}_\alpha = 1$ and the equations (11) imply $\Sigma_0^{m-1} w_\alpha \bar{w}_\alpha = 1$ for all t. The proof requires about two pages of computation. Since it is entirely mechanical, it is omitted.

The preceding result together with $p'T' \simeq Eh$ implies that $p'T'$ is homotopic to the $(2m-3)$-fold suspension of the Hopf map when m is even. It follows from §21.5 that $p'T'$ is essential.

The final statement of §24.3 is obtained as a corollary of the theorem below. Let Sp_n denote the *symplectic group* on n quaternionic variables (defined in §20.2). If each quaternionic variable is expressed in terms of its complex components $q_\alpha = z_{2\alpha} + jz_{2\alpha+1}$, each symplectic transformation of the q's is a unitary transformation of the z's. This follows since the scalar product $z \cdot z' = \Sigma_1^{2n} \bar{z}_\beta \cdot z'_\beta$ is the "complex" part of $q \cdot q' = \Sigma_1^n \bar{q}_\alpha \cdot q'_\alpha$. We may therefore regard Sp_n as a subgroup of U_{2n} which, in turn, is a subgroup of R_{4n}.

24.4. THEOREM. *For each $m > 0$, the bundle R_{2m} over S^{2m-1} is equivalent in its group, R_{2m-1}, to the unitary bundle U_m over S^{2m-1} whose group is $U_{m-1} \subset R_{2m-2}$. The bundle U_{2m} over S^{4m-1} is equivalent in its group, U_{2m-1}, to the symplectic bundle Sp_m over S^{4m-1} whose group is $Sp_{m-1} \subset U_{2m-2}$. Hence the bundle R_{4m} over S^{4m-1} is equivalent in R_{4m-1} to the bundle Sp_m over S^{4m-1} whose group is $Sp_{m-1} \subset R_{4m-4}$.*

Observe first that the projection p: $R_{2m} \to S^{2m-1}$ when restricted to U_m is the projection $U_m \to S^{2m-1}$ (the same base point was used). Let f be a local cross-section of U_{m-1} in U_m (see §7.4). Then f is also a local cross-section of R_{2m-1} in R_{2m}; for $U_m \cap R_{2m-1} = U_{m-1}$. Recall now the construction (§7.4) of a bundle structure in R_{2m} based on f. We chose the elements of R_{2m} as the indexing set. Consider the subset of coordinate functions with indices in U_m. Since U_m is transitive on S^{2m-1}, the corresponding coordinate neighborhoods cover S^{2m-1}. Hence the subset provides a bundle structure, and it is strictly equivalent to the original structure. However the coordinate transformations of the new bundle are precisely those which the construction of §7.4 assigns to the bundle U_m over S^{2m-1}. This proves the first assertion. The proof of the second is entirely similar. The transitivity of Sp_m on S^{4m-1} was noted in §20.2.

REMARK. The statement of §24.4 is somewhat awkward due to the comparison of bundles with different fibres. The reason for this is that, while R_{2m} is a principal bundle over S^{2m-1}, this is not true of the strictly equivalent structure with group U_{m-1}. The associated principal bundle of the latter is U_m over S^{2m-1}.

The groups of the bundles appearing in §24.4 are all arcwise connected. Hence any two characteristic maps for such a bundle are homotopic with base point fixed (see §§18.3–18.5). This yields

24.5. COROLLARY. *The characteristic map T_{2m}: $S^{2m-2} \to R_{2m-1}$ is homotopic in R_{2m-1} to the characteristic map T'_m: $S^{2m-2} \to U_{m-1}$; and T'_{2m}: $S^{4m-2} \to U_{2m-1}$ is homotopic in U_{2m-1} to a characteristic map T''_m: $S^{4m-2} \to Sp_{m-1}$ of the bundle Sp_m over S^{4m-1}.*

The final statement of §24.3 may now be proved. When $m = 2n - 1$ is odd, then T'_{m+1} is homotopic to T''_n. Since $Sp_{n-1} \subset U_{2n-2}$ and U_{2n-2} is the fibre of p': $U_{2n-1} \to S^{2m-1}$, the image of the homotopy under p' is a homotopy of $p'T'_{m+1}$ to a constant map.

24.6. The group $\pi_4(R_n)$.

THEOREM. *$\pi_4(R_n) = 0$ for $n \geqq 6$.*

The first part of the proof was given in §24.1. By §24.5, T_6 is homotopic in R_5 to T'_3: $S^4 \to U_2 \subset R_4$. By definition, the map σ of §22.6 maps S^3 topologically onto the symplectic group Sp_1. Since $Sp_1 \subset U_2$, the bundle U_2 over S^3 has a cross-section; and, therefore, U_2

is a product bundle (§8.4)

$$U_2 = S^3 \times U_1.$$

But $U_1 = S^1$, and $\pi_i(S^1) = 0$ for $i > 1$ (see §21.2). Therefore, by §17.8,

$$p_*: \quad \pi_i(U_2) \approx \pi_i(S^3) \qquad (i > 1),$$

where p is the projection $U_2 \to S^3$. By §24.3, pT'_3: $S^4 \to S^3$ is the non-zero element of $\pi_4(S^3)$. Since σ: $S^3 \to U_2$ is the cross-section, σ_* is inverse to p_*. Hence, the essential map h: $S^4 \to S^3$ followed by σ is homotopic in U_2 to T'_3. By 22.7, T'_3 represents β_4 in $\pi_4(R_3)$. This completes the proof.

24.7. Properties of the characteristic map T.

LEMMA. *For all n, the composition of the homomorphisms*

$$\pi_{n+1}(R_{n+1},R_n) \xrightarrow{\partial} \pi_n(R_n) \to \pi_n(R_n,R_{n-1})$$

is zero.

By §17.2, the second homomorphism is equivalent to p_*: $\pi_n(R_n) \to \pi_n(S^{n-1})$. It suffices to show that $p_*\partial = 0$. Since $\pi_1(S^0)$ and $\pi_2(S^1)$ are zero, the assertion is trivial for $n = 1$ and 2. When $n = 3$, $\pi_4(R_4,R_3) \approx \pi_4(S^3) = 2$ while $\pi_3(S^2) = \infty$; so, again, $p_*\partial = 0$.

Suppose $n \geqq 4$. Exactness of the homotopy sequence of (R_{n+1},R_n) implies that the image of ∂ is the kernel of $\pi_n(R_n) \to \pi_n(R_{n+1})$; and, by §23.2, the kernel is $T_{n+1*}\pi_n(S^{n-1})$. By §23.4, when n is odd, pT_{n+1} is homotopic to a constant. Then $p_*T_{n+1*} = 0$ and the result follows. When n is even, pT_{n+1} maps S^{n-1} on itself with degree 2. Let h be the essential map $S^n \to S^{n-1}$. Since $n \geqq 4$, h is a suspension. Therefore the left distributive law §23.8 applies, and it asserts that $pT_{n+1}h$ represents twice the element represented by h. Since $\pi_n(S^{n-1}) = 2$, $pT_{n+1}h$ represents zero. This completes the proof.

24.8. THEOREM. *If $n \equiv 0 \bmod 4$, then the characteristic map T_{n+2}: $S^n \to R_{n+1}$ is not homotopic to a map of S^n into R_{n-1}.*

Consider the diagram

$$\begin{array}{ccccc} & & \pi_n(R_n) & \xrightarrow{i_*} & \pi_n(R_{n+1}) \\ & & \downarrow j_* & & \downarrow k_* \\ \pi_{n+1}(R_{n+1},R_n) & \xrightarrow{\partial} & \pi_n(R_n,R_{n-1}) & \xrightarrow{m_*} & \pi_n(R_{n+1},R_{n-1}). \end{array}$$

The lower line is from the homotopy sequence of the triple (R_{n+1},R_n,R_{n-1}). By §24.7, $\partial = 0$. Exactness implies that the kernel of m_* is zero.

Let $q = (n + 2)/2$ so that q is odd. By §24.5, T_{n+2} is homotopic to T'_q: $S^n \to U_{q-1}$. Since $U_{q-1} \subset R_n$, T'_q represents an element α in $\pi_n(R_n)$ such that $i_*\alpha$ is represented by T_{n+2}. The composition

$$S^n \xrightarrow{T'} U_{q-1} \xrightarrow{l} R_n \xrightarrow{j} (R_n,R_{n-1}) \xrightarrow{p} S^{n-1}$$

where l is the inclusion map, and p is the projection, is just the map $p'T'$ shown in §24.3 to be essential. It follows that $j_*\alpha \neq 0$. Since the kernel of m_* is zero, we have $m_*j_*\alpha \neq 0$. Since $mj = ki$, it follows that $k_*i_*\alpha \neq 0$. Therefore kT_{n+2} represents a non-zero element of $\pi_n(R_{n+1},R_{n-1})$. This is equivalent to the desired conclusion.

24.9. LEMMA. *If n is even the element of $\pi_n(R_{n+1})$ represented by T_{n+2} is either zero or of order 2.*

Consider the diagram

$$\pi_{n+1}(R_{n+2}) \xrightarrow{k_*} \pi_{n+1}(R_{n+2},R_{n+1}) \xrightarrow{\partial} \pi_n(R_{n+1})$$

where k is the inclusion map. Since $n + 1$ is odd, $p'T_{n+3}$ has degree 2. Hence kT_{n+3} represents twice a generator of $\pi_{n+1}(R_{n+2},R_{n+1})$. By exactness, the kernel of ∂ contains at least the even elements. Therefore the image of ∂ (generated by T_{n+2}) is at most a cyclic group of order 2.

24.10. THEOREM. *If $n \equiv 0 \bmod 4$, then $\pi_n(R_{n+1})$ contains a cyclic group of order 2 whose non-zero element is represented by T_{n+2}.*

This is a corollary of the last two results.

24.11. Remarks. Further progress in the determination of the groups $\pi_i(R_n)$ awaits the determination of the groups $\pi_i(S^n)$. Assuming the truth of Pontrjagin's assertion that $\pi_5(S^3) = 0$, it has been shown [99] that $\pi_5(R_n) = 0$ when $n \neq 6$, and $\pi_5(R_6) = \infty$.

The important role in the calculations of $\pi_i(R_n)$ played by the characteristic maps of the orthogonal and unitary bundles over spheres suggests that a characteristic map

$$T''_m\colon\ S^{4m-2} \to Sp_{m-1}$$

for the symplectic bundle Sp_m over S^{4m-1} may be useful in future work (see §24.5). Such a map is readily obtained. The construction of §24.2 is followed in detail using quaternions in place of complex numbers. The term $z_\alpha\bar{z}_\beta/(1 + \bar{z}_m)$ in the matrix (2) must be written $q_\alpha(1 + \bar{q}_m)^{-1}\bar{q}_\beta$. Then (5) becomes

$$T''_{m+1}(x) = \|\delta^\alpha_\beta - 2q_\alpha(1 + q_m)^{-2}\bar{q}_\beta\|.$$

Perhaps it is worthwhile to summarize the main results concerning $\pi_i(R_n)$ in the form of a table. As usual ∞ means an infinite cyclic group, 2 means a cyclic group of order 2, and + means direct sum.

	R_2	R_3	R_4	R_5	R_6	$\cdots$	R_n
π_1	∞	2	2	2	2	$\cdots$	2
π_2	0	0	0	0	0	$\cdots$	0
π_3	0	∞	$\infty + \infty$	∞	∞	$\cdots$	∞
π_4	0	2	2 + 2	2	0	$\cdots$	0

ADDED IN PROOF. If we adjoin to the foregoing constructions the result that $\pi_5(S^3) = 2$ (see §21.7), we obtain easily the following:

$$\pi_5(R_2) = 0, \quad \pi_5(R_3) = 2, \quad \pi_5(R_4) = 2 + 2,$$
$$\pi_5(R_5) = 2, \quad \pi_5(R_6) = \infty, \quad \pi_5(R_n) = 0 \text{ for } n > 6.$$

§25. THE HOMOTOPY GROUPS OF MISCELLANEOUS MANIFOLDS

25.1. The unitary groups. The first four homotopy groups of the unitary groups are readily obtained from the results of §24. The group U_1 is, topologically, a 1-sphere; hence

$$\pi_i(U_1) = \pi_i(S^1) = \begin{cases} \infty & i = 1, \\ 0 & i > 1. \end{cases} \tag{1}$$

In the proof of §24.6, it is observed that U_2 is homeomorphic to $S^3 \times S^1$. Therefore, by §17.8,

$$\pi_1(U_2) = \pi_1(S^1) = \infty, \tag{2}$$

$$\pi_i(U_2) = \pi_i(S^3) = \begin{cases} 0 & i = 2, \\ \infty & i = 3, \\ 2 & i = 4. \end{cases} \tag{3}$$

The generator of $\pi_1(U_2)$ is the image of that of $\pi_1(U_1)$. A generator of $\pi_3(U_2)$ is represented by the map σ of §22.6. The generator of $\pi_4(U_2)$ is represented by σh where h: $S^4 \to S^3$ is essential.

25.2. THEOREM. *If $i < 2n$, $\pi_i(U_n) \approx \pi_i(U_{n+1})$. The group $\pi_{2n}(U_n)$ maps onto $\pi_{2n}(U_{n+1})$, and its kernel is a cyclic group with a generator represented by the characteristic map T'_{n+1}: $S^{2n} \to U_n$ (see §24.3).*

Since $U_{n+1}/U_n = S^{2n+1}$, we have $\pi_i(U_{n+1}, U_n) = 0$ for $i < 2n + 1$ (see §17.2). This and the usual exactness argument yield the first two conclusions. The final statement follows from §23.2.

25.3. THEOREM. *If n is even, T'_{n+1} represents a non-zero element of $\pi_{2n}(U_n)$.*

By §24.3, $p'T'_{n+1}$: $S^{2n} \to S^{2n-1}$ is essential; hence T'_{n+1} is essential

25.4. THEOREM. *For all n,* $\pi_1(U_n) \approx \pi_1(U_1) = \infty$. *For all n,* $\pi_2(U_n) = 0$. *For all* $n \geqq 2$, $\pi_3(U_n) \approx \pi_3(U_2) = \infty$. *For all* $n \geqq 3$, $\pi_4(U_n) = 0$.

The first three statements follow directly from §25.2. Since $\pi_4(U_2)$ has only one non-zero element, by §25.3, it is represented by T'_3. This and the last statement of §25.2 imply $\pi_4(U_3) = 0$, and hence $\pi_4(U_n) = 0$.

25.5. The sympletic groups. We have $Sp_1 = S^3$ (the group of unit quaternions). Since $Sp_{n+1}/Sp_n = S^{4n+3}$, the analog of §25.2 is

$$\pi_i(Sp_n) \approx \pi_i(Sp_{n+1}) \qquad \text{for } i \leqq 4n+1$$

and $\pi_{4n+2}(Sp_n)$ maps onto $\pi_{4n+2}(Sp_{n+1})$. We have therefore

$$\pi_i(Sp_n) \approx \pi_i(Sp_1) = \pi_i(S^3) \qquad \text{for } i \leqq 5.$$

Hence, for every n,

$$\pi_1(Sp_n) = \pi_2(Sp_n) = 0, \qquad \pi_3(Sp_n) = \infty, \qquad \pi_4(Sp_n) = 2.$$

25.6. The Stiefel manifolds. We turn our attention now to the Stiefel manifolds

$$V_{n,k} = R_n/R_{n-k}, \qquad k < n \qquad \text{(see §7.8)}$$

of orthogonal k-frames in n-space.

THEOREM. *The Stiefel manifold* $V_{n,k}$ *is arcwise connected, and* $\pi_i(V_{n,k}) = 0$ *for* $i < n - k$, *and*

$$\pi_{n-k}(V_{n,k}) = \begin{cases} \infty & \text{if } n-k \text{ is even, or } k = 1, \\ 2 & \text{if } n-k \text{ is odd, and } k > 1. \end{cases}$$

A generator of $\pi_{n-k}(V_{n,k})$ *is represented by a map* $f\colon\ S^{n-k} \to V_{n,k}$ *constructed as follows. Let* v_0 *be a fixed orthogonal* $(k-1)$*-frame in n-space. Let* S^{n-k} *be the unit sphere in the* $(n-k+1)$*-space orthogonal to* v_0. *Then f assigns to x in* S^{n-k} *the orthogonal k-frame consisting of the vector x followed by the vectors of* v_0.

Since $V_{n,k} = O_n/O_{n-k}$ and $k < n$, the first two statements repeat the lemma of §19.5. Set $h = n - k$, and consider the following section of the homotopy sequence of the triple (R_n, R_{h+1}, R_h):

$$\pi_{h+1}(R_n,R_{h+1}) \xrightarrow{\partial'} \pi_h(R_{h+1},R_h) \xrightarrow{j_*} \pi_h(R_n,R_h) \to \pi_h(R_n,R_{h+1}).$$

Since $h < n - k + 1$, the part already proved implies

$$\pi_h(R_n,R_{h+1}) \approx \pi_h(V_{n,k-1}) = 0.$$

Hence the exactness of the sequence implies that j_* is onto. Since

$$\pi_h(R_{h+1},R_h) \approx \pi_h(S^h) = \infty,$$

it remains to compute the kernel of j_* = the image of ∂'. When $k = 1$, we have $n = h + 1$, so $\pi_{h+1}(R_n,R_{h+1}) = 0$, and the image of ∂' is zero. Assume $k > 1$, and consider the diagram:

$$\begin{array}{lccc} \pi_{h+1}(R_{h+2},R_{h+1}) & & & \\ \downarrow j'_* & \searrow \partial & & l_* \\ & & \pi_h(R_{h+1}) & \longrightarrow \pi_h(R_{h+1},R_h) \\ & \nearrow \partial'' & & \\ \pi_{h+1}(R_n,R_{h+1}) & & & \end{array}$$

Replacing h by $h + 1$ in the preceding argument shows that j'_* is onto. By definition, $\partial' = l_*\partial''$. Hence the image of ∂' is the image of $l_*\partial$. As $R_{h+1}/R_h = S^h$, l_* is equivalent to p_*: $\pi_h(R_{h+1}) \longrightarrow \pi_h(S^h)$. By §23.5, the image of ∂ is a cyclic group and T_{h+2} represents a generator. By §23.4, pT_{h+2} has degree 0 or 2 according as $h + 1$ is odd or even. Hence the image of $p_*\partial$ is accordingly zero or the set of even elements. This holds equally for the image of $l_*\partial$ = image of ∂'. It follows that $\pi_h(V_{n,k}) \approx \pi_h(R_n,R_h)$ is, correspondingly, infinite cyclic or cyclic of order 2.

To show that f represents a generator, let p be the projection $(R_n,R_h) \longrightarrow (R_n/R_h,x_0) = (V_{n,k},x_0)$ where $x_0 = p(R_h)$. Let $p' = p|R_{h+1}$ so that p': $(R_{h+1},R_h) \longrightarrow (R_{h+1}/R_h,x_0) = (S^h,x_0)$. Then commutativity holds in the diagram

$$\begin{array}{ccc} \pi_h(R_{h+1},R_h) & \xrightarrow{j_*} & \pi_h(R_n,R_h) \\ \downarrow p'_* & & \downarrow p_* \\ \pi_h(S^h) & \xrightarrow{m_*} & \pi_h(V_{n,k}) \end{array}$$

where m is the inclusion map, and p'_*,p_* are isomorphisms. Since j_* is onto, so also is m_*. It follows that m: $S^h \longrightarrow V_{n,k}$ represents a generator of $\pi_h(V_{n,k})$. Now

$$p'': \quad V_{n,k} \longrightarrow V_{n,k-1} = R_n/R_{h+1}$$

is the projection which assigns to each k-frame the $(k - 1)$-frame obtained by omitting the first vector (see §7.7). Clearly, S^h is a fibre of p''. Since $V_{n,k-1}$ is arcwise connected, any one fibre may be deformed into any other. (Since $\pi_1(V_{n,k}) = 0$, a homotopy need not keep fixed a reference point.) The result follows now if we observe that f maps S^{n-k} topologically onto the fibre of p'' over v_0.

25.7. The complex Stiefel manifolds. Let $W_{n,k} = U_n/U_{n-k}$ be the Stiefel manifold of orthogonal k-frames in a complex n-space. The results analogous to §25.6, are

$$\pi_i(W_{n,k}) = \begin{cases} 0 & \text{for } i < 2n - 2k + 1, \\ \infty & \text{for } i = 2n - 2k + 1 \end{cases}$$

and, in the second case, any fibre of the projection $W_{n,k} \to W_{n,k-1}$ represents a generator. The proofs are omitted since they are similar and somewhat easier.

25.8. The Grassmann manifolds. Recall that the Grassmann manifold of k-planes in n-space is given by

$$M_{n,k} = O_n/O_k \times O'_{n-k} \qquad \text{(see §7.9)}.$$

If we set $V_{n,k} = O_n/O'_{n-k}$, then we have a bundle projection p: $V_{n,k} \to M_{n,k}$ whose fibre is O_k. By §25.5, every third term $\pi_i(V_{n,k})$ of the homotopy sequence of this bundle is zero up through $i = n - k - 1$. Since the homotopy sequence is exact, we have

THEOREM. *If $1 \leqq i < n - k$, then*

$$\Delta\colon \quad \pi_i(M_{n,k}) \approx \pi_{i-1}(O_k).$$

In particular, $\pi_1(M_{n,k})$ is cyclic of order 2. Its simply-connected covering space is $\tilde{M}_{n,k} = R_n/R_k \times R'_{n-k}$ (see §7.9).

In applying §25.8 when $k > n/2$, one first uses the homeomorphism of $M_{n,k}$ with $M_{n,n-k}$. Then §25.8 gives the more useful results

$$\pi_i(M_{n,k}) \approx \pi_{i-1}(O_{n-k}), \qquad 1 \leqq i < k.$$

§26. SPHERE BUNDLES OVER SPHERES

26.1. Bundles over S^1. Recall that a k-sphere bundle is one whose fibre is S^k and whose group is O_{k+1}. If the base space is an n-sphere, the classification theorem 18.5 reduces the enumeration of bundles to the computation of $\pi_{n-1}(O_{k+1})$ and the operations of $\pi_0(O_{k+1})$. Having computed some of these homotopy groups, we may interpret the results in terms of sphere bundles over spheres. This process, of course, is entirely mechanical. We shall supplement it with comments on the structures of the various bundle spaces obtained.

Taking $n = 1$, we have $\pi_0(O_{k+1}) = 2$. Thus there is one non-trivial k-sphere bundle over S^1. Since $\pi_0(O_1)$ maps onto $\pi_0(O_{k+1})$, it is equivalent in O_{k+1} to a bundle with group O_1. When $k = 0$, B is a circle and p is the double covering. When $k = 1$, B is the Klein bottle (§1.4). For $k > 1$, one might call B a generalized Klein bottle. It is constructed by forming the product of S^k with an interval and matching the ends under an orientation reversing transformation. The resulting bundle is always non-orientable; hence it is not equivalent to a product bundle in any group.

Since O_1 consists of two points, $\pi_i(O_1) = 0$ for $i > 0$. It follows that any 0-sphere bundle over S^n is a product bundle for all $n > 1$. This is also evident since $\pi_1(S^n) = 0$ (see §13.9).

26.2. Bundles over S^2. Consider now the 1-sphere bundles over S^2. Since $\pi_1(R_2) = \infty$, there is a bundle corresponding to each integer. Since $\pi_0(R_2) = 0$, no two of these are equivalent in R_2. However, under equivalence in O_2, the bundle corresponding to an integer m is equivalent to the one corresponding to $-m$; for α_0 reverses sign in $\pi_1(R_2)$ (see §16.10). We thus have one type of bundle for each $m \geqq 0$.

As shown §22.3, S^2 is the coset space of S^3 (the group of unit quaternions) by a 1-parameter subgroup S^1. Since $\pi_2(S^3) = \pi_1(S^3) = 0$, exactness of the homotopy sequence of this bundle implies that $\Delta\colon\ \pi_2(S^2) \approx \pi_1(S^1)$. Hence the bundle corresponds to $m = 1$ in the above classification. Now let H_m be a cyclic subgroup of S^1 of order $m > 1$. Let $B_m = S^3/H_m$, then $B_m \rightarrow S^3/S^1 = S^2$ is a bundle over S^2 with fibre and group S^1/H_m (see §7.4), and S^1/H_m is a 1-sphere. Since S^3 is connected and simply-connected, it follows that $\pi_1(B_m)$ is cyclic of order m. From $\pi_1(S^2) = 0$ and exactness of the homotopy sequence of the bundle it follows that $\pi_1(S^1/H_m)$ maps onto $\pi_1(B_m)$. Applying exactness again, it follows that Δ maps a generator of $\pi_2(S^2)$ onto m times a generator of $\pi_1(S^1/H_m)$. Therefore $B_m \rightarrow S^2$ corresponds to the integer m in the above classification.

Taking $m = 2$ we obtain the bundle $R_3 \rightarrow S^2$.

Any two of the spaces B_m $(m = 0, 1, \cdots)$ are topologically distinct since their fundamental groups are different. Hence no two are equivalent in any group of homeomorphisms of S^1.

The space B_m is just the "lens space" $(m,1)$ (see [85, p. 210]).

26.3. When $k > 1$, $\pi_1(R_{k+1}) = 2$; hence there is just one non-trivial k-sphere bundle over S^2. Since $\pi_1(R_2)$ maps onto $\pi_1(R_{k+1})$, this bundle is equivalent to a bundle with the group R_2. The latter is not unique but any two such are equivalent in R_3.

As an example, let $k = 2$. Let T map the equator of S^2 topologically onto the circle R_2 in R_3. Let $\mathfrak{B}$ denote the coordinate bundle over S^2 having S^2 as fibre, R_2 as group, and T as characteristic map (see §18.2). Since T represents a generator of $\pi_1(R_2)$, the equivalence class of $\mathfrak{B}$ in the larger group R_3 is the non-trivial class noted above.

Now R_2 leaves fixed a 0-sphere S^0 of S^2. Then S^0 determines a subbundle (see §6.5) $\mathfrak{B}_0$ of $\mathfrak{B}$ which is a product bundle $S^2 \times S^0$. It follows that $\mathfrak{B}$ admits a cross-section.

The equator S^1 of S^2 orthogonal to S^0 is mapped on itself by R_2.

It determines therefore a subbundle $\mathfrak{B}_1$ of $\mathfrak{B}$ which is a 1-sphere bundle over S^2. Since T is topological, $\mathfrak{B}_1$ is equivalent to the bundle $S^3 \to S^2$ corresponding to $m = 1$ in §26.2. So B_1 is a 3-sphere.

Let E_+, E_- denote the closed hemispheres into which S^1 divides S^2. Each is mapped on itself by R_2; hence they determine subbundles $\mathfrak{B}_+, \mathfrak{B}_-$ of $\mathfrak{B}$. The reflection through the plane of S^1 interchanges E_+ and E_-, and it commutes with R_2. Then $\mathfrak{B}_+$ and $\mathfrak{B}_-$ become equivalent bundles if their fibres are identified by the reflection. In this way, B is divided into two equivalent parts having the 3-sphere B_1 as their intersection.

An interpretation of B_+ can be given in terms of a mapping cylinder. Recall that, if $f\colon\ X \to Y$, the mapping cylinder of f is a space containing X and Y and, for each x in X, a line segment from x to $f(x)$. It is constructed by forming the product space $X \times I$ ($I = [0,1]$) and identifying $f^{-1}(y) \times 1$ with y for each y in Y.

The 2-cell E_+ may be regarded as the mapping cylinder of the map of S^1 into the midpoint x_+ of E_+. Let $g\colon\ S^2 \to B_+$ be the cross-section (of $\mathfrak{B}_0$) corresponding to x_+. If we identify the base space with the cross-section, it is seen that B_+ is just the mapping cylinder of the projection of the 3-sphere B_1 into S^2, i.e. B_+ is the mapping cylinder of the Hopf map $S^3 \to S^2$.

We give now a second description of B_+. Let $S^5 \to M$ be the fibering of the 5-sphere into great circles, given in §20.4, so that the 4-manifold M is the complex projective plane. Equating the last complex variable z_3 to zero defines an S^3 on S^5 whose image is a complex projective line S^2 in M, and $S^3 \to S^2$ is the Hopf map. Requiring z_3 to be real and non-negative defines a 4-cell E in S^5 whose boundary is S^3, and $E - S^3$ is mapped topologically onto $M - S^2$. Let E_1 be an open 4-cell in E such that $E - E_1$ is homeomorphic to $S^3 \times I$. Then the image of $E - E_1$ in M is seen to be the mapping cylinder of the Hopf map. It follows that B_+ is the space obtained by deleting from M a smoothly imbedded open 4-cell.

Now S^2 can be represented as a cell complex (of a more general type than defined in §19.1) consisting of one 0-cell E^0, and one 2-cell E^2 whose boundary is "pinched" to the point E^0. The image in M of the 4-cell E of S^5 is a 4-cell E^4 whose boundary has been pinched to S^2. We obtain thus a cellular decomposition of M consisting of E^0, E^2 and E^4. Using this the homology groups of M are readily described. Based on integer coefficients $H_0 = H_2 = H_4 = \infty$ and $H_1 = H_3 = 0$. By the Poincaré duality theorem, the generating 2-cycle of H_2 (represented by S^2) has a self intersection number of 1.

As shown above, B is constructed by taking two copies of M, deleting open 4-cells and matching the boundaries. This permits the derivation of the structure of the homology ring of B with the following results: $H_0 = H_4 = \infty$, $H_1 = H_3 = 0$ and $H_2 = \infty + \infty$. A base for H_2 is provided by the two cross-sections S^2_+ and S^2_- of $\mathcal{B}_0$ corresponding to the mid points x_+, x_- of E_+,E_- respectively. Then each of S^2_+ and S^2_- has a self-intersection number ± 1, and the intersection number of S^2_+ with S^2_- is zero (they are disjoint).

Since the homology ring of B differs from that of $S^2 \times S^2$, it follows that $\mathcal{B}$ is not equivalent to a product bundle in any group of homeomorphisms of S^2.

26.4. Bundles over S^3. From $\pi_2(O_{k+1}) = 0$ for all k, we obtain that any k-sphere bundle over S^3 is equivalent to the product bundle.

26.5. Bundles over S^4. From $\pi_3(R_2) = 0$ it follows that any 1-sphere bundle over S^4 is a product.

Since $\pi_3(R_3) = \infty$ and α_0 operates trivially, there are infinitely many classes of 2-sphere bundles over S^4 in natural correspondence with the integers.

Let $\mathcal{B}_m$ be the bundle corresponding to m and let $\tilde{\mathcal{B}}_m$ be the associated principal bundle. Choose a reference point y_0 in S^2, and let y_0: $\tilde{B}_m \to B_m$ be the associated principal map (see §8.12). Let x_0 be a reference point of S^4, and let S^2,R_3 be the fibres of $\mathcal{B}_m,\tilde{\mathcal{B}}_m$ over x_0. Then commutativity relations hold in the diagram

$$\begin{array}{ccccc} \pi_3(R_3) & \xleftarrow{\partial} & \pi_4(\tilde{B}_m,R_3) & & \\ & & & \searrow^{\tilde{p}_*} & \\ \downarrow g_* & & \downarrow f_* & & \pi_4(S^4) \\ & & & \nearrow_{p_*} & \\ \pi_3(S^2) & \xleftarrow{\partial} & \pi_4(B_m,S^2) & & \end{array}$$

where f,g are maps induced by y_0. By definition of the index m, $\partial\tilde{p}_*^{-1}$ maps a generator of $\pi_4(S^4)$ into m times a generator of $\pi_3(R_3)$. Now g is just the standard projection $R_3 \to S^2$. Since π_2 and π_3 of the fibre S^1 of g are zero, the exactness of the homotopy sequence implies that g_* in the diagram is an isomorphism. This and commutativity imply that $\Delta = \partial p_*^{-1}$ maps a generator of $\pi_4(S^4)$ into m times a generator of $\pi_3(S^2)$. Since the section

$$\pi_4(S^4) \xrightarrow{\Delta} \pi_3(S^2) \to \pi_3(B_m) \to \pi_3(S^4)$$

of the homotopy sequence of $\mathcal{B}_m$ is exact, and the end term is zero, it follows that $\pi_3(B_m)$ is cyclic of order m. This proves that $|m| \neq |n|$ implies that B_m and B_n are not homeomorphic. Hence $\mathcal{B}_m,\mathcal{B}_n$ are not equivalent in any group of homeomorphisms of S^2.

By §18.7, $\mathcal{B}_m$ and $\mathcal{B}_{-m}$ are weakly equivalent; hence B_m and B_{-m} are homeomorphic.

26.6. Since $\pi_3(R_4) = \infty + \infty$ (see §§22.6–22.7), there are an infinity of 3-sphere bundles over S^4. We obtain a doubly indexed family of bundles if we assign to $m\alpha_3 + n\beta_3$ the bundle $\mathcal{B}_{m,n}$. These are not all pairwise distinct since α_0 is non-trivial. Referring to (13) and (15) of §22.7, $\mathcal{B}_{m,n}$ and $\mathcal{B}_{m+n,-n}$ are equivalent.

The bundles $\mathcal{B}_{m,0}$ are reducible to the group R_3. Since R_3 leaves fixed a 0-sphere on S^3, the fixed points provide subbundles which are cross-sections; hence all of these bundles have cross-sections.

The bundles $\mathcal{B}_{0,n}$ are reducible to the symplectic group Sp_1; for the map σ, representing β_3, maps S^3 topologically onto Sp_1. We will show that $\mathcal{B}_{0,1}$ or $\mathcal{B}_{0,-1}$ is equivalent to the bundle $S^7 \to S^4$ given in §20.4. The latter bundle is principal since S^3 is its fibre and group. From $\pi_4(S^7) = \pi_3(S^7) = 0$, it follows that $\Delta\colon\ \pi_4(S^4) \approx \pi_3(S^3)$. Hence the characteristic map represents a generator of $\pi_3(S^3)$. It therefore represents either β_3 or $-\beta_3$ in R_4.

Since we may regard $\mathcal{B}_{0,n}$ as a principal bundle with group $Sp_1 = S^3$, it follows that Δ maps a generator of $\pi_4(S^4)$ into n times a generator of $\pi_3(S^3)$. Since $\pi_3(S^4) = 0$, exactness of the homotopy sequence implies that $\pi_3(B_{0,n})$ is cyclic of order n. Thus, if $|m| \neq |n|$, the spaces $B_{0,m}, B_{0,n}$ are not homeomorphic.

Having shown that $\mathcal{B}_{m,0}$ admits a cross-section, §17.7 yields

$$\pi_i(B_{m,0}) \approx \pi_i(S^4) + \pi_i(S^3).$$

Then $\pi_3(B_{m,0}) = \infty$. This shows that $B_{m,0}$ and $B_{0,n}$ are not homeomorphic if $n \neq 0$.

It also shows that $B_{m,0}$ and $B_{n,0}$ $(n \neq m)$ cannot be distinguished by their homotopy groups. Neither can they be distinguished by their homology rings. The first non-trivial homotopy group of $B_{m,0}$ is π_3 and it is infinite cyclic. By the Hurewicz theorem 15.10, the first non-trivial homology group is $H_3(B_{m,0}) = \infty$. By duality in the 7-manifold $B_{m,0}$, $H_4 = \infty$, $H_5 = H_6 = 0$; and generators of H_3 and H_4 can be selected having an intersection number 1. Therefore $B_{m,0}$ has both the homology structure and the homotopy group structure of the product $S^4 \times S^3$.

By §18.7, $\mathcal{B}_{m,n}$ and $\mathcal{B}_{-m,-n}$ are weakly equivalent; hence $B_{m,n}$ and $B_{-m,-n}$ are homeomorphic.

It would be of interest to decide whether $B_{m,0}$ is homeomorphic to $S^4 \times S^3$. Perhaps a simpler problem along the same lines is to decide whether the non-trivial 3-sphere bundle over S^2 is homeomorphic to $S^2 \times S^3$. (See App. sect. 6.)

26.7. Since $\pi_3(R_{k+1}) = \infty$ for $k > 3$, there are an infinity of k-sphere bundles over S^4. All such are equivalent to bundles with the group Sp_1; for $\pi_3(R_{k+1})$ is generated by the image of β_3. Since Sp_1 acts only in the linear space of the first four coordinates, it leaves point-wise fixed a $(k - 4)$-sphere on S^k. Hence all of these bundles admit cross-sections. Therefore their homotopy groups are isomorphic to those of $S^4 \times S^k$.

When $k = 4$, the situation is similar to that of the non-trivial 2-sphere bundle over S^2 (see §26.3). One of the bundles corresponding to a generator of $\pi_3(R_5)$ is a union of two copies of the mapping cylinder of the projection $S^7 \to S^4$. An argument, using the projective space of three homogeneous quaternion variables, leads to the conclusion that the only non-trivial homology group is $H_4 = \infty + \infty$, and generators can be chosen having self-intersections of ± 1 and a crossed intersection of 0. Hence the bundle space is not homeomorphic to $S^4 \times S^4$.

26.8. Bundles over S^5. Since $\pi_4(R_{k+1}) = 0$ when $k = 1$ and $k > 4$, all k-sphere bundles over S^5 are product bundles unless $k = 2$, 3 or 4.

From $\pi_4(R_3) = 2$ there is just one non-trivial 2-sphere bundle over S^5. Let $\mathfrak{B}$ denote this bundle and $\tilde{\mathfrak{B}}$ its principal bundle. Choose a principal map $\tilde{B} \to B$ determined by some point y_0 of S^2 (§8.12). It induces a homomorphism of the homotopy sequence of $\tilde{\mathfrak{B}}$ into that of $\mathfrak{B}$. Consider the portion

$$\begin{array}{ccccccc} \pi_5(S^5) & \xrightarrow{\Delta} & \pi_4(R_3) & \longrightarrow & \cdots & & \\ \downarrow f_* & & \downarrow g_* & & & & \\ \pi_5(S^5) & \xrightarrow{\Delta'} & \pi_4(S^2) & \longrightarrow & \pi_4(B) & \longrightarrow & \pi_4(S^5) \end{array}$$

where f is the identity map, and g is the map of the fibres determined by y_0. By the definition of $\mathfrak{B}$, Δ maps a generator of $\pi_5(S^5)$ onto the non-zero element of $\pi_4(R_3)$. Now g_* is an isomorphism (see similar argument in §26.5). Since f_* is the identity, it follows that Δ' is onto. From $\pi_4(S^5) = 0$ and exactness it follows that $\pi_4(B) = 0$. This implies that B is not homeomorphic to $S^5 \times S^2$, for $\pi_4(S^5 \times S^2) \approx \pi_4(S^2) = 2$.

26.9. Since $\pi_4(R_4) = 2 + 2$, generated by α_4, β_4 and $\alpha_0(\alpha_4) = \alpha_4$ and $\alpha_0(\beta_4) = \alpha_4 + \beta_4$ (see §22.7), it follows that there are two distinct non-trivial 3-sphere bundles over S^5. Let $\mathfrak{B}_1$ correspond to α_4 and $\mathfrak{B}_2$ to β_4.

The bundle $\mathfrak{B}_1$ is reducible to the group R_3. It therefore admits a cross-section. Hence

$$\pi_i(B_1) \approx \pi_i(S^5 \times S^3) \approx \pi_i(S^5) + \pi_i(S^3).$$

By the Hurewicz theorem, $H_1 = H_2 = 0$ and $H_3 = \infty$. Duality gives $H_5 = \infty$, $H_6 = H_7 = 0$. Furthermore it can then be shown [101, Th. 1] that $H_4 = 0$. It follows that B_1 has the homotopy-homology structure of $S^5 \times S^3$.

The bundle $\mathcal{B}_2$ is reducible to the symplectic group Sp_1. As such it is a principal bundle. It follows that Δ maps a generator of $\pi_5(S^5)$ onto the non-zero element of $\pi_4(S^3)$. This and exactness imply that $\pi_4(B_2) = 0$. Since $\pi_4(S^5 \times S^3) \approx \pi_4(S^3) = 2$, it follows that B_2 is not homeomorphic to $S^5 \times S^3$ or to B_1.

26.10. From $\pi_4(R_5) = 2$, it follows that there is one non-trivial 4-sphere bundle over S^5. Its group is reducible to Sp_1; hence it admits a cross-section. It has therefore the homotopy groups of $S^5 \times S^4$. The Hurewicz theorem and duality show that it has the homology ring of $S^5 \times S^4$. (See App. sect. 6.)

§27.* The Tangent Bundle of S^n

27.1. The characteristic map. We have seen in §12.9 that the tangent bundle of any differentiable orientable n-manifold is equivalent to a bundle with group R_n, and the latter has an associated subbundle called the tangent $(n - 1)$-sphere bundle consisting of tangent vectors of length 1. In the case of the n-sphere S^n, we have shown in §7.7 that the tangent sphere bundle is the Stiefel manifold $V_{n+1,2}$ over $V_{n+1,1} = S^n$. Since these manifolds are coset spaces of R_{n+1}, §8.15 provides the following result:

27.2. Theorem. *The principal bundle of the tangent sphere bundle of S^n is the bundle R_{n+1} over $R_{n+1}/R_n = S^n$.*

27.3. Corollary. *The characteristic map for a normal form of the tangent bundle of S^n is T_{n+1}: $S^{n-1} \to R_n$ (see §23.4).*

27.4. Corollary. *The tangent bundle of S^n is equivalent to a product bundle if and only if the map T_{n+1}: $S^{n-1} \to R_n$ is homotopic to a constant.*

27.5. Representations as products. In §8.6 it is shown that the tangent bundles of S^1, S^3, and S^7 are product bundles. When n is even, §23.4 asserts that $p'T_{n+1}$ is not homotopic to a constant; hence T_{n+1} is not homotopic to a constant. If $n > 1$ and $n \equiv 1 \bmod 4$, §24.8 asserts that T_{n+1} is not homotopic to a constant. This yields

Theorem. *The tangent bundles of S^1, S^3, and S^7 are equivalent to product bundles. If n is even, or if $n > 1$ and $n \equiv 1 \bmod 4$, then the tangent bundle of S^n is not equivalent to a product bundle.*

The first unsettled case is the tangent bundle of S^{11}. G. W. Whitehead has proved [100] that, if $n > 3$ and $n \equiv 3 \bmod 8$, then the tangent

* See App. sect. 7.

bundle of S^n is not a product bundle. His proof assumes the unproved assertion of Pontrjagin that $\pi_5(S^3) = 0$ (see §21.7).

27.6. THEOREM. *The following properties of S^n are equivalent:* (i) *the tangent bundle of S^n is equivalent in R_n to a bundle with group R_k (regarded as a subgroup of R_n),* (ii) *T_{n+1} is homotopic in R_n to a map of S^{n-1} into R_k, and* (iii) *S^n admits a continuous field of tangent $(n-k)$-frames (see §7.7).*

Let $\mathcal{B}$ denote the tangent bundle of S^n. Suppose (i) holds and $\mathcal{B}'$ is the equivalent bundle. Let T': $S^{n-1} \to R_k$ be a characteristic map for $\mathcal{B}'$. By 18.3, $T \simeq aT'a^{-1}$. But R_n is connected and a can be joined to e by a curve. It follows that $aT'a^{-1} \simeq T'$; and (ii) is proved.

Suppose (ii) holds, and $T \simeq T'$ where $T'(S^{n-1}) \subset R_k$. If $x_0 \in S^{n-1}$ and $T'(x_0) \neq e$, we join $T'(x_0)$ to e by a path in R_k and multiply T' by the inverses of the elements of the path. This deforms T' in R_k into a map T'': $(S^{n-1},x_0) \to (R_k,e)$. By §18.2, there is a bundle with T'' as characteristic map and group R_k. The homotopy $T \simeq T''$ may not leave x_0 at e. Since π_1 of a group operates trivially on π_{n-1} (see §16.9), there exists a homotopy leaving x_0 at e. Then §18.3 implies that T and T'' determine bundles equivalent in R_n; hence (ii) implies (i).

According to §9.4, condition (i) is equivalent to the condition that the associated bundle $\mathcal{B}'$ of $\mathcal{B}$ with fibre R_n/R_k admits a cross-section. But $V_{n+1,n-k+1} = R_{n+1}/R_k$ is a bundle over S^n with fibre R_n/R_k. Its principal bundle is R_{n+1} over S^n (see §9.1). Hence it is equivalent to $\mathcal{B}'$. In §7.7 it is shown that $V_{n+1,n-k+1}$ may be regarded as the bundle of $(n-k)$-frames tangent to S^n. Thus (i) is equivalent to (iii).

27.7. Fields of tangent vectors. A field of k-frames on a manifold determines k vector fields (the field of first vectors, of second vectors, etc.) and at each point the vectors are independent. Conversely if k vector fields, independent at each point, are given, the usual orthogonalization process leads to a field of k-frames. A set of k vector fields independent at each point is called a *k-field.* Its existence is equivalent to that of a field of k-frames.

In §23 and §24, we have obtained information concerning the property (ii) of §27.6. These yield corresponding results concerning properties (i) and (iii) as follows.

27.8. THEOREM. *If n is even, the tangent bundle of S^n is not equivalent to a bundle in R_{n-1}, and S^n does not admit a continuous 1-field.*

By §23.4, $p'T_{n+1}$ is not homotopic to a point. Since $p'(R_{n-1})$ is a point, T_{n+1} is not deformable into R_{n-1}.

27.9. THEOREM. *If $n = 2m + 1$, then the tangent bundle of S^n is equivalent to a bundle with the group $U_m \subset R_{n-1}$, and S^n admits a continuous 1-field.*

27.10. THEOREM. *If $n = 4m + 3$, then the tangent bundle of S^n is equivalent to a bundle with the group $Sp_m \subset R_{n-3}$, and S^n admits a continuous 3-field.*

Both of these theorems follow from §24.4.

27.11. THEOREM. *If $n = 4m + 1$ and $n > 1$, then the tangent bundle of S^n is not equivalent to a bundle with group R_{n-2}, and S^n does not admit a continuous 2-field.* (See App. sect. 7.)

This follows from §24.8.

27.12. THEOREM. *If $n = 8m + 7$, then T_{n+1} is homotopic to a map of S^{n-1} into R_{n-7}, the tangent bundle of S^n is equivalent to a bundle with the group R_{n-7}, and S^n admits a continuous 7-field.*

The proof differs from the preceding because the Cayley numbers do not form a group. We shall prove the final statement, the others will follow by §27.6.

The case $m = 0$ has already been proved (§27.5). Let $V^i(x)$ ($x \in S^7$, $i = 1, \cdots, 7$) be a set of continuous tangent vector fields over S^7, each vector of length 1, and at each point the vectors are orthogonal. Let E^8 be the euclidean 8-space having S^7 as the sphere $|x| = 1$. Extend V^i over E^8 by setting

$$V^i(x) = |x| V^i\left(\frac{x}{|x|}\right), \qquad x \in E^8,\ x \neq 0,$$

and $V^i(0) = 0$. Then V^i is continuous over E^8, and, at each $x \neq 0$, the seven vectors are orthogonal. Let $V^8(x)$ be the vector x. Then $V^1, \cdots, V^8$ are orthogonal.

Let E^{8m+8} be the product of $(m + 1)$-factors E^8; then $w \in E^{8m+8}$ is a set $(x_1, \cdots, x_{m+1})$ of points of E^8. Let $W^i(w)$ be the vector in E^{8m+8} whose component in the jth factor E^8 is $V^i(x_j)$ for $j = 1, \cdots, m + 1$. Clearly, the vectors $W^1, \cdots, W^8$ are orthogonal at each point. Since $W^8(w) = w$, it follows that, for w in S^{8m+7}, (i.e. $|w| = 1$), the vectors $W^1(w), \cdots, W^7(w)$ are tangent to S^{8m+7} and are of unit length. This completes the proof.

The preceding results determine the maximum k for which S^n admits a continuous k-field for $n \leqq 10$. S^{11} admits at least a 3-field. Whether it admits a 4-field is not known (see the remark and reference following §27.5).

27.13. Fields of tangent hyperplanes. We shall discuss now the problem of constructing over S^n a continuous field of tangent k-dimensional hyperplanes (briefly: a continuous field of k-planes). Note that, if $k = 0$ or n, there is no problem, the field exists and is unique. Furthermore the problems for k and $n - k$ are equivalent under the obvious duality.

The configuration of a k-plane tangent at a point of S^n is clearly equivalent to that of a (parallel) k-plane E^k through the origin and a unit vector x orthogonal to E^k (from the origin to the point of tangency). If $k < n$ (assumed hereafter), the group R_{n+1} is obviously transitive on the set of such pairs (E^k,x). Let E_0^k be the space of the first k coordinates, and let $x_{n+1} = (0, \cdots, 0, 1)$. Let H be the subgroup of R_{n+1} which leaves x_{n+1} fixed and maps E_0^k on itself. It follows that the coset space

$$W_{n,k} = R_{n+1}/H$$

may be identified with the manifold of k-planes tangent to S^n. Since H leaves x_{n+1} fixed, $H \subset R_n$. Therefore $W_{n,k}$ is a bundle over $R_{n+1}/R_n = S^n$ and the projection $W_{n,k} \to S^n$ induced by the inclusion of cosets coincides with the projection which assigns to a tangent k-plane the point of tangency.

The group H has two components determined by the sign of the transformation restricted to E_0^k. The component of e in H is the direct product $R_k \times R'_{n-k}$ where R'_{n-k} is the subgroup of R_n leaving E_0^k pointwise fixed. Define

$$\tilde{W}_{n,k} = R_{n+1}/R_k \times R'_{n-k}.$$

We call $\tilde{W}_{n,k}$ the manifold of oriented k-planes tangent to S^n. Since both $W_{n,k}$ and $\tilde{W}_{n,k}$ are connected, the natural projection

$$\tilde{W}_{n,k} \to W_{n,k}$$

is a 2-fold covering.

27.14. Theorem. *The n-sphere S^n admits a continuous field of tangent k-planes if and only if the characteristic map T_{n+1}: $S^{n-1} \to R_n$ is homotopic in R_n to a map of S^{n-1} into $R_k \times R'_{n-k}$.*

Both $W_{n,k}$ and S^n are coset spaces of R_{n+1}. Therefore the bundle R_{n+1} over S^n is the principal bundle of $W_{n,k}$ over S^n. By §9.4, $W_{n,k} \to S^n$ admits a cross-section if and only if the bundle $R_{n+1} \to S^n$ is equivalent in R_n to a bundle with group H. According to §18.3, the latter condition is equivalent to deforming T_{n+1} into a map T': $S^{n-1} \to H$. Omitting the trivial case $n = 1$, S^{n-1} is connected; hence $T'(S^{n-1})$ must lie in one of the two components of H. If we multiply H by each point of a curve in R_n connecting the two components, we obtain a homotopy of H which interchanges the components. Therefore deformability into H is equivalent to deformability into the component $R_k \times R'_{n-k}$ of H. This completes the proof.

Incidentally the argument shows that S^n admits a field of k-planes if and only if it admits a field of oriented k-planes.

27.15. THEOREM. *If $2k \leqq n$, then the inclusion map of $R_k \times R'_{n-k}$ in R_n is homotopic in R_n to a map of $R_k \times R'_{n-k}$ into R'_{n-k}.*

Since $2k \leqq n$, R'_{n-k} contains a subgroup R'_k conjugate to R_k. Suppose $a \varepsilon R_n$ and $aR_ka^{-1} = R'_k$. Let a_t $(0 \leqq t \leqq 1)$ be a curve in R_n from e to a. Then the required homotopy

$$h\colon\ R_k \times R'_{n-k} \times I \to R_n$$

is given by $h(r,r',t) = a_t r a_t^{-1} r'$.

As a corollary, the contractibility of T_{n+1} into $R_k \times R'_{n-k}$ is equivalent to contractibility into R'_{n-k}. Applying §27.6 and §27.7, we have

27.16. THEOREM. *If $2k \leqq n$, then S^n admits a continuous field of tangent k-planes if and only if it admits a continuous k-field.*

Since k independent vectors span a k-plane, any k-field determines a field of k-planes. Thus the "if" part of the above result is trivial. It should be noted that the proof does *not* show that any field of k-planes is determined in this way by some k-field. When $k = 2$, this is the case because $\pi_{n-1}(R_2) = 0$.

27.17. THEOREM. *For at least the following values of n and k, the n-sphere admits a continuous field of tangent k-planes:*

(i) *n even, $k = 0$ and $k = n$,*
(ii) *n odd, $k = 0,1,n - 1$ and n,*
(iii) *$n \equiv 3 \bmod 4$, $0 \leqq k \leqq 3$ and $n - 3 \leqq k \leqq n$,*
(iv) *$n \equiv 7 \bmod 8$, $0 \leqq k \leqq 7$ and $n - 7 \leqq k \leqq n$.*

If, in each case, we delete the latter half of the values of k, then the theorem has been proved with k-frame in place of k-plane (§§27.9, 27.10, 27.12). Since a k-frame determines a k-plane, the theorem follows for the restricted values of k. The results for the remaining k values follow by duality.

27.18. THEOREM. *The n-sphere does not admit a continuous field of k-planes if n is even and $1 \leqq k \leqq n - 1$, or if $n \equiv 1 \bmod 4$ and $2 \leqq k \leqq n - 2$.* (See App. sect. 7.)

When $2k \leqq n$ we have only to apply §27.17, §27.8, and §27.11. The cases $2k > n$ follow from duality.

The preceding two theorems settle the existence question for fields of k-planes over S^n for all $n \leqq 10$ and $0 \leqq k \leqq n$. The first unsettled case is $n = 11$ and $k = 4$.

§28. ON THE NON-EXISTENCE OF FIBERINGS OF SPHERES BY SPHERES

28.1. Necessary relations on the homotopy groups. In §20, S^{2n-1} is represented as an $(n - 1)$-sphere bundle over S^n for the values 2, 4, and 8 of n. The case $n = 1$ must also be included since the double covering of a circle by a circle is a 0-sphere bundle. These bundles were extremely useful in computing some of the homotopy groups of spheres. More such bundles are needed. This raises the question: For what values of n and k can S^{n+k} be a k-sphere bundle over S^n?

Assuming $n > 0$, any k-sphere in S^{n+k} is contractible to a point. Therefore §17.10 gives (for such a bundle)

$$\pi_i(S^n) \approx \pi_{i-1}(S^k) + \pi_i(S^{n+k}).$$

This implies that π_{n-1} is the first non-zero homotopy group of S^k; hence $k = n - 1$. Thus we have

28.2. THEOREM. *If S^m is representable as a bundle over S^n with a sphere as fibre (and any group of the fibre), then $m = 2n - 1$ and the fibre is an S^{n-1}.*

This result restricts greatly the possibilities. We proceed to derive further restrictions.

28.3. On $\pi_k(B)$ for any k-sphere bundle.

THEOREM. *Let $\mathfrak{B}$ be a k-sphere bundle and S_0^k a fibre. If k is even, the inclusion map i: $S_0^k \to B$ maps $\pi_k(S_0^k)$ isomorphically into $\pi_k(B)$. If $k > 1$ and $k \equiv 1 \bmod 4$, then the image of $\pi_k(S_0^k)$ under i_* is either infinite cyclic or finite cyclic of even order.*

By §17.13, the kernel of i_* is contained in the image of h'_*: $\pi_k(G_0) \to \pi_k(S_0^k)$. Now $G = R_{k+1}$, and h' is topologically equivalent to the bundle map $R_{k+1} \to S^k$. (This is proved by using associated maps $R_{k+1} \to G_0$, $S^k \to S_0^k$ as in §8.16.) The image of $\pi_k(R_{k+1})$ in $\pi_k(S^k)$ is the kernel of Δ: $\pi_k(S^k) \to \pi_{k-1}(R_k)$. Now T_{k+1} represents a generator α of the image of Δ. When k is even, §23.5 asserts that α is not of finite order. This implies that the image of $\pi_k(R_{k+1})$ in $\pi_k(S^k)$ is zero. Hence the kernel of i_* is zero. If $k \equiv 1 \bmod 4$, §24.10 asserts that α is of order 2. Hence $\pi_k(R_{k+1})$ maps onto the "even" elements of $\pi_k(S^k)$. This implies that the kernel of i_* is a subgroup of the even elements; and the theorem is proved.

28.4. COROLLARY. *Under the above hypotheses, the fibre S_0^k is not contractible to a point in B.*

28.5. COROLLARY. *If $k > 0$* and *k is even, or if $k > 1$ and $k \equiv 1 \bmod 4$, then the m-sphere ($m > k$) is not the bundle space of any k-sphere bundle.*

This follows from the preceding corollary since $k < m$ implies $\pi_k(S^m) = 0$.

28.6. COROLLARY. *If S^{2n-1} is an $(n - 1)$-sphere bundle over S^n, then n is 1 or 2 or $n \equiv 0 \bmod 4$.* (See App. sect. 8.)

It is a reasonable conjecture that the conclusion of the corollary should read "n is a power of 2." Compare this with §20.7.

28.7. Non-existence with G bigger than O_n. It is to be noted that §28.2 holds without restriction on the group of the fibre while §28.6 specifies the group O_n. A part of §28.6 holds without this restriction, namely:

THEOREM. *If S^{2n-1} is a bundle over S^n with fibre S^{n-1} and any group G of homeomorphisms of S^{n-1}, then n is 1, or n is even.*

Let y_0 be a reference point of the fibre S^{n-1}. Define h': $G \to S^{n-1}$ by $h'(g) = g \cdot y_0$. Let $\mathfrak{B}$ denote the given bundle $S^{2n-1} \to S^n$, let S_0^{n-1} be an $(n-1)$-sphere equator of S^n, and suppose $\mathfrak{B}$ is in normal form relative to S_0^{n-1}. Let T: $S_0^{n-1} \to G$ be the characteristic map. We proceed to show that the composition

$$S_0^{n-1} \xrightarrow{T} G \xrightarrow{h'} S^{n-1}$$

has degree ± 1.

Let $\tilde{\mathfrak{B}}$ be the associated principal bundle (defined as in §8.1 so as to be in normal form). Let $x_1 \varepsilon S^n$ be the pole of S_0^{n-1} in V_1 (see §18.1). let S_1^{n-1} and G_1 be the fibres over x_1, let

$$\xi = \phi_{1,x_1}, \qquad \tilde{\xi} = \tilde{\phi}_{1,x_1},$$
$$y_1 = \xi(y_0), \qquad g_1 = \tilde{\xi}(e).$$

Now y_0 determines a principal map

$$h\colon\ (\tilde{B},G_1,g_1) \to (S^{2n-1},S_1^{n-1},y_1)$$

given by $h(\tilde{b}) = P(\tilde{b},y)$ (see §8.10). Let $h_1 = h|G_1$. Then we have the diagram:

$$\begin{array}{ccccc}
\pi_{n-1}(G) & \xrightarrow{\tilde{\xi}_*} & \pi_{n-1}(G_1) & \xleftarrow{\tilde{\Delta}} & \pi_n(S^n) \\
\downarrow h'_* & & \downarrow h_{1*} & \swarrow \Delta & \\
\pi_{n-1}(S^{n-1}) & \xrightarrow{\xi_*} & \pi_{n-1}(S_1^{n-1}) & &
\end{array}$$

and the commutativity relations $\Delta = h_{1*}\tilde{\Delta}$, $h_{1*}\tilde{\xi}_* = \xi_* h'_*$ are readily verified.

Since $\pi_i(S^{2n-1}) = 0$ for $i = n$ and $n - 1$ (assuming $n > 1$), the exactness of the homotopy sequence of $\mathfrak{B}$ implies that Δ is an isomorphism. By §18.4, $\tilde{\xi}T$ represents a generator of the image of $\tilde{\Delta}$. Hence $h_1\tilde{\xi}T$ generates the image of Δ, namely $\pi_{n-1}(S_1^{n-1})$. But $h_1\tilde{\xi}T = \xi h'T$, and ξ_* is an isomorphism. Therefore $h'T$ represents a generator of $\pi_{n-1}(S^{n-1})$, i.e. $h'T$ has degree ± 1.

Define the map

$$f\colon\ S_0^{n-1} \times S^{n-1} \to S^{n-1}$$

by

$$f(x,y) = T(x)\cdot y.$$

When $x = x_0$, $T(x_0) = e$, so $f(x_0,y) = y$. When $y = y_0$, $f(x,y_0) = h'T(x)$. It follows that f maps cross-sections $x \times S^{n-1}$ with degree 1 and cross-sections $S_0^{n-1} \times y$ with degree 1 (when S_0^{n-1} is suitably oriented).

We will show that a map f of degree 1 in each cross-section is impossible when n is odd. This requires the use of "cup" products of cohomology classes. Integer coefficients are used throughout. The cohomology group $H^{n-1}(S^{n-1})$ is infinite cyclic; let u be a generator. The group $H^{n-1}(S_0^{n-1} \times S^{n-1})$ is a free group on two generators, say v and w, and their product $v \smile w$ generates $H^{2n-2}(S_0^{n-1} \times S^{n-1})$. Furthermore $v \smile v = 0$ and $w \smile w = 0$.

Since f has degree 1 on each cross-section, the signs of v and w can be chosen so that

$$f^*u = v + w$$

under the induced homomorphism $f^*\colon\ H^{n-1}(S^{n-1}) \to H^{n-1}(S_0^{n-1} \times S^{n-1})$. Now $u \smile u = 0$, and f^* preserves products; hence

$$(v + w) \smile (v + w) = 0.$$

Applying the distributive law and using $v \smile v = w \smile w = 0$, we obtain

$$v \smile w = -w \smile v.$$

Now the general law for commuting cohomology classes v,w of dimensions p,q, respectively, is

$$v \smile w = (-1)^{pq} w \smile v.$$

Setting $p = q = n - 1$, and comparing the two formulas gives $(-1)^{n-1} = -1$. Therefore n is even. This completes the proof.

28.8. A related problem is the following: Is S^m a bundle over S^n for some fibre Y? Since the fibres must be contractible in S^m, §17.10 gives

$$\pi_i(S^n) \approx \pi_{i-1}(Y) + \pi_i(S^m), \qquad i \geqq 2.$$

This requires that the homotopy groups of S^n be at least as big as those of S^m. As $n < m$, this is no restriction for $i < m$. Taking $i = m$, a necessary condition is that $\pi_m(S^n)$ contain an infinite cyclic group. Since $\pi_{n+1}(S^n)$ does not when $n > 2$, we have

If $n > 2$, S^{n+1} is not the space of any bundle over S^n.

Since $\pi_4(S^2) = 2$, we have

S^4 is not the space of any bundle over S^2.

Part III. The Cohomology Theory of Bundles

§29. THE STEPWISE EXTENSION OF A CROSS-SECTION

29.1. Extendability when Y is q-connected. We turn now to the application of cohomology theory to the problem of constructing a cross-section of a bundle. It will be assumed throughout that the base space of the bundle $\mathfrak{B}$ is a finite complex K as defined in §19.1. The problem of constructing a cross-section is a special case of that of extending a cross-section f already given over a subspace of K. We assume that the subspace is a subcomplex L.

If L does not contain all of the 0-dimensional skeleton K^0 of K, then f can be extended continuously over $L \cup K^0$ by defining $f(x)$ in Y_x arbitrarily for each vertex x not in L.

Assuming that f is given on $L \cup K^0$, consider the problem of extending over $L \cup K^1$. If f can be extended over $L \cup K^1$ then, for each 1-cell σ of K not in L, the cross-section $f|\dot\sigma$ ($\dot\sigma$ = boundary of σ) can be extended over σ. Conversely, a set of extensions over the individual 1-cells form an extension over $L \cup K^1$; for the interiors of the 1-cells are disjoint open sets of $L \cup K^1$.

Let σ be a 1-cell not in L, and $\mathfrak{B}_\sigma$ the part of $\mathfrak{B}$ over σ. Since σ is contractible on itself to a point, $\mathfrak{B}_\sigma$ is a product bundle (§11.6). Therefore we have a bundle map

(1) $$\phi_\sigma\colon\ \sigma \times Y \rightarrow \mathfrak{B}_\sigma$$

and a map

(2) $$p_\sigma\colon\ B_\sigma \rightarrow Y$$

such that

(3) $$\phi_\sigma(p(b),p_\sigma(b)) = b \qquad \text{for } b \in B_\sigma.$$

The composition of $f|\dot\sigma$ and p_σ is a map $f_{\dot\sigma}\colon\ \dot\sigma \rightarrow Y$. If f extends over σ, then $p_\sigma f|\sigma$ extends $f_{\dot\sigma}$ over σ. Conversely, if $f_{\dot\sigma}$ extends over σ to a map f_σ, then $f(x) = \phi_\sigma(x,f_\sigma(x))$ extends f over σ. Since σ is a 1-cell, the extendability of $f_{\dot\sigma}$ is just the question of whether the images of the two vertices of σ can be joined by a curve in Y. Thus, *if Y is arcwise connected, each $f_{\dot\sigma}$ can be extended, and the cross-section f on $L \cup K^0$ can be extended over $L \cup K^1$.*

Suppose, in general, a cross-section f is given on $L \cup K^q$, and we consider the problem of extending it over $L \cup K^{q+1}$. Again the problem reduces to extending, over each $(q+1)$-cell σ of K not in L, the cross-section $f|\dot{\sigma}$. Since σ is contractible, we may choose a product representation as in (1) and a projection (2) satisfying (3). Setting $f_{\dot{\sigma}} = p_\sigma[f|\dot{\sigma}]$, we reduce the problem to extending $f_{\dot{\sigma}}$: $\dot{\sigma} \rightarrow Y$ to a map f_σ: $\sigma \rightarrow Y$. Again there is a blanket assumption which permits the extension, namely, the homotopy group $\pi_q(Y,y_0) = 0$ for each base point y_0 (if Y is arcwise connected, it suffices to impose this for a single base point, and we write $\pi_q(Y) = 0$).

We say that a space Y is *q-connected* ($q \geqq 0$) if it is arcwise connected and $\pi_i(Y) = 0$ for $i = 1, \cdots, q$.

With this definition, we may summarize the preceding argument in

29.2. Theorem. *If f is a cross-section of the part of $\mathfrak{B}$ over a subcomplex L of K, then f may be extended over $L \cup K^0$. If Y is q-connected, then f may be extended over $L \cup K^{q+1}$.*

29.3. Corollary. *If Y is q-connected, and $K - L$ has dimension $\leqq q + 1$, then any cross-section over L may be extended over K.*

Note that a solid space (§12.1) is q-connected for every q. The above corollary is therefore a sharpening of §12.2.

29.4. The obstruction cocycle. The stepwise extension of a cross-section does not lead to an interesting situation until a dimension q is reached for which $\pi_q(Y)$ is not zero. Suppose then that f is given on $L \cup K^q$. For any $(q + 1)$-cell σ the preceding construction yields a map $f_{\dot{\sigma}}$: $\dot{\sigma} \rightarrow Y$ whose extendability over σ is equivalent to that of f. Assuming $\pi_q(Y) \neq 0$, we meet with an *obstruction* to extending f. We propose to measure this obstruction. The next few articles are devoted to the measuring procedure. Roughly it runs as follows. An orientation of σ induces one of $\dot{\sigma}$; the latter and $f_{\dot{\sigma}}$ determine an element of $\pi_q(Y)$ denoted by $c(f,\sigma)$. This function of oriented $(q + 1)$-cells proves to be a cocycle. If f is altered on the q-cells of $K - L$, the cocycle varies by a coboundary. We thus arrive at a cohomology class of K mod L which gives a precise measure of the obstruction. Its vanishing is necessary and sufficient for $f|L \cup K^{q-1}$ to be extendable over $L \cup K^{q+1}$.

There are several difficulties. The first of these concerns the base point of $\pi_q(Y)$. This is eliminated by assuming that Y is q-simple (§16.5), so that any map of an oriented q-sphere in Y determines a unique element of $\pi_q(Y,y_0)$ for any y_0. That this assumption is not too restrictive is shown by §16.9 and §16.11. The first asserts that any group space is q-simple for all q. The second asserts the same for a coset space of a Lie group by a connected subgroup. Also $\pi_1 = 0$

implies q-simple for each q. Thus, the hypothesis is satisfied if we are dealing with a principal bundle, or with a sphere bundle, or with most of the bundles associated with a sphere bundle. It is probable that the hypothesis can be avoided since it has been avoided in a similar situation.

In the case $q = 1$, simplicity means that $\pi_1(Y)$ is abelian (§16.4). Thus, in every case, $\pi_q(Y)$ is abelian. The case $q = 0$ will be included by defining $\pi_0(Y)$ to be the reduced 0-dimensional singular homology group $H_0(Y)$, i.e. only such 0-cycles are used as have a coefficient sum of zero. Then a map of the boundary of an oriented 1-cell σ into Y determines an element of $\pi_0(Y)$ whose vanishing is necessary and sufficient for the extendability of the map over σ. In the case $Y = G$, this convention supersedes the convention $\pi_0(G) = G/G_e$ of §16.10.

There is a second and more serious difficulty. The map $f_{\dot{\sigma}}$ depends on the choice of the product representation (1). A different choice may lead to an entirely different element of $\pi_q(Y)$. As will be shown there are several different assumptions which eliminate this difficulty, e.g. G is connected, or $\pi_1(K) = 0$, or that the characteristic class χ: $\pi_1(K) \rightarrow G/G_e$ is zero. But this would also eliminate important cases, e.g. tensor bundles over non-orientable manifolds. Moreover, it is possible to circumvent the difficulty without restrictions. This is accomplished by an altered procedure, and an elaboration of the cohomology theory to be used. We discuss now the altered procedure.

29.5. *The cross-section $f|\dot{\sigma}$ can be extended to a cross-section over σ if and only if it can be extended to a map of σ into B_σ.*

Half of the assertion is trivial. Suppose f': $\sigma \rightarrow B_\sigma$ is an extension of $f|\dot{\sigma}$, but is not necessarily a cross-section. Choose a product representation of B_σ as in (1), (2) and (3) of §29.1. Define

$$f''(x) = \phi_\sigma(x, p_\sigma f'(x)), \qquad x \in \sigma.$$

Then $f''|\dot{\sigma} = f'|\dot{\sigma} = f|\dot{\sigma}$, and $pf''(x) = x$. Hence f'' is a cross-section extending $f|\dot{\sigma}$.

29.6. *If Y_x is the fibre over a point x of σ, then the inclusion map of Y_x in B_σ induces an isomorphism $\pi_q(Y_x) \approx \pi_q(B_\sigma)$. Therefore $f|\dot{\sigma}$ is homotopic in B_σ to a map $f_{\dot{\sigma}}$: $\dot{\sigma} \rightarrow Y_x$. The homotopy class of $f_{\dot{\sigma}}$ in Y_x depends only on that of $f|\dot{\sigma}$ in B_σ; and $f_{\dot{\sigma}}$ is extendable to f_σ: $\sigma \rightarrow Y_x$ if and only if $f|\dot{\sigma}$ is extendable to a cross-section over σ.*

This is a trivial consequence of (i) the existence of a product representation $\sigma \times Y$ for B_σ, (ii) the vanishing of all homotopy groups of σ, and (iii) the result §17.8 on the homotopy groups of a product space.

A direct visualization is provided by choosing a contraction of σ on itself to the point x and picturing a covering homotopy which contracts B_σ into the fibre Y_σ. The latter deforms $f|\dot{\sigma}$ into $f_{\dot{\sigma}}$. If the contraction of $\dot{\sigma}$ over σ into x is chosen to sweep out each point of $\sigma - x$ just once (e.g. a radial contraction), then the covering homotopy of $f|\dot{\sigma}$ sweeps out an extension of $f|\dot{\sigma}$ to a single-valued continuous cross-section over $\sigma - x$. We thus obtain an extension with a singularity at x; and $f_{\dot{\sigma}}$ is clearly a measure of this singularity.

29.7. We choose now, in each $(q + 1)$-cell σ, a reference point x_σ, and denote by Y_σ the fibre over x_σ. Choose, for each $(q + 1)$-cell σ, an orientation; and denote by σ the oriented cell, and by $\dot{\sigma}$ the oriented boundary. If f is a cross-section over $L \cup K^q$, we define $c(f,\sigma)$ to be the element of $\pi_q(Y_\sigma)$ given by 29.6 applied to $f|\dot{\sigma}$. (We are assuming that Y is q-simple.) Then we have

29.8. *A cross-section f over $L \cup K^q$ is extendable over $L \cup K^{q+1}$ if and only if $c(f,\sigma) = 0$ for each $(q + 1)$-cell σ.*

Denote by $c(f)$ the function of σ given by $c(f)(\sigma) = c(f,\sigma)$. We call $c(f)$ *the obstruction cocycle of f*. The terminology anticipates showing that $c(f)$ is a cochain, in some sense, and proving that its coboundary is zero. A cochain in the usual sense is a function assigning to oriented cells elements of an abelian group—the same group for each cell. In the present case, the values of $c(f)$ lie in different groups. It is true that they are all isomorphic, but there is no natural unique isomorphism between any two of them. Thus we meet, in a different form, the second of the difficulties described in §29.4. As promised there, we circumvent the difficulty by broadening the notion of cochain so as to allow functions such as $c(f)$. This requires a broadening of the related concepts of cohomology theory to which the next two articles are devoted.

§30. Bundles of Coefficients

30.1. Definitions. By a *bundle of groups* is meant a bundle with a fibre Y which is a group, and the group G acts as automorphisms of Y. As observed in §6.6, we can define, in each fibre Y_x, a group structure so that each admissible map $Y \to Y_x$ is an isomorphism.

By a *bundle of coefficients* (for homology or cohomology groups) is meant a bundle of groups where the fibre is an abelian group, written additively, and the group of the bundle is totally disconnected. The fibre will be denoted by π, and the group of the bundle (acting as automorphisms of π) by Γ.

Since Γ is totally disconnected, the results of §13 may be applied. In particular, any curve C from x_0 to x_1 in X determines an isomorphism $C^\#\colon\ \pi_1 \approx \pi_0$ which depends only on the homotopy class of C. If C' is a curve from x_1 to x_2, then $(CC')^\# = C^\# C'^\#$. The bundle is determined up to an equivalence by its characteristic homomorphism χ: $\pi_1(X) \to \Gamma$ (X connected and locally connected).

The bundle of coefficients is called *simple* if it is a product bundle. According to §13.7, this happens if and only if χ is zero, i.e. each closed path operates as the identity.

Examples of bundles of coefficients are easily obtained. According to §13.8 we need only choose a group π and a homomorphism χ of $\pi_1(X)$ into the automorphism group of π to obtain one such. But certain bundles of coefficients arise naturally from other bundles.

30.2. The associated bundle $\mathfrak{B}(\pi_q)$. The example of most importance to us is the following. Let $\mathfrak{B} = \{B,p,X,Y,G\}$ be a bundle such that Y is q-simple. Then each fibre Y_x is also q-simple and the groups $\pi = \pi_q(Y)$ and $\pi_x = \pi_q(Y_x)$ are defined without reference to a base point in Y_x (e.g. as the group of homotopy classes of maps of a q-sphere into Y_x). Let Π be the union of the sets π_x for all x in X. Define ρ: $\Pi \to X$ so that ρ maps π_x into x for each x. We proceed to define a topology and bundle structure in Π.

For each coordinate neighborhood V_j of $\mathfrak{B}$ define

$$\psi_j\colon\quad V_j \times \pi \to \rho^{-1}(V_j) \tag{1}$$

by

$$\psi_j(x,\alpha) = (\phi_{j,x})_*\alpha, \tag{2}$$

i.e. $\psi_j(x,\alpha)$ is the image of α under the isomorphism $\pi \to \pi_x$ induced by $\phi_{j,x}$: $Y \to Y_x$. Define

$$\rho_j\colon\quad \rho^{-1}(V_j) \to \pi \tag{3}$$

by

$$\rho_j(\beta) = (p_j|Y_x)_*(\beta) \qquad \text{for } \beta \in \pi_x. \tag{4}$$

It is easily seen that $\rho\psi_j(x,\alpha) = x$, $\rho_j\psi_j(x,\alpha) = \alpha$ and $\psi_j(\rho(\beta),\rho_j(\beta)) = \beta$. It follows that (1) is a 1-1 map. If we set $\gamma_{ji}(x) = \rho_j\psi_{i,x}$, we have

$$\gamma_{ji}(x) = g_{ji}(x)_*. \tag{5}$$

Thus we have most of the elements of a bundle structure. As yet we do not have a topology in Π, a topology in π, and a group of transformations of π. The last is readily obtained. Each g in G induces an automorphism g_* of π. Let H be the subgroup which acts as the identity in π. Set $\Gamma = G/H$ and let η: $G \to \Gamma$ be given by $\eta(g) = g_*$. Then Γ is an automorphism group of π, and by (5) the coordinate transformations of the projected bundle are in Γ.

It remains to assign topologies to π, Γ, and Π. We give to π and Γ

the discrete topologies. This choice leads to difficulties in the most general situation since we must show that (5) is continuous in x. We therefore make the restriction: *let X be locally arcwise connected.* Then, if $x_1 \varepsilon V_i \cap V_j$, there exists a neighborhood N of x_1 such that, for each x_2 in N there is a curve C in $V_i \cap V_j$ from x_1 to x_2. Since g_{ji} is continuous, it maps C into a curve in G from $g_{ji}(x_1)$ to $g_{ji}(x_2)$. Then $g_{ji}(x_1)$ and $g_{ji}(x_2)$ are homotopic maps of Y on itself. Therefore they induce the same homomorphism of π. This shows that γ_{ji} is constant over N, hence continuous over $V_i \cap V_j$.

There is an alternative restriction which is equally effective: *let the component G_e of e in G be arcwise connected and open in G.* It follows quickly that $G_e \subset H$, and G/G_e is discrete with the coset space topology. Since G/H is a coset space of G/G_e, it too has the discrete topology as a coset space of G. This implies that η: $G \to G/H$ is continuous when G/H is discrete. Then the continuity of g_{ji} implies that of γ_{ji}.

Finally we topologize Π so that the 1-1 maps ψ_j are homeomorphisms. There is just one way of doing this. For each α in Π we select a j such that $\rho(\alpha) \varepsilon V_j$, and define neighborhoods of α to be the images of neighborhoods of $(\rho(\alpha),\rho_j(\alpha))$ under ψ_j. That this yields a topology, under which ρ and each ψ_j,ρ_j are continuous, follows quickly from the continuity of γ_{ji}.

The bundle of coefficients so constructed is denoted by $\mathcal{B}(\pi_q)$. It is defined whenever Y is q-simple, and X is locally arcwise connected. Since π is discrete, the bundle is a covering space of X (§14).

30.3. Lemma. *If C is a curve from x_1 to x_2 in X, and Y_{x_2} is translated along C^{-1} into Y_{x_1} in $\mathcal{B}$, then the induced isomorphism of π_{x_2} into π_{x_1} coincides with the unique isomorphism $C^{\#}$ obtained by translating π_{x_2} along C^{-1} into π_{x_1} in the bundle $\mathcal{B}(\pi_q)$.*

The proof is obtained quickly by first observing that it holds when C lies in some V_j, and then noting that any curve is a composition of a finite number of such curves.

30.4. Theorem. *Let $\mathcal{B}$ be a bundle whose fibre Y is q-simple, and whose base space X is arcwise connected and arcwise locally connected. Suppose also that the component G_e of e in G is arcwise connected. If the characteristic class χ: $\pi_1(X) \to G/G_e$ of B is trivial, then the bundle $\mathcal{B}(\pi_q)$ is a product bundle.*

The hypothesis on χ means that, if a fibre Y_0 is translated around a closed curve C into itself, the resulting self map is homotopic to the identity. By §30.3, this implies that $C^{\#}$ is the identity map of $\pi_q(Y_0)$. Hence the characteristic class χ: $\pi_1(X) \to \Gamma$ of $\mathcal{B}(\pi_q)$ is also trivial. By §13.7, it must be a product bundle.

It is to be noted that formula (5) implies that $\mathcal{B}(\pi_q)$ is the bundle weakly associated with $\mathcal{B}$ relative to η: $G \to \Gamma$.

30.5. LEMMA. *Let h: $\mathcal{B} \to \mathcal{B}'$ be a bundle map where the common fibre Y is q-simple, and the base spaces X,X' are locally arcwise connected. If $x \in X$ and $x' = \bar{h}(x)$, let h_{x*} be the isomorphism $\pi_q(Y_x) \approx \pi_q(Y_{x'})$ induced by h_x. Then the maps h_{x*}, for all x in X, define a bundle map κ: $\mathcal{B}(\pi_q) \to \mathcal{B}'(\pi_q)$.*

If $x \in V_j \cap \bar{h}^{-1}(V'_k)$, and $x' = \bar{h}(x)$, we have

$$(\phi'^{-1}_{k,x'})_* h_{x*} \phi_{j,x*} = (\phi'^{-1}_{k,x'} h_x \phi_{j,x})_* = \bar{g}_{kj}(x)_*.$$

The continuity of $\bar{g}_{kj}(x)_*$ in x is proved precisely as in the case of (5) above. By §2.6, there is a unique bundle map κ: $\mathcal{B}(\pi_q) \to \mathcal{B}'(\pi_q)$ corresponding to $\bar{h}$ and the $\bar{g}_{kj}(x)_*$. A brief glance at the construction of κ in §2.6 reveals that $\kappa_x = h_{x*}$ for each x; and the lemma is proved.

30.6. It is to be noted that the preceding construction of $\mathcal{B}(\pi_q)$ can be carried through with the homology group $H_q(Y)$ replacing $\pi_q(Y)$. The only properties of $\pi_q(Y)$ used in the discussion are (i) it is an abelian group, (ii) any g in G induces an automorphism g_* of π, (iii) $(gg')_* = g_*g'_*$, and (iv) g_* depends only on the homotopy class of g. Since $H_q(Y)$ has the same four properties, the weakly associated bundle $\mathcal{B}(H_q)$ can be defined in the same fashion.

The cohomology group $H^q(Y)$ satisfies all save (iii) which is replaced by $(gg')^* = g'^*g^*$. If we set $g^\# = (g^{-1})^*$, then $g \to g^\#$ is a homomorphism of G onto a group Γ of automorphisms of H^q. With $g^\#$ in place of g^*, the four properties hold, and we may define $\mathcal{B}(H^q)$.

30.7. The bundle of homotopy groups. Another important example of a bundle of coefficients is provided by the homotopy groups $\pi_q(X,x)$ where X is arcwise connected, arcwise locally connected, and semi-locally 1-connected (§13.8).

We set $\pi_x = \pi_q(X,x)$, and let Π denote the union of the groups π_x for all x in X. Define ρ: $\Pi \to X$ by $\rho(\pi_x) = x$.

Let x_0 be a reference point, and let the fibre π be π_{x_0}. As shown in §16.4, $\pi_1(X,x_0)$ acts as a group of automorphisms of π. We define Γ to be the factor group of $\pi_1(X,x_0)$ by the subgroup which acts as the identity automorphism of π. We give to π and Γ the discrete topology.

For each x in X there exists a neighborhood V of x which is arcwise connected and such that any closed curve in V is homotopic to a point in X. For such a neighborhood V and any $\alpha \in \pi_x$, we define the *neighborhood* $V(\alpha)$ in Π to be the set of elements obtained by deforming α along curves in V, i.e. if C is a curve from x' to x in V, then $C^\#\alpha$ is in $V(\alpha)$.

Clearly ρ maps $V(\alpha)$ onto V. In fact, under the topology defined in Π by these neighborhoods, ρ *maps* $V(\alpha)$ *topologically onto* V. Suppose C_1, C_2 are two curves in V from x to x'. Then $C_1C_2^{-1}$ is contractible in X, and this implies that $C_1^\# = C_2^\#$. It follows that $\rho | V(\alpha)$ is 1-1. Since, for each V and α, ρ maps $V(\alpha)$ into V, ρ is continuous. If β is in $V(\alpha)$, and $V'(\beta) \subset V(\alpha)$, it is easily seen that $V' \subset V$. Hence $(\rho | V(\alpha))^{-1}$ maps V' into $V'(\beta)$. This proves that $\rho | V(\alpha)$ is topological.

We have thereby shown that ρ: $\Pi \to X$ is a covering in the sense of §14.1. By §14.3, it may be given a bundle structure. One checks readily that the characteristic class of the bundle is the natural map $\pi_1(X, x_0) \to \Gamma$.

§31. Cohomology Groups Based on a Bundle of Coefficients

31.1. Introductory remarks. This generalization of cohomology theory was given first by Reidemeister [81]. He called the bundle of coefficients an Überdeckung. Subsequently, an extensive survey was made by the author [87]. In the latter, the bundle of coefficients was called a *system of local coefficients.* Although their definitions differ, it is easily proved that, in a connected and semi-locally 1-connected space (e.g. a complex), a system of local coefficients is a bundle of coefficients.

The following treatment is restricted to cohomology. One may also treat homology theory with coefficients in a bundle. A reader, familiar with the parallelism between cohomology and homology, will be able to state and prove the corresponding facts about the latter. But these will not be used in the sequel.

We will assume that the reader is familiar with certain basic material concerning ordinary homology theory with integer coefficients in a complex. All such can be found in the book of Lefschetz [64]. We could avoid this and achieve greater simplicity if we dealt only with simplicial complexes and simplicial maps. But the needs of subsequent articles demand the use of the cell complex (of §19.1) and arbitrary continuous maps.

It is not generally realized that the satisfactory use of cell complexes in homology theory presupposes the theorem on the invariance of the homology groups. For example, if the cell complex K consists of a single n-cell and its faces, then, for $q > 0$, $H_q(K) = 0$ is a consequence of the invariance theorem; but, to my knowledge, is not provable in any other way. If K is a simplex, the fact is directly deducible from the definition of H_q in terms of cycles and boundaries. This difference accounts for the preferred treatment accorded simplicial complexes.

But one rarely computes the homology groups of spaces from simplicial decompositions. The number of simplexes required can be impractically large. For example a simplicial division of a torus requires a minimum of 42 elements. A cell decomposition with 16 elements can be given. In higher dimensions the discrepancy is greater. An n-simplex has $2^n - 1$ elements. A cellular decomposition of an n-cell need have only $2n + 1$ elements.

Still fewer cells are needed if one allows cell complexes in which identifications occur on the boundaries of the cells. These are frequently used to compute ordinary homology groups. For the generalization to be given, it seems to be necessary that the closed cells be at least simply connected. However we shall adhere to the definition of §19.1.

31.2. Cochains, cocycles and cohomology. Let K be a finite cell complex and let $X = |K|$ be the space of K. Let $\mathcal{B} = \{\Pi,\rho,X,\pi,\Gamma\}$ be a bundle of coefficients over X (§30.1).

For each $q \geqq 0$ and each q-cell σ of K, we choose a reference point x_σ in σ, and denote by π_σ the fibre of $\mathcal{B}$ over x_σ. We call π_σ *the coefficient group of* σ.

A *q-cochain of K with coefficients in* $\mathcal{B}$ is a function c which attaches to each oriented q-cell σ an element $c(\sigma)$ of π_σ and satisfies $c(-\sigma) = -c(\sigma)$ where $-\sigma$ denotes the orientation opposite to that of σ. The q-cochain c is said to be zero on a subcomplex L if $c(\sigma) = 0$ for each σ in L. We add cochains by adding functional values:

$$(1) \qquad (c_1 + c_2)(\sigma) = c_1(\sigma) + c_2(\sigma).$$

It follows that the q-cochains form an additive abelian group denoted by $C^q(K;\mathcal{B})$. Those which are zero on L form a subgroup $C^q(K,L;\mathcal{B})$.

If we choose a fixed orientation of each cell, then a choice of one element from π_σ for each q-cell σ determines a unique q-cochain which on each oriented σ has the prescribed value in π_σ. It follows that $C^q(K,L;\mathcal{B})$ is isomorphic to the direct sum $\Sigma\pi_\sigma$ for q-cells σ in $K - L$.

Choose now for each cell σ a fixed reference orientation and let σ also denote the oriented cell. If σ is a q-cell and is a face of the $(q + 1)$-cell τ (written: $\sigma < \tau$), let $[\sigma:\tau] = \pm 1$ denote the incidence number of σ and τ. If σ is a q-face of the $(q + 2)$-cell ξ, and τ, τ' are the two $(q + 1)$-cells such that $\sigma < \tau < \xi$, $\sigma < \tau' < \xi$, then we have the usual relation

$$(2) \qquad [\sigma;\tau][\tau:\xi] + [\sigma:\tau'][\tau':\xi] = 0.$$

For each relation $\sigma < \tau$, we choose a curve C in τ from x_τ to x_σ and denote by $w_{\sigma\tau}$ the isomorphism $C^\#$ of π_σ onto π_τ. Note that w is independent of the choice of C since τ is simply-connected. It follows that $\sigma < \tau < \xi$ implies

$$(3) \qquad w_{\tau\xi}w_{\sigma\tau} = w_{\sigma\xi}.$$

For any q-cochain c we define its *coboundary* δc in $C^{q+1}(K;\mathcal{B})$ by

$$(4) \qquad \delta c(\tau) = \Sigma[\sigma:\tau]w_{\sigma\tau}(c(\sigma))$$

where τ is a $(q+1)$-cell and the sum is extended over all q-faces σ of τ. Since each w is a homomorphism, it follows from (1) that δ is a homomorphism. If c is zero on L, it is clear that δc is zero on L. Hence

$$(5) \qquad \delta\colon \quad C^q(K,L;\mathcal{B}) \to C^{q+1}(K,L;\mathcal{B}).$$

If the orientation of σ is reversed, both $[\sigma:\tau]$ and $c(\sigma)$ change sign and $\delta c(\tau)$ remains unchanged. If we reverse that of τ, both sides of (4) change sign. It follows that δ is independent of the choice of the reference orientations.

If we calculate $\delta\delta c$ from (4), apply (3) and then (2), we arrive at the basic relation

$$(6) \qquad \delta\delta = 0.$$

The kernel of (5), denoted by $Z^q(K,L;\mathcal{B})$, is called *the group of q-cocycles of K mod L with coefficients in* $\mathcal{B}$. The image of (5), denoted by $B^{q+1}(K,L;\mathcal{B})$ is called *the group of* $(q + 1)$*-coboundaries of K mod L*. By (6), we have

$$Z^q(K,L;\mathcal{B}) \supset B^q(K,L;\mathcal{B}).$$

We define *the qth cohomology group of K mod L with coefficients in* $\mathcal{B}$ by

$$H^q(K,L;\mathcal{B}) = Z^q(K,L;\mathcal{B})/B^q(K,L;\mathcal{B}).$$

To have a proper definition for all $q \geqq 0$, we define $C^q(K,L;\mathcal{B}) = 0$ when $K - L$ has no q-cells, and we set $B^0(K,L;\mathcal{B}) = 0$.

The cohomology groups are independent of the choice of the base points x_σ. For suppose x'_σ is a second set of choices. Choose a path in σ from x'_σ to x_σ and use it to define an isomorphism $u_\sigma\colon \ \pi_\sigma \to \pi'_\sigma$. Use these to map each old cochain into a new one. This gives an isomorphism of the old group of cochains onto the new one. If $\sigma < \tau$, then $w'_{\sigma\tau}u_\sigma$ and $u_\tau w_{\sigma\tau}$ are both isomorphisms of π_σ onto π'_τ induced by traversing curves in τ; so they must be equal. From this it follows that the isomorphisms of the old cochains onto the new ones commute with δ. Hence they induce isomorphisms of the respective cohomology groups.

31.3. Simple coefficients. Whenever the bundle $\mathcal{B}$ of coefficients is a product bundle, the cohomology groups $H^q(K,L;\mathcal{B})$ reduce in a natural way to the ordinary cohomology groups $H^q(K,L;\pi)$. This is proved as follows.

Choose a bundle map μ: $\mathcal{B} \to \pi$ which exists since $\mathcal{B}$ is a product. Then μ maps each π_σ isomorphically onto π, and in such a way that $\mu w_{\sigma\tau}|\pi_\sigma = \mu|\pi_\sigma$ for each $\sigma < \tau$. Using μ we obtain, quickly, isomorphisms $C^q(K,L;\mathcal{B}) \approx C^q(K,L;\pi)$ for each q which commute with δ. They induce therefore isomorphisms of the cohomology groups.

31.4. The Kronecker index. In the case of ordinary cochains with coefficients in π, one has the notion of a Kronecker index. Let c be a p-cochain with coefficients in π, and let $z = \Sigma_{i=1}^r a_i\sigma_i$ be a p-chain with integer coefficients. Then the *Kronecker index* $c \cdot z$ in π is defined by

$$c \cdot z = \sum_{i=1}^{r} a_i c(\sigma_i). \tag{7}$$

Clearly, $c \cdot z$ is bilinear. Furthermore it is readily shown that

$$(\delta c) \cdot z = c \cdot (\partial z) \tag{8}$$

when c is a $(p - 1)$-cochain and z is a p-chain.

Passing to the case of a bundle $\mathcal{B}$ of coefficients, let $c \in C^p(K;\mathcal{B})$, and let z be a p-chain with ordinary integer coefficients. We note that (7) has no meaning because the various $c(\sigma_i)$ lie in different groups. It is possible to bring them together into the same group by translating along curves of X into a single fibre. If $\mathcal{B}$ is not a product, the result will depend on the choice of the curves. Thus, we must abandon a Kronecker index in the usual sense.

However, if the chain z lies on a subcomplex E of K such that $\mathcal{B}|E$ is a product bundle, then the terms of (7) can be accumulated in a single fibre, in just one way, by using curves in E. Adding them in this fibre determines a unique value of $c \cdot z$. It is easily seen that the values of $c \cdot z$ obtained in the various fibres of $\mathcal{B}|E$ correspond to one another under translation along curves in E. We are thus led to a bilinear operation, called the Kronecker index, which pairs $C^p(K;\mathcal{B})$ and $C_p(E)$ to the cross-sections of $\mathcal{B}|E$.

Since $\mathcal{B}|E$ is a product, a cochain c' on E is an ordinary cochain, and $c' \cdot z$ is definable as usual. A cochain c of K with coefficients in $\mathcal{B}$ determines a cochain c' on E by restricting c to cells of E. It is clear that $c \cdot z = c' \cdot z$ under this correspondence. It follows that (8) holds for the extended Kronecker index.

Any closed cell σ of K and its faces form a subcomplex E such that $\mathcal{B}|E$ is a product. In this sense, we have

$$c(\sigma) = c \cdot \sigma. \tag{9}$$

And (8) gives

$$\delta c(\sigma) = c \cdot \partial\sigma. \tag{10}$$

Thus, we always have a "local" Kronecker index with the usual properties.

31.5. Carrier of a mapping. A *carrier* for a continuous function $h\colon\ K \to K'$ of one cell complex into another is a function assigning to each cell σ of K a closed subcomplex E_σ of K' such that $h(\sigma) \subset E_\sigma$ and $\sigma < \tau$ implies $E_\sigma \subset E_\tau$. A carrier for a map $h\colon\ (K,L) \to (K',L')$ is required to satisfy the additional condition that $E_\sigma \subset L'$ when σ is in L.

The intersection of two carriers of h is again a carrier. If, for each σ, E_σ is the smallest closed subcomplex containing $h(\sigma)$, then $\{E_\sigma\}$ is called the minimal carrier. It is contained in every carrier.

A carrier $\{E_\sigma\}$ of h is called *solid* if, for each σ, E_σ is contractible to a point (see §12.1). This implies that E_σ is connected, simply-connected, and the homology groups $H_q(E_\sigma)$, with integer coefficients, are zero for $q = 1, 2, \cdots$. The last means that each q-cycle on E_σ is a boundary. Since E_σ is connected, a 0-cycle of the form $v_1 - v_2$, where v_1,v_2 are vertices of E_σ, bounds a 1-chain on E_σ.

31.6. Chain homomorphisms and homotopies. Let $\{E_\sigma\}$ be a solid carrier for h. Then *there exist chain homomorphisms (integer coefficients)*

$$h_\#\colon\ C_q(K) \to C_q(K'), \qquad q = 0, 1, \cdots, \tag{11}$$

such that $h_\#$ carries a vertex into a vertex,

$$h_\#\sigma \subset E_\sigma, \qquad \text{and} \qquad \partial h_\# = h_\#\partial. \tag{12}$$

The proof proceeds by induction. To each vertex v of K we assign a vertex $h_\# v$ of E_v. We then extend $h_\#$ to all 0-chains by the requirement of linearity. Assuming that (11) is defined for $q < p$, we choose a base for $C_p(K)$ consisting of one orientation of each p-cell. If σ is a base element, then (12) implies that $h_\#\partial\sigma$ is a $(p - 1)$-cycle on E_σ. Since E_σ is solid, we can choose a p-chain on E_σ whose boundary is $h_\#\partial\sigma$ and denote it by $h_\#\sigma$. We then extend $h_\#$ to all p-chains by the requirement of linearity. This completes the general step of the induction.

Any two chain homomorphisms $h_\#,h'_\#$, having the same solid carrier $\{E_\sigma\}$, are chain homotopic, i.e. there exist homomorphisms

$$D\colon\ C_q(K) \to C_{q+1}(K'), \qquad q = 0, 1, \cdots, \tag{13}$$

such that

$$D\sigma \subset E_\sigma, \qquad \text{and} \qquad \partial Dz = h'_\# z - h_\# z - D\partial z \tag{14}$$

for any chain z.

This again is proved by induction. We define Dv, for a vertex v, to be a 1-chain of E_v whose boundary is $h'_\# v - h_\# v$. Assuming D

defined for $q < p$, then, for any base element σ of $C_p(K)$, by applying (14) with $z = \partial\sigma$, we find that $h'_{\#}\sigma - h_{\#}\sigma - D\partial\sigma$ is a p-cycle on E_σ. We let $D\sigma$ be a $(p + 1)$-chain of E_σ which has it for boundary.

31.7. Induced homomorphisms. Let $\mathfrak{B},\mathfrak{B}'$ be bundles of coefficients over K,K', and let h: $\mathfrak{B} \rightarrow \mathfrak{B}'$ be a bundle map such that the induced map $\bar{h}$: $(K,L) \rightarrow (K',L')$ has a solid carrier $\{E_\sigma\}$. Select a chain homomorphism $h_{\#}$ as above.

Let c' be an element of $C^q(K',L';\mathfrak{B}')$, and let σ be an oriented q-cell. Since E_σ is simply connected, $\mathfrak{B}'|E_\sigma$ is a product bundle; hence, we can form the Kronecker index $c' \cdot h_{\#}\sigma$, and it is a cross-section of $\mathfrak{B}'|E_\sigma$ (see §31.4). Let h_σ denote the isomorphism of the coefficient group π_σ onto the fibre of $\mathfrak{B}'$ over $\bar{h}(x_\sigma)$. Then we define $h^{\#}c'$ by

(15) $$h^{\#}c'(\sigma) = h_\sigma^{-1}(c' \cdot h_{\#}\sigma).$$

It is readily checked that $h^{\#}$ is a homomorphism

(16) $$h^{\#}: \quad C^q(K',L';\mathfrak{B}') \rightarrow C^q(K,L;\mathfrak{B}).$$

If c' is a $(q - 1)$-cochain of K', and σ is a q-cell of K, then (12) and (8) yield

$$\begin{aligned}(h^{\#}\delta c')(\sigma) &= h_\sigma^{-1}(\delta c' \cdot h_{\#}\sigma) = h_\sigma^{-1}(c' \cdot \partial h_{\#}\sigma) \\ &= h_\sigma^{-1}(c' \cdot h_{\#}\partial\sigma) = h^{\#}c' \cdot \partial\sigma = (\delta h^{\#}c')(\sigma).\end{aligned}$$

This proves

(17) $$\delta h^{\#} = h^{\#}\delta.$$

It follows that $h^{\#}$ carries cocycles into cocycles and coboundaries into coboundaries, thereby inducing a homomorphism

(18) $$h^*: \quad H^q(K',L';\mathfrak{B}') \rightarrow H^q(K,L;\mathfrak{B}).$$

31.8. The uniqueness of h^* for proper maps. The definition of h^* depends on the choice of $\{E_\sigma\}$, and on the choice of $h_{\#}$. Consider first the latter choice. Let $h_{\#},h'_{\#}$ be any two such choices; then a chain homotopy (13) may be chosen. Define a corresponding *cochain homotopy*

(19) $$D: \quad C^q(K',L';\mathfrak{B}') \rightarrow C^{q-1}(K,L;\mathfrak{B})$$

by using the local Kronecker index:

(20) $$Dc'(\sigma) = h_\sigma^{-1}(c' \cdot D\sigma).$$

Using (14) we obtain

(21) $$\delta Dc' = h'^{\#}c' - h^{\#}c' - D\delta c'.$$

This implies, for a cocycle c', that $h'^{\#}c'$ and $h^{\#}c'$ belong to the same cohomology class. Thus we have proved that h^* is independent of the choice of $h_{\#}$.

If we had started with two maps h and h' having the common solid carrier $\{E_\sigma\}$, the foregoing argument proves that $h^* = h'^*$

In general, h^* depends on the choice of the solid carrier. We shall say that h is a *proper* map if the minimal carrier is solid. Then we can define h^* using the minimal carrier. Since any solid carrier contains the minimal carrier, any h^* coincides with the one assigned to the minimal carrier. *Thus, for proper maps, h^* is unique and can be constructed from any solid carrier.*

It is to be noted that any inclusion map $(K,L) \to (K',L')$ is proper where K is a subcomplex of K' and L is a subcomplex of L'. The minimal carrier of σ consists of σ and its faces.

If K,K' are simplicial complexes, and $\bar{h}$ is simplicial, then $\bar{h}(\sigma)$ is a simplex and is therefore solid. Thus simplicial maps are proper.

There is a wide class of proper maps for which the $h_{\#}$ assigned to the minimal carrier is unique. The map $\bar{h}$: $K \to K'$ is called *cellular* if, for each q, $\bar{h}$ maps the q-skeleton K^q of K into K'^q. Suppose that $\bar{h}$ is both proper and cellular. For any q-cell σ of K, the minimal carrier E_σ is a subcomplex of K'^q, and is thereby q-dimensional. Then $H_q(E_\sigma) = 0$ means $Z_q(E_\sigma) = 0$; for, the absence of $(q+1)$-cells in E_σ means $B_q(E_\sigma) = 0$. But $Z_q(E_\sigma)$ is the kernel of ∂: $C_q(E_\sigma) \to C_{q-1}(E_\sigma)$. Since this kernel is zero, there can be at most one q-chain of E_σ whose boundary is $h_{\#}\partial\sigma$. Thus, at each stage of the inductive construction of $h_{\#}$, the choice is unique:

If $\bar{h}$: $K \to K'$ is both proper and cellular, then there is just one $h_{\#}$ associated with the minimal carrier.

31.9. Subdivision. Let (K',L') be a subdivision of (K,L), i.e. their spaces coincide, and each cell of K (L) is the union of the cells of K' (L') which it contains. We assert that *the inclusion maps*

$$(22) \qquad h\colon\ (K,L) \to (K',L'), \qquad h'\colon\ (K',L') \to (K,L)$$

are proper. In the case of h, the minimal carrier of σ is the subdivision in K' of the complex composed of σ and its faces. Denote this by $Sd\,\sigma$. Since any subdivision of a cell is solid, h is proper. If σ' is in K', its interior lies in the interior of just one cell σ of K; then its minimal carrier $E_{\sigma'}$ consists of σ and its faces. Thus h' is proper.

Using these carriers, choose $h_{\#}$ and $h'_{\#}$ and let $h^{\#},h'^{\#}$, as in (16), be defined accordingly. Let i be the identity map of (K,L) and let $i_{\#},i^{\#}$ be identity maps of chains and cochains. Now $i_{\#}$ and $h'_{\#}h_{\#}$ have in common the minimal solid carrier $\{E_\sigma\}$ where E_σ consists of σ and its faces. Since i is cellular, the uniqueness statement of 31.8 yields $i_\# = h'_\# h_\#$. It follows that $i^* = h^* h'^*$.

Letting i' denote the identity map of (K',L'), we find that $i'_\#$ and $h_\# h'_\#$ have the common solid carrier $\{Sd\ E_{\sigma'}\}$. Hence they are chain homotopic. As in §31.8, this implies $i'^* = h'^* h^*$. Thus we have proved

INVARIANCE UNDER SUBDIVISION: *If (K',L') is a subdivision of (K,L), then the inclusion maps of (22) are proper, and they induce isomorphisms*

$$H^q(K,L;\mathfrak{B}) \approx H^q(K',L';\mathfrak{B}).$$

31.10. The h^* of a general map. It is important to define h^* when h is not proper. The procedure is to factor h into the composition of three proper maps

$$(23) \qquad (K,L) \xrightarrow{i} (K_1,L_1) \xrightarrow{h'} (K'_1,L'_1) \xrightarrow{i'} (K',L')$$

where (K_1,L_1) is a simplicial subdivision of (K,L), (K'_1,L'_1) is a simplicial subdivision of (K',L'), i and i' are inclusion maps and $h'(x) = h(x)$ for each x.

The first regular subdivision of K is simplicial. It is constructed, inductively, by introducing one new vertex on each σ and subdividing σ into the join of this vertex with the subdivision of the boundary of σ. Thus, simplicial subdivisions can be found. Arbitrarily fine subdivisions can be found by repeated barycentric subdivisions of the regular subdivision.

The existence of (23) is proved as follows. Let (K'_1,L'_1) be any simplicial subdivision of (K',L'). The open stars of vertices of K'_1 cover K'. Choose a simplicial subdivision of (K_1,L_1) of (K,L) so fine that the image of each simplex of K_1 lies in the open star of a vertex of K'_1. If $h(\sigma)$ lies in the star of v, then each closed simplex of K'_1 which meets $h(\sigma)$ has v as a vertex. Therefore their union is contractible to v. But this union is the minimal carrier of $h(\sigma)$. It follows that h' is proper. By §31.9, i and i' are proper.

Assuming now that h: $\mathfrak{B} \to \mathfrak{B}'$ is a map of bundles of coefficients, we choose a factorization (23), and define h^* by

$$(24) \qquad h^* = i^* h'^* i'^*.$$

Various facts must now be proved to insure that this definition of h^* is satisfactory. We list these without proofs since the proofs are simple applications of the cochain homotopy construction of §31.8.

31.11. *The h^* of (24) is independent of the choice of the factorization (23).*

31.12. *If h is a proper map, the h^* defined in §31.7 coincides with the h^* of (24).*

31.13. *If h: $\mathcal{B} \to \mathcal{B}'$ and h': $\mathcal{B}' \to \mathcal{B}''$, then $(h'h)^* = h^*h'^*$.*

31.14. *The identity map $\mathcal{B} \to \mathcal{B}$ induces the identity map of $H^q(K,L;\mathcal{B})$.*

If h: $\mathcal{B} \to \mathcal{B}'$ induces a homeomorphism $\bar{h}$: $(K,L) \to (K',L')$, the last two propositions imply that h^* is an isomorphism. This is a precise formulation of the statement: *$H^q(K,L;\mathcal{B})$ is a topological invariant.*

31.15. *If h_0,h_1: $\mathcal{B} \to \mathcal{B}'$ are homotopic maps, then $h_0^* = h_1^*$.*

Let h: $\mathcal{B} \times I \to \mathcal{B}'$ be the homotopy. Let h'_0,h'_1: $\mathcal{B} \to \mathcal{B} \times I$ be defined by $h'_0(x) = (x,0)$ and $h'_1(x) = (x,1)$. Then $h_i = hh'_i$ $(i = 0,1)$. By §31.13, it suffices to prove that $h_0'^* = h_1'^*$. But, relative to the product complex $K \times I$ (see §19.1), h'_0 and h'_1 have the common solid carrier $E_\sigma = \sigma \times I$. We may therefore choose $h'_{0\#} = h'_{1\#}$, and the desired result follows.

31.16. In the factorization (23), the map i is cellular as well as proper while h' and i' need not be cellular. Suppose the subdivision K_1 of K is chosen so fine that, for any vertex v of K_1, the image of the star of v lies in the star of some vertex v' of K'_1. As is well known (the simplicial approximation theorem), setting $k(v) = v'$ determines a unique simplicial map k: $(K_1,L_1) \to (K'_1,L'_1)$ such that $k(x)$ lies in the closure of the smallest simplex containing $h(x)$. Then $k \simeq h$, and the minimal carrier of k is contained in that of h'. The latter implies that $h'_\#$ can be chosen to be the unique $k_\#$ (k is cellular and proper). Then $h'^* = k^*$.

If K'_1 is not K', then i' is not cellular; however its inverse i'' is cellular, and i''^* is the inverse of i'^*. Then (24) becomes

$$h^* = i^*k^*(i''^*)^{-1}. \tag{24'}$$

We have proved:

Any h^* can be factored into the form (24′) where i,k and i'' are proper cellular maps.

This result enables us to extend to the general h^* those of its properties proved when h is proper and cellular.

31.17. The coboundary operator. Let $\mathcal{B}$ be a bundle of coefficients over (K,L) and let i and j be the inclusion maps

$$L \xrightarrow{i} K \xrightarrow{j} (K,L).$$

These maps are proper and cellular, so $i_\#$ and $j_\#$ are unique. These determine homomorphisms

$$(25) \qquad 0 \to C^q(K,L;\mathfrak{B}) \xrightarrow{j^\#} C^q(K;\mathfrak{B}) \xrightarrow{i^\#} C^q(L;\mathfrak{B}) \to 0.$$

Direct interpretations are as follows. $C^q(K,L;\mathfrak{B})$ is the subgroup of $C^q(K;\mathfrak{B})$ consisting of cochains which are zero on L; and $j^\#$ is the inclusion map. For any c in $C^q(K;\mathfrak{B})$, $i^\#c$ is the cochain of L obtained by restricting c to cells of L. The image of $i^\#$ is $C^q(L;\mathfrak{B})$ since a function of cells of L can be extended to a function of cells of K by assigning values to the cells of $K - L$. It follows that the sequence (25) is exact.

Consider now the diagram

$$(26) \qquad \begin{array}{ccccccccc} 0 \to & C^q(K,L;\mathfrak{B}) & \xrightarrow{j^\#} & C^q(K;\mathfrak{B}) & \xrightarrow{i^\#} & C^q(L;\mathfrak{B}) & \to 0 \\ & \downarrow\delta & & \downarrow\delta & & \downarrow\delta & \\ 0 \to & C^{q+1}(K,L;\mathfrak{B}) & \xrightarrow{j^\#} & C^{q+1}(K;\mathfrak{B}) & \xrightarrow{i^\#} & C^{q+1}(L;\mathfrak{B}) & \to 0. \end{array}$$

If c is in $Z^q(L;\mathfrak{B})$, i.e. $\delta c = 0$, we choose an extension c' of c in $C^q(K;\mathfrak{B})$. Then

$$i^\#\delta c' = \delta i^\# c' = \delta c = 0.$$

Therefore $\delta c'$ lies in $C^{q+1}(K,L;\mathfrak{B})$. Since $\delta\delta c' = 0$, $\delta c'$ is a cocycle. If, also, $i^\#c'' = c$. Then $\delta(c' - c'') = \delta c' - \delta c''$, and $c' - c''$ is zero on L. Thus the cohomology class in $H^{q+1}(K,L;\mathfrak{B})$ of $\delta c'$ is independent of the extension c'. We obtain thus a unique homomorphism $Z^q(L;\mathfrak{B}) \to H^{q+1}(K,L;\mathfrak{B})$. It is easily proved that it carries $B^{q+1}(L;\mathfrak{B})$ into zero. It thereby induces a homomorphism

$$(27) \qquad \delta\colon\ H^q(L;\mathfrak{B}) \to H^{q+1}(K,L;\mathfrak{B}), \qquad q = 0, 1, \cdots.$$

31.18. The cohomology sequence. Associated with the bundle $\mathfrak{B}$ of coefficients over (K,L) is the infinite sequence of groups and homomorphisms

$$(28) \qquad \cdots \to H^{q-1}(L;\mathfrak{B}) \xrightarrow{\delta} H^q(K,L;\mathfrak{B}) \xrightarrow{j^*} H^q(K;\mathfrak{B}) \xrightarrow{i^*} H^q(L;\mathfrak{B}) \to \cdots$$

It is called *the cohomology sequence of* (K,L) *with coefficients in* $\mathfrak{B}$.

The cohomology sequence is exact.

As an example we shall prove exactness at the terms of the form $H^q(K,L;\mathfrak{B})$; the proofs for the other two cases are left to the reader. Let c be a $(q-1)$-cocycle of L representing $\bar{c}$ in $H^{q-1}(L;\mathfrak{B})$. To find $\delta\bar{c}$, we choose an extension c' of c in $C^{q-1}(K;\mathfrak{B})$. By definition, $\delta c'$ in $Z^q(K,L;\mathfrak{B})$ represents $\delta\bar{c}$. Since $j^\#$ is an inclusion, $j^\#\delta c' = \delta c'$

regarded as an element of $Z^q(K;\mathcal{B})$. But $\delta c'$ is a coboundary. Hence $\delta c'$ represents zero in $H^q(K;\mathcal{B})$. Thus $j^*\delta\bar{c} = 0$, and we have proved that the kernel of j^* contains the image of δ.

Now, let $\bar{c}$ in $H^q(K,L;\mathcal{B})$ be such that $j^*\bar{c} = 0$. Let c in $Z^q(K,L;\mathcal{B})$ represent $\bar{c}$. Then $j^\# c = c$ in $Z^q(K;\mathcal{B})$ represents $j^*\bar{c} = 0$. Hence $c = \delta c'$ where $c' \varepsilon C^{q-1}(K;\mathcal{B})$. Let $c_1 = i^\# c'$. Then $\delta c_1 = \delta i^\# c' = i^\# \delta c' = i^\# j^\# c = 0$. Therefore c_1 is a $(q-1)$-cocycle of L. If $\bar{c}_1$ is its class in $H^{q-1}(L;\mathcal{B})$, we have $\delta\bar{c}_1 = \bar{c}$. This proves that the kernel of j^* is contained in the image of δ; and completes the proof of the exactness of (28) at the term $H^q(K,L;\mathcal{B})$.

31.19. Commutativity. Let $\mathcal{B},\mathcal{B}'$ be bundles of coefficients over (K,L), (K',L') respectively, and let h: $\mathcal{B} \to \mathcal{B}'$. Then h restricted to $\mathcal{B}|L$ is a bundle map h_1: $\mathcal{B}|L \to \mathcal{B}'|L'$. We obtain the diagram

$$\begin{array}{ccc} H^{q+1}(K',L';\mathcal{B}') & \xrightarrow{h^*} & H^{q+1}(K,L;\mathcal{B}) \\ \uparrow\delta & & \uparrow\delta \\ H^q(L';\mathcal{B}') & \xrightarrow{h_1^*} & H^q(L;\mathcal{B}). \end{array} \tag{29}$$

We assert that

$$h^*\delta = \delta h_1^*. \tag{30}$$

In view of the definition (24), it suffices to prove (30) when h is a proper map. Let h_2 denote h regarded as a map of $\mathcal{B}$ over K into $\mathcal{B}'$ over K'. Using the minimal carrier, we select an $h_\#$ satisfying (11) and (12). Then we set $h_{2\#} = h_\#$ and $h_{1\#} = h_\#|C_q(L)$. Passing to the associated cochain maps, as in (15) and (16), we obtain the commutativity relations

$$h_2^\# j'^\# = j^\# h^\#, \qquad h_1^\# i'^\# = i^\# h_2^\#.$$

Let c be a representative cocycle of $\bar{c}$ in $H^q(L';\mathcal{B}')$. Extend c to c' in $C^q(K';\mathcal{B}')$. Then, by definition, $h^\#\delta c'$ represents $h^*\delta\bar{c}$. Using (17), $h^\#\delta c' = \delta h_2^\# c'$. But $i^\# h_2^\# c' = h_1^\# i'^\# c' = h_1^\# c$ represents $h_1^*\bar{c}$. Therefore $\delta h_2^\# c'$ represents $\delta h_1^*\bar{c}$, and the proof is complete.

31.20. The 0-dimensional group. When K is a connected complex, the ordinary cohomology group $H^0(K;\pi)$ is isomorphic to π. This is not the case for $H^0(K;\mathcal{B})$ when $\mathcal{B}$ is not a product bundle. The structure of this group is obtained as follows. Since $B^0(K;\mathcal{B}) = 0$, we have

$$H^0(K;\mathcal{B}) = Z^0(K;\mathcal{B}).$$

We must interpret the condition for a 0-cochain c to be a cocycle. Let σ be an edge with vertices A and B so that $\partial\sigma = B - A$. We can suppose that $x_\sigma = A$. Then

$$\delta c(\sigma) = w_{B\sigma}c(B) - c(A).$$

Thus, $\delta c(\sigma) = 0$ is equivalent to the statement: $c(B)$ translates along σ into $c(A)$. Then $\delta c = 0$ is equivalent to the statement: for any two vertices A and B, translation along a curve in K^1 from B to A carries $c(B)$ into $c(A)$. Since any curve in K from A to B is homotopic to one in K^1, we may replace K^1 by K in the statement. Then, translating $c(A)$, say, to the various points of K provides a uniquely defined cross-section of $\mathfrak{B}$. It follows that the 0-cocycles are in 1-1 correspondence with the cross-sections of $\mathfrak{B}$. Since the fibre π is discrete, the cross-sections are in 1-1 correspondence with the elements of π pointwise invariant under the operations of $\chi(\pi_1(K))$ on π. *Thus,* $H^0(K;\mathfrak{B})$ *is isomorphic to the subgroup of* π *pointwise invariant under* $\chi(\pi_1(K))$.

31.21. The ordinary homology and cohomology theory of complexes has been extended to spaces other than complexes by two distinct methods: the Čech method based on coverings and their nerves, and the method based on singular simplexes. One would expect to find corresponding generalizations for homology and cohomology theory with coefficients in bundles. These do exist. We shall have no need of them in the sequel since our work is restricted to complexes. For a discussion of these matters see [87].

§32. THE OBSTRUCTION COCYCLE

32.1. The proof of $\delta c(f) = 0$. Let $\mathfrak{B}$ be a bundle over the cell complex K with a fibre Y which is q-simple, let L be a subcomplex of K, and let f be a cross-section of $\mathfrak{B}|L \cup K^q$. According to §30.2, the groups $\pi_q(Y_x)$ form a bundle $\mathfrak{B}(\pi_q)$ of coefficients over K. Let $c(f)$ be the obstruction cocycle of f as defined in §§29.7–29.8. By §31.2, $c(f)$ *is a* $(q + 1)$*-cochain of* K *mod* L *with coefficients in* $\mathfrak{B}(\pi_q)$. To prove that it is a cocycle requires the use of a "homotopy addition theorem." The latter relates the addition of spherical cycles to the addition of the corresponding elements of homotopy groups. It is used in the proof of the Hurewicz isomorphism (§15.10), and is a consequence of it. We avoid the addition theorem by assuming §15.10 and deriving the following consequence.

32.2. LEMMA. *If* K *is* $(q - 1)$*-connected (§29.1), then the natural homomorphism* $\pi_q(K^q) \to H_q(K^q)$ *is an isomorphism, and* $H_q(K^q)$ *coincides with the group of* q*-cycles* $Z_q(K)$*; hence*

$$\pi_q(K^q) \approx Z_q(K).$$

The relation $\pi_i(K^q) \approx \pi_i(K)$, $i < q$, holds in any complex. (This follows quickly from the well-known homotopy approximation theorem:

If a complex of dimension $\leqq q$ is mapped into K, the map is homotopic to a map into K^q, leaving fixed any points already mapped into K^q.)

Since K is $(q-1)$-connected, it follows that K^q is $(q-1)$-connected. Then the Hurewicz theorem, §15.10, gives $\pi_q(K^q) \approx H_q(K^q)$. Since $C_{q+1}(K^q) = 0$, we have $B_q(K^q) = 0$; hence $H_q(K^q) = Z_q(K^q) = Z_q(K)$.

32.3. Lemma. *If K is $(q-1)$-connected, and the bundle $\mathfrak{B}$ over K is equivalent to $K \times Y$, and if f is a cross-section defined over K^q, then $c(f)$ is a coboundary in K.*

Let p' be the projection $K \times Y \to Y$, and let $f' = p'f$. In this case $\mathfrak{B}(\pi_q)$ is a product bundle, and we may identify each $\pi_q(Y_x)$ with $\pi_q(Y)$. Then $c(f,\sigma)$ is the element of $\pi_q(Y)$ represented by $f'|\dot{\sigma}$. Letting $C_p(K)$ denote the group of p-chains of K with integer coefficients, we have the diagram

$$C_{q+1}(K) \xrightarrow{\partial} Z_q(K) \xleftarrow{\psi} \pi_q(K^q) \xrightarrow{f'_*} \pi_q(Y)$$

where ψ is the isomorphism of §32.2. Then, for any oriented $(q+1)$-cell σ,

$$c(f,\sigma) = f'_*\psi^{-1}\partial\sigma.$$

Since $C_{q-1}(K)$ is a free abelian group, the kernel of ∂: $C_q(K) \to C_{q-1}(K)$, namely $Z_q(K)$, is a direct summand of $C_q(K)$. Therefore the homomorphism $f'_*\psi^{-1}$ extends to a homomorphism

$$d\colon \quad C_q(K) \to \pi_q(Y).$$

As a function on the oriented q-cells of K, we have $d \varepsilon C^q(K;\pi_q(Y))$. Then the relation $c(f,\sigma) = d(\partial\sigma)$, and the general relation $d(\partial\sigma) = (\delta d)(\sigma)$ imply $\delta d = c(f)$.

32.4. Theorem. *The obstruction cochain $c(f)$ is a cocycle.*

We must show that $\delta c(f)$ is zero on any $(q+2)$-cell ζ. Let K' be the subcomplex consisting of ζ and its faces. Let $c' = c(f)|K'$. Then $\delta c'$ has the same value on ζ as does $\delta c(f)$. If $\mathfrak{B}'$ is $\mathfrak{B}|K'$, and if $f' = f|K'^q$, then it is clear that $c' = c(f')$. Now K' is a cell; hence it is $(q-1)$-connected, and $\mathfrak{B}' = K' \times Y$. Then, by §32.3, $c' = \delta d$ is a coboundary. Since $\delta\delta d = 0$ for any cochain, we have $\delta c' = 0$, and the theorem is proved.

32.5. Homotopies of cross-sections. If f is a cross-section of the bundle $\mathfrak{B} = \{B,p,X,Y,G\}$, a *homotopy* of f is a map F: $X \times I \to B$ such that $pF(x,t) = x$ for all t and $F(x,0) = f(x)$. If we define f_t: $X \to B$ by $f_t(x) = F(x,t)$, then f_t is a cross-section. We call F a *homotopy* of f_0 into f_1; and f_0 and f_1 are said to be *homotopic*.

32.6. LEMMA. *Let $\mathcal{B}$ be a bundle over K, and let f_0,f_1 be homotopic cross-sections of $\mathcal{B}|L \cup K^q$. Then $c(f_0) = c(f_1)$.*

For any $(q + 1)$-cell σ, the homotopy $f_0 \simeq f_1$ induces a homotopy $f_0|\dot{\sigma} \simeq f_1|\dot{\sigma}$ in B_σ. Therefore they determine the same element of $\pi_q(B_\sigma)$; hence, by §29.6, the same element of π_σ.

32.7. Invariance of $c(f)$ under mappings.

LEMMA. *Let $\mathcal{B},\mathcal{B}'$ be bundles over complexes (K,L), (K',L') respectively, and let h: $\mathcal{B} \to \mathcal{B}'$ be a bundle map which induces a proper cellular map $\bar{h}$: $(K,L) \to (K',L')$. Let k: $\mathcal{B}(\pi_q) \to \mathcal{B}'(\pi_q)$ be the induced map of the coefficient bundles (§30.5), and let*

$$h^{\#}:\quad C^{q+1}(K',L';\mathcal{B}'(\pi_q)) \to C^{q+1}(K,L;\mathcal{B}(\pi_q))$$

denote the unique cochain homomorphism induced by k (§31.8). Let f' be a cross-section of $\mathcal{B}'|L' \cup K'^q$, and let f be the cross-section of $\mathcal{B}|L \cup K^q$ induced by f' and h (§2.11). Then

$$c(f) = h^{\#}c(f').$$

We must prove that the two sides have the same value on any $(q + 1)$-cell σ of K. Since f is induced by h, the statement to be proved is a commutativity relation. It is a consequence of a rather large number of trivial commutativity relations. Let E_σ be the minimal carrier of σ, E^q_σ its q-skeleton, and τ a $(q + 1)$-cell of E_σ. We obtain the diagram

$$\begin{array}{ccccccccc}
C_{q+1}(\sigma) & \xrightarrow{\partial} & Z_q(\dot{\sigma}) & = & H_q(\dot{\sigma}) & \xleftarrow{\psi} & \pi_q(\dot{\sigma}) & \xrightarrow{f_{\dot{\sigma}*}} & \pi_q(Y_\sigma) \\
\downarrow h_{\#} & & \downarrow h_{\#} & & \downarrow \bar{h}_* & & \downarrow \bar{h}_* & & \downarrow h_{1*} \quad \searrow^{h_\sigma} \\
C_{q+1}(E_\sigma) & \xrightarrow{\partial} & Z_q(E^q_\sigma) & = & H_q(E^q_\sigma) & \xleftarrow{\psi'} & \pi_q(E^q_\sigma) & \xrightarrow{f'_*} & \pi_q(B'_\sigma) \xleftarrow{i_*} \pi_q(Y_0) \\
\uparrow k_{\#} & & \uparrow k_{\#} & & \uparrow k_* & & \uparrow k_* & & \uparrow j_* \quad \nearrow_{w_\tau} \\
C_{q+1}(\tau) & \xrightarrow{\partial} & Z_q(\dot{\tau}) & = & H_q(\dot{\tau}) & \xleftarrow{\psi''} & \pi_q(\dot{\tau}) & \xrightarrow{f'_{\dot{\tau}*}} & \pi_q(Y_\tau)
\end{array}$$

We have denoted by B'_σ the part of B' over E_σ. The fibre of $\mathcal{B}$ over x_σ is Y_σ, Y_0 is the fibre of $\mathcal{B}'$ over $\bar{h}(x_\sigma)$, and Y_τ is the fibre over x'_τ. Restricting h to the domain Y_σ and range B'_σ gives h_1. The maps i, j and k are inclusions, h_σ is induced by h_{x_σ}: $Y_\sigma \to Y_0$, and w_τ is induced by translating Y_τ along a curve in E_σ to Y_0.

Observe first that commutativity holds in each square and triangle of the diagram. This is trivial for the six squares on the left and the upper right triangle. Since $f_{\dot{\sigma}} \simeq f|\dot{\sigma}$, and $hf = f'\bar{h}$, we obtain $h_1 f_{\dot{\sigma}} \simeq f'\bar{h}|\dot{\sigma}$. This implies commutativity in the upper right square. The same for the lower right square follows from $f'_{\dot\tau} \simeq f'|\dot\tau$. Since translation of Y_τ along a curve of E_σ keeps Y_τ in B'_σ, commutativity holds in the lower right triangle.

The fact that ψ and ψ'' are isomorphisms has already been used in defining $c(f)$ and $c(f')$. Perhaps the only non-trivial point of this proof is that ψ' is an isomorphism onto. This follows from §32.2 and the assumption that E_σ is solid. Thus, we may reverse the arrows of ψ, ψ', ψ'' and commutativity still holds.

In the two triangles on the right, all maps are isomorphisms onto. This follows since the contractibility of E_σ implies that B'_σ can be contracted into Y_0.

By definition,

$$c(f',\tau) = f'_{\dot\tau *}\psi''^{-1}\partial\tau.$$

Applying the definition (§31.4) of the Kronecker index in E_σ,

$$c(f')\cdot\tau = w_\tau f'_{\dot\tau *}\psi''^{-1}\partial\tau.$$

Identifying $C_{q+1}(\tau)$ with a subgroup of $C_{q+1}(E_\sigma)$ under $k_\#$, and using commutativity in the lower half of the diagram, we obtain

$$c(f')\cdot\tau = i_*^{-1}f'_*\psi'^{-1}\partial\tau.$$

Since both sides are additive, the last relation holds with τ replaced by any chain in $C_{q+1}(E_\sigma)$, in particular, by $h_\#\sigma$. It follows, from §31.7, that

$$h^\# c(f')(\sigma) = h_\sigma^{-1}(c(f')\cdot h_\#\sigma) = h_\sigma^{-1}i_*^{-1}f'_*\psi'^{-1}\partial h_\#\sigma.$$

Using commutativity in the upper half of the diagram, the right side of the last equation reduces to $f_{\dot\sigma *}\psi^{-1}\partial\sigma$. But this is the definition of $c(f,\sigma)$; and the proof is complete.

§33. The Difference Cochain

33.1. Motivation. Recall that the obstruction cocycle $c(f)$ is met after a stepwise extension to $L \cup K^q$ of a cross-section given on L. We will show that an alteration of the extension over the q-cells alters $c(f)$ by a coboundary. For this purpose we must introduce the difference cochain associated with two different extensions over the q-cells. Now two maps f_0,f_1 of a q-cell τ which agree on $\dot\tau$ determine, in a natural way, a map of a q-sphere, and this, in turn, determines an element $d(f_0,f_1,\tau)$ of $\pi_q(Y_\tau)$. In this way we obtain a cochain $d(f_0,f_1)$ in $C^q(K,L;\mathcal{B}(\pi_q))$. If we were to adopt this direct and intuitive definition of the difference cochain, we would be required to give it a formal treatment as extensive as that of $c(f)$. Fortunately $d(f_0,f_1)$ is essentially an obstruction cochain on the product complex $K \times I$. We shall use this fact to define d, and then its properties will follow quickly from those of $c(f)$. The cost of this procedure is that we must digress to consider the product complex and cross-products of cochains.

33.2. Products of chains and cochains. As in §19.1, the product $K \times K_1$ of two cell complexes is a cell complex whose cells are the products $\sigma \times \tau$ of cells of K and K_1. Using ordinary chains with integer coefficients, we have the well-known result that, for all p and q, there exists a bilinear pairing of $C_p(K)$ and $C_q(K_1)$ to $C_{p+q}(K \times K_1)$, denoted by $\times$, with the following properties: if σ and τ are oriented cells, then $\sigma \times \tau$ is an orientation of their product cell, and

$$(1) \qquad \partial(\sigma \times \tau) = \partial\sigma \times \tau + (-1)^p \sigma \times \partial\tau, \qquad p = \dim \sigma.$$

We wish to extend this result to cochains with coefficients in bundles. Because of the limited application we will not consider the most general situation. We shall suppose that the bundle $\mathfrak{B}'$ of coefficients over $K \times K_1$ is the one induced by the projection $K \times K_1 \to K$ and a bundle $\mathfrak{B}$ of coefficients over K. The coefficients for K_1 will be ordinary integers, denoted by J. The reference point for $\sigma \times \tau$ will be the point (x_σ, y_τ). Then the coefficient group $\pi_{\sigma\times\tau}$ is naturally isomorphic to π_σ under the projection $\mathfrak{B}' \to \mathfrak{B}$. The pairing of π_σ and J to $\pi_{\sigma\times\tau}$ is the ordinary multiplication of a group element by an integer followed by the inverse of the isomorphism $\pi_{\sigma\times\tau} \approx \pi_\sigma$. Then, if

$$u \in C^p(K;\mathfrak{B}), \qquad v \in C^q(K_1;J),$$

we define $u \times v \in C^{p+q}(K \times K_1;\mathfrak{B}')$ by setting $u \times v = 0$ on all cells $\sigma \times \tau$ unless $\dim \sigma = p$ and $\dim \tau = q$, in which case

$$(2) \qquad u \times v(\sigma \times \tau) = u(\sigma)v(\tau).$$

It is easily seen that $u \times v$ is bilinear. If we agree that $u(\sigma) = 0$ whenever $\dim \sigma \neq \dim u$, then (2) defines $u \times v$ on any product cell.

It is important to prove the analog of (1):

$$(3) \qquad \delta(u \times v) = \delta u \times v + (-1)^p u \times \delta v.$$

We must evaluate both sides of (3) on a product cell $\sigma \times \tau$. This cell and its faces form a simply-connected subcomplex of $K \times K_1$ which is the product of the analogous subcomplexes for σ and τ. This reduces the proof to the case of simple coefficients. For any cochain w and cell ζ, we have the basic relation $(\delta w)(\zeta) = w(\partial\zeta)$. Using this repeatedly, we have

$$\begin{aligned}(\delta(u \times v))(\sigma \times \tau) &= u \times v(\partial(\sigma \times \tau)) \\ &= u \times v(\partial\sigma \times \tau + (-1)^r \sigma \times \partial\tau), \qquad r = \dim \sigma \\ &= u(\partial\sigma)v(\tau) + (-1)^r u(\sigma)v(\partial\tau) \\ &= \delta u(\sigma)v(\tau) + (-1)^r u(\sigma)\delta v(\tau) \\ &= (\delta u \times v)(\sigma \times \tau) + (-1)^r(u \times \delta v)(\sigma \times \tau).\end{aligned}$$

The last term is zero unless $\dim u = \dim \sigma$; hence replacing $(-1)^r$ by $(-1)^p$ leaves it unaltered. Then (3) follows.

33.3. The complex $K \times I$. Let $\mathfrak{B}$ be a bundle over (K,L). For convenience of notation, let

$$\mathfrak{B}^{\square} = \mathfrak{B} \times I, \qquad K^{\square} = K \times I,$$
$$L^{\square} = (K \times 0) \cup (L \times I) \cup (K \times 1).$$

We regard I as a complex composed of two 0-cells $\bar{0}$ and $\bar{1}$ and the 1-cell $\bar{I}$. We also let $\bar{0},\bar{1}$ stand for the generating 0-cochains of $C^0(I)$ (integer coefficients); and $\bar{I}$ will denote a generator of $C^1(I)$ chosen so that

(4) $$\delta\bar{0} = -\bar{I}, \qquad \delta\bar{1} = \bar{I}.$$

If $d \in C^p(K,L;\mathfrak{B}(\pi_q))$, it is readily checked that $d \times \bar{I}$ is zero on $L^{\square}$. Hence

$$d \times \bar{I} \in C^{p+1}(K^{\square},L^{\square};\ \mathfrak{B}^{\square}(\pi_q)).$$

Since $\tau \to \tau \times I$ is a 1-1 correspondence between the p-cells of $K - L$ and the $(p+1)$-cells of $K^{\square} - L^{\square}$, it follows that $d \to d \times \bar{I}$ is an isomorphism

(5) $$C^p(K,L;\mathfrak{B}(\pi_q)) \approx C^{p+1}(K^{\square},L^{\square};\mathfrak{B}^{\square}(\pi_q)).$$

Since $\delta\bar{I} = 0$, (3) implies that the isomorphisms (5) commute with δ:

(6) $$\delta(d \times \bar{I}) = \delta d \times \bar{I}.$$

Therefore (5) induces

(7) $$H^p(K,L;\mathfrak{B}(\pi_q)) \approx H^{p+1}(K^{\square},L^{\square};\mathfrak{B}^{\square}(\pi_q)).$$

33.4. Definition. Let f_0,f_1 be cross-sections of the part of $\mathfrak{B}$ over $L \cup K^q$, let $f_0 = f_1$ on L, and let k be a homotopy (as in §32.5)

$$f_0|L \cup K^{q-1} \simeq f_1|L \cup K^{q-1} \qquad \text{relative to } L.$$

The *associated* cross-section F of the part of $\mathfrak{B}^{\square}$ $(= \mathfrak{B} \times I)$ over $L^{\square} \cup K^{\square q}$ is defined by

(8) $$\begin{aligned} F(x,0) &= (f_0(x),0), \qquad F(x,1) = (f_1(x),1), \\ F(x,t) &= (k(x,t),t) \qquad \text{for } x \in L \cup K^{q-1},\ t \in I.\end{aligned}$$

Then an obstruction cocycle

$$c(F) \in C^{q+1}(K^\square, L \times I; \mathcal{B}^\square(\pi_q))$$

is defined. It coincides with $c(f_0) \times \bar{0}$ on $K \times 0$ and with $c(f_1) \times \bar{1}$ on $K \times 1$. Hence

$$c(F) - c(f_0) \times \bar{0} - c(f_1) \times \bar{1} \in C^{q+1}(K^\square, L^\square; \mathcal{B}^\square(\pi_q)).$$

Using the isomorphism (5), we define the *deformation cochain*

$$d(f_0,k,f_1) \in C^q(K,L;\mathcal{B}(\pi_q))$$

by

$$(9) \qquad d(f_0,k,f_1) \times \bar{I} = (-1)^{q+1}\{c(F) - c(f_0) \times \bar{0} - c(f_1) \times \bar{1}\}.$$

Whenever $f_0 = f_1$ on $L \cup K^{q-1}$ and $k(x,t) = f_0(x)$ for all t, we abbreviate $d(f_0,k,f_1)$ by $d(f_0,f_1)$ and call it the *difference cochain*.

33.5. The coboundary formula.

THEOREM. *Under the hypotheses of §33.4,*

$$\delta d(f_0,k,f_1) = c(f_0) - c(f_1).$$

If we apply δ to both sides of (9), use (6), (3), (4), and the fact that $c(F)$, $c(f_0)$, $c(f_1)$ are cocycles (§32.7), we obtain

$$(\delta d(f_0,k,f_1)) \times \bar{I} = c(f_0) \times \bar{I} - c(f_1) \times \bar{I}.$$

Since the operation $u \to u \times \bar{I}$ is an isomorphism of cochains (see (5)), the theorem follows.

33.6. Invariance under mappings.

THEOREM. *Let f_0,k,f_1 be as in §33.4. Let h: $\mathcal{B}' \to \mathcal{B}$ induce a proper cellular map $\bar{h}$: $(K',L') \to (K,L)$, and let f'_0,f'_1,k' be the cross-sections and homotopy induced by h and f_0,f_1,k. Then*

$$h^\# d(f_0,k,f_1) = d(f'_0,k',f'_1).$$

The proof is entirely mechanical so we only sketch its outline. Define $h^\square$: $\mathcal{B}'^\square \to \mathcal{B}^\square$ by $h^\square(b',t) = (h(b'),t)$. Then verify that (i) $h^{\square\#}(d \times \bar{I}) = (h^\# d) \times \bar{I}$, and (ii) the cross-section F' associated with f'_0,k',f'_1 is induced by $h^\square$ and F. Now apply $h^{\square\#}$ to both sides of (9) and use §32.7 to provide $h^{\square\#}c(F) = c(F')$, and

$$h^{\square\#}(c(f_i) \times \bar{\imath}) = (h^\# c(f_i)) \times \bar{\imath} = c(f'_i) \times \bar{\imath}, \qquad i = 0,1.$$

This gives $(h^\# d(f_0,k,f_1)) \times \bar{I} = d(f'_0,k',f'_1) \times \bar{I}$, and this, in view of (5), proves the theorem.

33.7. The addition formula.

THEOREM. *Let f_0,f_1,f_2 be cross-sections defined on $L \cup K^q$ which coincide on L. Let $f_i' = f_i|L \cup K^{q-1}$ $(i = 0,1,2)$. Let k,k' be homotopies*

$$k:\ f_0' \simeq f_1', \qquad k':\ f_1' \simeq f_2', \qquad \text{relative to } L,$$

and let k'': $f_0' \simeq f_2'$ rel. L *be their composition* (i.e. $k''(x,t) = k(x,2t)$ *for* $0 \leqq t \leqq 1/2$, $k''(x,t) = k'(x,2t-1)$ *for* $1/2 \leqq t \leqq 1$). *Then*

$$d(f_0,k'',f_2) = d(f_0,k,f_1) + d(f_1,k',f_2).$$

To prove that the two sides coincide on a q-cell τ, it suffices to restrict attention to the subcomplex consisting of τ and its faces. If §33.6 is applied to the inclusion map of this subcomplex, we thereby reduce the proof to the case where K consists of a q-cell τ and its faces. Then $\mathfrak{B} = K \times Y$, $\mathfrak{B}^\square = K \times I \times Y$, and we have projections p': $\mathfrak{B} \to Y$ and $p'^\square$: $\mathfrak{B}^\square \to Y$. The coefficient bundles are product bundles, and all coefficient groups may be identified with $\pi_q(Y)$ under these projections.

Let F_0 be the cross-section associated with f_0,k,f_1; F_1 with f_1,k',f_2; and F with f_0,k'',f_2. Let I' denote the complex obtained by dividing I into two subintervals $I_0 = [0,1/2]$ and $I_1 = [1/2,1]$. Extend F to a cross-section F' over $(K \times I')^q$ by setting $F'(x,1/2) = (f_1(x),1/2)$. Let

$$\sigma = \tau \times I, \qquad \sigma_0 = \tau \times I_0, \qquad \sigma_1 = \tau \times I_1.$$

Then the chain $\sigma_0 + \sigma_1$ is the subdivision of σ. Hence $\partial\sigma_0 + \partial\sigma_1$ is the subdivision of $\partial\sigma$. Applying §32.2, it follows that $\partial\sigma$ represents in $\pi_q((K \times I')^q)$ the sum of the elements represented by $\partial\sigma_0$ and $\partial\sigma_1$. Taking images under $p'^\square F'$, we obtain

$$c(F,\sigma) = c(F',\sigma_0) + c(F',\sigma_1).$$

Defining g: $\mathfrak{B} \times I_0 \to \mathfrak{B} \times I$ by $g(b,t) = (b,2t)$, then $gF' = F_0\bar{g}$, and this implies $c(F',\sigma_0) = c(F_0,\sigma)$. Similarly $c(F',\sigma_1) = c(F_1,\sigma)$. Therefore

$$c(F,\sigma) = c(F_0,\sigma) + c(F_1,\sigma). \tag{10}$$

Since f_i $(i = 0,1,2)$ is defined on all of K, $c(f_i) = 0$. The theorem follows now from (10), (9) and the isomorphism (5).

33.8. LEMMA. *We have $d(f_0,k,f_1) = 0$ if and only if k can be extended to a homotopy*

$$f_0|L \cup K^q \simeq f_1|L \cup K^q.$$

Now $d = 0$ if and only if $c(F)$ is zero on each $\tau \times I$; and this occurs if and only if F is extendable over each $\tau \times I$. But the latter is equivalent to the extendability of k.

33.9. LEMMA. *If f_0 is a cross-section of the part of $\mathfrak{B}$ over $L \cup K^q$, and*

$$d \in C^q(K,L;\mathfrak{B}(\pi_q)),$$

then $f_0|L \cup K^{q-1}$ may be extended to a cross-section f_1 defined on $L \cup K^q$ such that

$$d(f_0,f_1) = d.$$

For each q-cell τ, we shall extend $f_0|\dot{\tau}$ to a cross-section f_1 over τ so that the value of $d(f_0,f_1)$ on τ is $d(\tau)$. This reduces the proof to the case where K consists of a q-cell τ and its faces. We may therefore suppose that $\mathfrak{B} = K \times Y$, and $\mathfrak{B}^\square = K \times I \times Y$. Let p': $\mathfrak{B}^\square \to Y$ be the natural projection. Define

$$F(x,0) = (f_0(x),0) \text{ for } x \in \tau; \qquad F(x,t) = (f_0(x),t) \text{ for } x \in \dot{\tau}.$$

Then F is defined on $E = \tau \times 0 \cup \dot{\tau} \times I$. Let g: $(\tau \times I)^\cdot \to Y$ represent $(-1)^{q+1}\, d(\tau) \in \pi_q(Y)$. Since E is a q-cell, there exists a homotopy k of $g|E$ into $p'F$ (shrinking E to a point deforms both maps into constant maps, and the two resulting image points can be connected by a curve in Y, for Y is q-simple). By §16.2, the homotopy $k|(\tau \times 1)^\cdot$ can be extended to a homotopy of $\tau \times 1$. Then g is homotopic to a map g_1 such that $g_1|E = p'F$; and g_1 represents $(-1)^{q+1}\, d(\tau)$. Using the representation $\mathfrak{B} = K \times Y$, define $f_1(x) = (x,g_1(x,1))$. It follows immediately that $d(f_0,f_1)(\tau) = d(\tau)$.

§34. EXTENSION AND DEFORMATION THEOREMS

34.1. Extensions of cross-sections. We put together now the results of the preceding articles. It is assumed that $\mathfrak{B}$ is a bundle over the cell complex K, L is a subcomplex, and Y is q-simple.

34.2. THEOREM. *Let f be a cross-section of $\mathfrak{B}|L \cup K^{q-1}$, and let f be extendable over $L \cup K^q$. Then the set $\{c(f')\}$ of $(q+1)$-dimensional obstruction cocycles of all such extensions f' of f forms a single cohomology class*

$$\bar{c}(f) \in H^{q+1}(K,L;\mathfrak{B}(\pi_q));$$

and f is extendable over $L \cup K^{q+1}$ if and only if $\bar{c}(f) = 0$.

Let f_0, f_1 be two extensions of f over $L \cup K^q$. Then §33.5 gives $\delta d(f_0,f_1) = c(f_0) - c(f_1)$; hence $c(f_0)$ and $c(f_1)$ belong to the same cohomology class.

Let f_0 be an extension of f over $L \cup K^q$, and c a cocycle in the cohomology class of $c(f_0)$. Then there is a q-cochain d such that $\delta d = c(f_0) - c$. By §33.9, there is an extension f_1 of f over $L \cup K^q$ such that $d(f_0,f_1) = d$. It follows from §33.5, that $c(f_1) = c$.

If f is extendable over $L \cup K^{q+1}$ and f' is such an extension, then $f_0 = f'|L \cup K^q$ is extendable over $L \cup K^{q+1}$, so $c(f_0) = 0$, and $\bar{c}(f) = 0$.

If $\bar{c}(f)$ is zero, the part already proved provides an extension f' over $L \cup K^q$ such that $c(f') = 0$. Then f' is extendable over $L \cup K^{q+1}$. This completes the proof.

34.3. COROLLARY. *If f is a cross-section of $\mathfrak{B}|L \cup K^q$, then $f|L \cup K^{q-1}$ is extendable over $L \cup K^{q+1}$ if and only if $c(f)$ is a coboundary in $K - L$.*

34.4. COROLLARY. *If $\mathfrak{B}$ is a bundle over (K,L) and, for each $q = 1, 2, \cdots, \dim (K - L)$, Y is $(q - 1)$-simple and $H^q(K,L;\mathfrak{B}(\pi_{q-1})) = 0$, then any cross-section f of $\mathfrak{B}|L$ can be extended to a full cross-section of $\mathfrak{B}$. In particular, if $H^q(K;\mathfrak{B}(\pi_{q-1})) = 0$ for $q = 1, 2, \cdots, \dim K$, then $\mathfrak{B}$ has a cross-section.*

If f is extendable to $L \cup K^q$, the assumption $H^{q+1} = 0$ and the preceding corollary imply that f is extendable to $L \cup K^{q+1}$. The result follows by induction.

34.5. Homotopies of cross-sections. Suppose now that f_0,f_1 are two cross-sections of $\mathfrak{B}$, and $f_0|L = f_1|L$. And let the problem be to construct a homotopy

$$k: \ f_0 \simeq f_1, \qquad \text{relative } L \text{ (see §32.5).}$$

Defining $\mathfrak{B}^\square$, $K^\square$ and $L^\square$ as in §33.3, and setting

$$f(x,0) = (f_0(x),0), \qquad f(x,1) = (f_1(x),0),$$
$$f(x,t) = (f_0(x),t) \qquad \text{for } x \in L,\ t \in I,$$

we obtain a cross-section of $\mathfrak{B}^\square|L^\square$. If the homotopy k exists, then $f'(x,t) = (k(x,t),t)$ is an extension of f to a full cross-section. Conversely if f' is such an extension of f, the x-coordinate, $k(x,t)$, of $f'(x,t)$ is the required homotopy. Thus *the homotopy problem is equivalent to an extension problem.* Using this equivalence, the preceding results of this article yield the following three propositions concerning the homotopy problem. The proofs are omitted since they are obvious formal translations. We note that, if

$$k: \ f_0|L \cup K^{q-1} \simeq f_1|L \cup K^{q-1} \qquad \text{relative } L,$$

then $d(f_0,k,f_1)$ is a cocycle; for, by §33.5, $\delta d = c(f_0) - c(f_1)$ and both obstructions are zero since f_0,f_1 are full cross-sections.

34.6. THEOREM. *Let f_0,f_1 be two cross-sections of $\mathfrak{B}$ which coincide on L, and let k be a homotopy*

$$k: \ f_0|L \cup K^{q-2} \simeq f_1|L \cup K^{q-2} \qquad \textit{relative } L$$

which is extendable to a homotopy

$$k': \quad f_0|L \cup K^{q-1} \simeq f_1|L \cup K^{q-1} \qquad \textit{relative } L.$$

Then the set $\{d(f_0,k',f_1)\}$ *of deformation cocycles of all such extensions* k' *forms a single cohomology class*

$$\bar{d}(f_0,k,f_1) \in H^q(K,L;\mathfrak{B}(\pi_q)),$$

and k *is extendable to a homotopy*

$$f_0|L \cup K^q \simeq f_1|L \cup K^q \qquad \textit{relative } L$$

if and only if $\bar{d}(f_0,k,f_1) = 0$.

34.7. COROLLARY. *If* k *is a homotopy*

$$k: \quad f_0|L \cup K^{q-1} \simeq f_1|L \cup K^{q-1} \qquad \textit{relative } L,$$

then $k|(L \cup K^{q-2}) \times I$ *is extendable to a homotopy* $f_0|L \cup K^q \simeq f_1|L \cup K^q$ *if and only if* $d(f_0,k,f_1)$ *is a coboundary in* $K - L$.

34.8. THEOREM. *If* $\mathfrak{B}$ *is a bundle over* (K,L), *and, for each* $q = 0$, $1, \cdots$, dim $(K - L)$, Y *is* q*-simple and* $H^q(K,L;\mathfrak{B}(\pi_q)) = 0$, *then any two cross-sections of* $\mathfrak{B}$, *equal on* L, *are homotopic relative to* L.

34.9. Extension of a homotopy. We prove now a homotopy extension theorem which will provide a reinterpretation of these results.

THEOREM. *Let* $\mathfrak{B}$ *be a bundle over* (K,L), f *a cross-section of* $\mathfrak{B}$, *and* F': $L \times I \to B$ *a homotopy of* $f' = f|L$. *Then* F' *can be extended to a homotopy* F: $K \times I \to B$ *of* f.

We order the cells of $K - L$ in a finite sequence so that no cell precedes any of its faces. The extension of F' to F is carried out a cell at a time in the prescribed order. For the extension over a particular cell σ, we need use only the part of $\mathfrak{B}$ over σ. In this way we reduce the proof to the case where K is a q-cell σ and its faces, and L is the collection of proper faces of σ. Choose, then, a product representation

$$\phi: \quad \sigma \times Y \to \mathfrak{B}, \qquad p': \quad B \to Y$$

with the usual properties. Define

$$h: \quad (\sigma \times 0) \cup (\dot{\sigma} \times I) \to Y$$

by

$$h(x,t) = \begin{cases} p'F'(x,t), & x \in \dot{\sigma}, \\ p'f(x), & x \in \sigma,\ t = 0. \end{cases}$$

According to §16.2, there is a retraction r of $\sigma \times I$ into $\sigma \times 0 \cup \dot{\sigma} \times I$. Define

$$F(x,t) = \phi(x,hr(x,t)).$$

Then F is the desired extension of F'.

34.10. THEOREM. *Let f_0, f_1 be two cross-sections of $\mathfrak{B}$ which coincide on $L \cup K^{q-1}$, then there exists a homotopy*

$$F: \quad f_0 \simeq f_1', \qquad \text{relative } L \cup K^{q-2},$$

such that

$$f_1' = f_1 \qquad \text{on} \qquad L \cup K^q$$

if and only if the difference cocycle $d(f_0,f_1)$ is a coboundary in $K - L$.

Taking $k(x,t) = f_0(x)$ for $x \in L \cup K^{q-1}$, then §34.7 states that there exists a homotopy

$$k': \quad f_0|L \cup K^q \simeq f_1|L \cup K^q \quad \text{relative } L \cup K^{q-2}$$

if and only if $d(f_0,f_1)$ is a coboundary. If k' exists, then §34.9 provides the extension F. If F exists, then $F|(L \cup K^q) \times I$ is a homotopy k'. Thus k' exists if and only if F exists.

§35. THE PRIMARY OBSTRUCTION AND THE CHARACTERISTIC COHOMOLOGY CLASS

35.1. Assumption on the dimension q. The result of §34 on the obstruction to an extension can be summarized as follows. If, in the stepwise process of extending a cross-section, we meet with a non-zero obstruction $c(f)$, then it is a cocycle, and it may be varied within its cohomology class by altering the choice of the extension at the last step. If the class of $c(f)$ is zero, the alteration of the last step can be chosen so that the next step of the extension is possible.

The weakness of this result is only too apparent if one asks the question: Suppose the class of $c(f)$ is not zero, can one alter the choice of the extension over the last *two* steps so as to make the next step of the extension possible? If not, what can be accomplished by redefining over three steps, etc.? A few special results have been achieved in this direction. A redefinition over two stages can alter the cohomology class of $c(f)$, usually by some kind of "product" of lower dimensional classes (see [89]). The general problem is highly interesting and much research remains to be done. (See App. sect. 11.)

There is a special case however where the results of article 34 are fully satisfactory. We turn to this now.

Throughout this article we shall let q denote the least integer such that $\pi_q(Y) \neq 0$.

We continue the convention that $\pi_0(Y)$ is the reduced 0th homology group with integer coefficients (in the singular sense). Thus $q = 0$ is possible. If $q > 0$, then Y is arcwise connected, and the condition $\pi_q \neq 0$ is independent of the base point. If $q = 1$, we assume that π_1 is abelian. If $q > 1$, then $\pi_1 = 0$. Thus, in all cases, Y is q-simple.

It is to be noted that q and π_q are effectively computable, at least for triangulable spaces Y; for, by the Hurewicz theorem 15.10, q is the dimension of the first non-vanishing homology group, and $\pi_q(Y) \approx H_q(Y)$.

35.2. LEMMA. *Any cross-section f of $\mathfrak{B}|L$ is extendable to a cross-section of $\mathfrak{B}|L \cup K^q$. If f_1,f_2 are any two such extensions then $c(f_1) - c(f_2)$ is a coboundary in $K - L$.*

Since Y is $(q-1)$-connected, the first statement follows from §29.2. If we apply §34.8 to $\mathfrak{B}|L \cup K^{q-1}$, we obtain a homotopy

$$k: \quad f_1|L \cup K^{q-1} \simeq f_2|L \cup K^{q-1} \qquad \text{relative } L.$$

Then $d(f_1,k,f_2)$ is defined and by §33.5 its coboundary is $c(f_1) - c(f_2)$.

In view of the lemma, we can state:

35.3. Definition. If $\mathfrak{B}$ is a bundle over (K,L), q is the least integer such that $\pi_q(Y) \neq 0$, and f is a cross-section of $\mathfrak{B}|L$, then the cohomology class of the obstruction $c(f')$, where f' is any extension of f over $L \cup K^q$, is called the *primary obstruction* to the extension of f. It is denoted by $\bar{c}(f)$ and is an element of $H^{q+1}(K,L;\mathfrak{B}(\pi_q))$. In the special case that L is vacuous, the cohomology class of $c(f')$ is denoted by $\bar{c}(\mathfrak{B})$ and is called *the characteristic cohomology class of* $\mathfrak{B}$. It is the primary obstruction to the construction of a cross-section.

35.4. The vanishing of $\bar{c}(f)$.

THEOREM. *The primary obstruction $\bar{c}(f)$ is an invariant of the homotopy class of f. Its vanishing is a necessary and sufficient condition for f to be extendable over $L \cup K^{q+1}$.*

Let f' be an extension of f to $L \cup K^q$, and let a homotopy $f \simeq f_1$ be given. By §34.9, the homotopy is extendable to $f' \simeq f_1'$. Then, by §32.4, $c(f') = c(f_1')$. Hence $\bar{c}(f) = \bar{c}(f_1)$, and the first assertion is proved.

If f is extendable over $L \cup K^{q+1}$ and f' is such an extension, then, by §29.8, $c(f'|L \cup K^q) = 0$. Hence $\bar{c}(f) = 0$. Conversely, if $\bar{c}(f) = 0$, and f' is an extension of f to $L \cup K^q$, then $c(f')$ is a coboundary in $K - L$, and, by §34.3, $f\,|L \cup K^{q-1}$ is extendable over $L \cup K^{q+1}$.

35.5. COROLLARY. *The vanishing of $\bar{c}(\mathfrak{B})$ is a necessary and sufficient condition for the existence of a cross-section over K^{q+1}.*

35.6. Invariance under mappings.

LEMMA. *Let $\mathfrak{B},\mathfrak{B}'$ be bundles over X,X', respectively, let h_0, h_1 be homotopic maps $\mathfrak{B} \rightarrow \mathfrak{B}'$, and let f' be a cross-section of $\mathfrak{B}'$. Then the cross-sections f_0,f_1 of $\mathfrak{B}$, induced by h_0,f' and h_1,f' respectively, are homotopic.*

Let h: $\mathfrak{B} \times I \rightarrow \mathfrak{B}'$ be a homotopy of h_0 into h_1. Let f be the cross-section of $\mathcal{B} \times I$ induced by h,f'. Then f has the form $f(x,t) = (k(x,t),t)$, and k is the required homotopy.

35.7. THEOREM. *Let $\mathcal{B},\mathcal{B}'$ be bundles over (K,L), (K',L') respectively, let h: $\mathcal{B} \to \mathcal{B}'$ induce a map $\bar{h}$: $(K,L) \to (K',L')$, and let*

$$h^*: \quad H^{q+1}(K',L';\mathcal{B}'(\pi_q)) \to H^{q+1}(K,L;\mathcal{B}(\pi_q))$$

be the induced homomorphism of cohomology groups. Let f' be a cross-section of $\mathcal{B}'|L'$, and f the induced cross-section of $\mathcal{B}|L$. Then

$$h^*\bar{c}(f') = \bar{c}(f).$$

By §31.16, h is homotopic to a map which can be factored into the form (24′) of §31.16. By §35.6, this alters f by a homotopy, and, by §35.4, $\bar{c}(f)$ is unchanged. Since h^* is unchanged by a homotopy (§31.15), we can suppose that h itself factors into $i''^{-1}ki$. Now i''^* is an isomorphism (§31.9). If we can show that i''^* carries the primary obstruction into itself, the same will hold for $(i''^*)^{-1}$. Thus, it suffices to prove the theorem in the special case that $\bar{h}$ is a proper cellular map.

Choose an extension of f' to a cross-section f_1' of $\mathcal{B}'|L' \cup K^q$. Since $\bar{h}$ is cellular, f_1' and $\bar{h}$ induce a cross-section f_1 of $\mathcal{B}|L \cup K^q$ which extends f. Then $c(f_1')$, $c(f_1)$ are cocycles representing $\bar{c}(f')$,$\bar{c}(f)$ respectively. By §32.7, $h^\#c(f_1') = c(f_1)$; and the theorem is proved.

35.8. COROLLARY. *If, in §35.7, h induces a homeomorphism of (K,L) onto (K',L'), then the conclusion asserts the topological invariance of the primary obstruction. In particular, if (K,L), (K',L') are two cellular decompositions of the same space and subspace, $\mathcal{B} = \mathcal{B}'$ and $h =$ the identity, it follows that $\bar{c}(f)$ is independent of the choice of the cell complex used to compute it.*

35.9. The generalized $\bar{c}(f)$. There is a useful generalization of the primary obstruction. Let us replace the assumption $\pi_i(Y) = 0$ for $i < q$ by the following weaker conditions on Y and (K,L) jointly:

(1) Y is i-simple for $i = 1, \cdots, q-1$.

(2) $H^{i+1}(K,L;\mathcal{B}(\pi_i)) = 0$ for $i = 0, 1, \cdots, q-1$.

(3) $H^i(K,L;\mathcal{B}(\pi_i)) = 0$ for $i = 0, 1, \cdots, q-1$.

(It is understood that H^0 is the *reduced* cohomology group.)

In the development of the primary obstruction, the only place, where the assumption $\pi_i(Y) = 0$ for $i < q$ was used, occurred in the proof of §35.2. It was used once to obtain an extension of the cross section f over L to $L \cup K^q$. But §34.4 states that (1) and (2) insure this. The assumption was used again to obtain a homotopy connecting two such extensions restricted to $L \cup K^{q-1}$. The homotopy was provided by §34.8. But (1) and (3) are the hypotheses of §34.8. The assumption was not used again except, possibly, for a tacit use in §35.6 where $h\colon \mathcal{B} \to \mathcal{B}'$ implies $h^*\bar{c}(f') = \bar{c}(f)$. But here we would require that (1), (2) and (3) hold for both (K,L) and (K',L'). Thus we have

35.10. THEOREM. *All of the preceding theorems of §35 hold if the restriction $\pi_i(Y) = 0$ for $i < q$ is replaced by the weaker condition (1) above, and only such bundles $\mathcal{B}$ over (K,L) are considered as satisfy conditions (2) and (3) above.*

There are important special cases where the weaker conditions hold. Thus (1) holds if Y is a group (e.g. for principal bundles), or if $\pi_1(Y) = 0$. Conditions (2) and (3) will hold if K is a $(q + 1)$-sphere and $L = 0$.

35.11. Bundles over spheres. Let $\mathcal{B}$ be a bundle over the $(q + 1)$-sphere S, and let Y be q-simple. Then $\bar{c}(\mathcal{B})$ in $H^{q+1}(S;\pi_q(Y))$ is defined. We suppose that $q > 0$ so that $\mathcal{B}(\pi_q)$ is a product bundle. Then, also, we can form the Kronecker index of $\bar{c}(\mathcal{B})$ with each homology class (integer coefficients, see §31.4) and obtain a homomorphism $H_{q+1}(S) \to \pi_q(Y_0)$ where Y_0 is the fibre over $x_0 \in S$. Defining Δ as in §17.3 and letting ψ be the Hurewicz isomorphism (§15.10) we have the diagram

$$\begin{array}{ccc} & H_{q+1}(S) & \\ \psi \nearrow & & \searrow \bar{c}(\mathcal{B}) \\ & \Delta & \\ \pi_{q+1}(S) & \longrightarrow & \pi_q(Y_0). \end{array} \tag{4}$$

35.12. THEOREM. *Under the above hypotheses, we have*

$$\bar{c}(\mathcal{B})\cdot\psi(\alpha) = -\Delta\alpha, \qquad \alpha \in \pi_{q+1}(S).$$

Let K be cellular decomposition of S whose $(q + 1)$-cells consist of the two hemispheres E_1,E_2 into which S is divided by a great q-sphere S'. We suppose $x_0 \in S'$. Orient E_1,E_2 so that

$$\partial E_1 = -\partial E_2. \tag{5}$$

Then $E_1 + E_2$ is a cycle representing a generator u of $H_{q+1}(S)$. It suffices to prove the theorem for the generator $\alpha = \psi^{-1}u$. Let f be a cross-section of $\mathcal{B}|E_2$, and let $f' = f|S'$. Then $c(f')$ is a cocycle representing $\bar{c}(\mathcal{B})$, and

$$c(f',E_2) = 0 \tag{6}$$

since f' is extendable over E_2.

Choose a homotopy k shrinking E_1 over itself into x_0, and then extend k to a homotopy of S (see §16.2). Then k deforms the identity into a map g_1 which maps (E_2,S') on (S,x_0) with degree 1. Hence $g_1|(E_2,S')$ represents α. Cover k by a homotopy k' of f into a map f_1: $(E_2,S') \to (B,Y_0)$. Since $pf_1 = g_1$, it follows that $f_1|S'$ represents $\Delta\alpha$ when S' is oriented so as to be positively incident to E_2. On the other hand $k'|S'$ deforms f' over $B|E_1$ into $f_1|S'$. Then, by definition, $f_1|S'$ represents $c(f',E_1)$ when S' is oriented so as to be positively incident to E_1. So, by (5),

$$c(f',E_1) = -\Delta\alpha. \tag{7}$$

Combining (6) and (7) gives

$$c(f')\cdot(E_1 + E_2) = -\Delta\alpha,$$

and the theorem is proved.

§36. The Primary Difference of Two Cross-sections

36.1. Assumption on the dimension q. We consider again the problem of the homotopy classification of cross-sections. Let $\mathfrak{B}$ be a bundle over (K,L), and let f_0,f_1 be cross-sections of $\mathfrak{B}$ which coincide on L. The problem is to construct a homotopy $f_0 \simeq f_1$ relative to L. As shown in §34.5, this is equivalent to an extension problem in $\mathfrak{B}^\square = \mathfrak{B} \times I$. We make the same assumption as in §35, namely, *q is the least integer such that $\pi_q(Y) \neq 0$.* Then the results of §35 on the extension problem yield corresponding results for the homotopy problem. We state these now. The proofs of the first few propositions are omitted since they are entirely mechanical.

36.2. Lemma. *If f_0,f_1 are two cross-sections of $\mathfrak{B}$ which coincide on L, then there exists a homotopy*

$$k\colon\ f_0|L \cup K^{q-1} \simeq f_1|L \cup K^{q-1} \qquad \textit{relative } L.$$

If k,k' are two such homotopies, then $d(f_0,k,f_1) - d(f_0,k',f_1)$ is a coboundary in $K - L$.

36.3. Definition. The conclusion of the preceding lemma asserts that the cohomology class of $d(f_0,k,f_1)$ depends only on f_0,f_1. This class, denoted by

$$\bar{d}(f_0,f_1) \in H^q(K,L;\mathfrak{B}(\pi_q))$$

is called the *primary difference* of f_0 and f_1.

36.4. The vanishing of $\bar{d}(f_0,f_1)$.

Theorem. *The primary difference $\bar{d}(f_0,f_1)$ is an invariant of the homotopy classes relative to L of f_0 and f_1. Its vanishing is a necessary and sufficient condition for*

$$f_0|L \cup K^q \simeq f_1|L \cup K^q \qquad \textit{relative } L.$$

Using §34.10, we can restate the last proposition:

36.5 COROLLARY. *The vanishing of* $\bar{d}(f_0,f_1)$ *is a necessary and sufficient condition for the existence of a homotopy* $f_0 \simeq f_1'$ *relative* L *such that* $f_1' = f_1$ *on* $L \cup K^q$.

36.6. The addition formula.

THEOREM. *Let* f_0,f_1,f_2 *be three cross-sections of* $\mathfrak{B}$ *which coincide on* L, *then*

$$\bar{d}(f_0,f_2) = \bar{d}(f_0,f_1) + \bar{d}(f_1,f_2).$$

This follows directly from §33.7.

36.7. The coboundary formula.

THEOREM. *Let* f_0,f_1 *be two cross-sections of* $\mathfrak{B}|L$. *Then* $\bar{d}(f_0,f_1)$ *is defined and is in* $H^q(L;\mathfrak{B}(\pi_q))$. *Under the coboundary operator*

$$\delta\colon\ H^q(L;\mathfrak{B}(\pi_q)) \to H^{q+1}(K,L;\mathfrak{B}(\pi_q)) \qquad \text{(see §31.6)},$$

we have

$$\delta\bar{d}(f_0,f_1) = \bar{c}(f_0) - \bar{c}(f_1).$$

By §35.2, we can choose extensions f_0',f_1' of f_0,f_1 over $L \cup K^q$. Applying §36.2 (with $L = 0$) we obtain a homotopy

$$k'\colon\ f_0'|K^{q-1} \simeq f_1'|K^{q-1}.$$

Now apply §33.5 (with $L = 0$) and we obtain

$$(1) \qquad \delta d(f_0',k',f_1') = c(f_0') - c(f_1').$$

Let k denote the homotopy k' restricted to L^{q-1}. By definition, $d(f_0,k,f_1)$ is a cocycle in the class $\bar{d}(f_0,f_1)$. Furthermore, $d(f_0',k',f_1')$ is an extension of $d(f_0,k,f_1)$ to a cochain of K. Hence $\delta d(f_0',k',f_1')$ is a cocycle in the class $\delta\bar{d}(f_0,f_1)$. Since $c(f_i')$ $(i = 0,1)$ is a cocycle in the class $\bar{c}(f_i)$, the theorem follows from (1) above.

36.8. Invariance under mappings.

LEMMA. *Let* f_0,f_1 *be cross-sections of* $\mathfrak{B}$ *which coincide on* L. *Define* $\mathfrak{B}^\square(= \mathfrak{B} \times I)$, $K^\square$ *and* $L^\square$ *as in §33.3. Let the cross-section* F *of the part of* $\mathfrak{B}^\square$ *over* $L^\square$ *be given by*

$$F(x,0) = (f_0(x),0), \quad F(x,1) = (f_1(x),1), \quad F(x,t) = (f_0(x),t) \quad \text{for } x \in L.$$

Then, under the isomorphism (7) of §33.3, we have

$$\bar{d}(f_0,f_1) \times \bar{I} = (-1)^{q+1}\bar{c}(F).$$

Using a homotopy k given by §36.2, and noting that $c(f_0) = c(f_1) = 0$, the result follows from (9) of §33.4.

36.9. THEOREM. *Let* $h\colon\ \mathfrak{B} \to \mathfrak{B}'$ *induce a map* $h\colon\ (K,L) \to (K',L')$. *Let* f_0',f_1' *be cross-sections of* $\mathfrak{B}'$ *which agree on* L'; *and let* f_0,f_1

be the cross-sections of $\mathfrak{B}$ *induced by* h. *Then*

$$h^*\bar{d}(f_0',f_1') = \bar{d}(f_0,f_1).$$

Construct F as in §36.8. In a similar way, construct the cross-section F' of $\mathfrak{B}'^{\square}|L'^{\square}$. Define $h^{\square}$: $\mathfrak{B}^{\square} \rightarrow \mathfrak{B}'^{\square}$ by $h^{\square}(b,t) = (h(b),t)$. It is obvious that F is induced by F' and $h^{\square}$. By §35.7, we have $h^{\square *}\bar{c}(F') = \bar{c}(F)$. The conclusion of §36.8 provides

$$(1) \qquad h^{\square *}(\bar{d}(f_0',f_1') \times \bar{I}) = \bar{d}(f_0,f_1) \times \bar{I}.$$

If we can show that

$$(2) \qquad h^{\square *}(\bar{d} \times \bar{I}) = (h^*\bar{d}) \times \bar{I},$$

the theorem will follow from (1) and the fact that $d \rightarrow d \times \bar{I}$ is an isomorphism. If $\bar{h}$ is a proper cellular map, we have noted in the proof of §33.6 that $h^{\square\#}(d \times \bar{I}) = (h^{\#}d) \times I$; and this implies (2) for this case. For the general case, we apply the factorization (24′) of §31.16. Since i, k, and i'' are proper cellular maps, the general case of (2) follows from the special case.

36.10. Corollary. *When* $\bar{h}$ *is a homeomorphism, the conclusion of §36.9 asserts the topological invariance of the primary difference. When* (K,L), (K',L') *are two cellular decompositions of the same space and subspace,* $\mathfrak{B} = \mathfrak{B}'$ *and* $h =$ *the identity, it asserts that* $\bar{d}(f_0,f_1)$ *is independent of the cellular decomposition used to compute it.*

36.11. The generalized primary difference. In §35.9 it is noted that the assumption on Y and q can be replaced by weaker conditions in defining the primary obstruction. The same obtains for the primary difference. Since the latter is just a primary obstruction in $(K^{\square},L^{\square})$ (see §36.8), it is enough for the weakened conditions of §35.9 to hold in the bundle $\mathfrak{B}^{\square}$. Using the isomorphisms (7) of §33.3, these conditions translate into

(1) Y is i-simple for $i = 1, \cdots, q-1$.

(2) $H^i(K,L;\mathfrak{B}(\pi_i)) = 0$ for $i = 0, 1, \cdots, q-1$.

(3) $H^{i-1}(K,L;\mathfrak{B}(\pi_i)) = 0$ for $i = 1, \cdots, q-1$.

These conditions suffice for all the results of this article except §36.7. Here the primary difference lies in L; so we impose (1), (2), and (3) with L in place of (K,L). In addition the theorem involves primary obstructions in (K,L). Hence we impose conditions (1), (2), and (3) of §35.9. The two sets of conditions suffice to prove §36.7.

§37. EXTENSIONS OF FUNCTIONS, AND THE HOMOTOPY CLASSIFICATIONS OF MAPS

37.1. Assumption on the dimension q. We continue with the assumption that q is the least integer such that $\pi_q(Y) \neq 0$, and we will apply the primary difference to the problems of extending a cross-section, and of the homotopy classification of cross-sections.

37.2. First extension theorem. *Let* $\dim (K - L) \leqq q + 1$. *Let* $\mathcal{B}$ *be a bundle over* (K,L) *which has a cross-section* f_0. *Then, for each* $d \in H^q(K,L;\mathcal{B}(\pi_q))$, *there exists an extension* f_1 *of* $f_0|L$ *to a full cross-section such that*

$$\bar{d}(f_0,f_1) = d.$$

Let d' be a cocycle in the class d. According to §33.9, $f_0|L \cup K^{q-1}$ extends to a cross-section f_1 over $L \cup K^q$ such that $d(f_0,f_1) = d'$. By §33.5,

$$c(f_0|L \cup K^q) - c(f_1) = \delta d(f_0,f_1) = \delta d' = 0,$$

for d' is a cocycle. Since $f_0|L \cup K^q$ is extendable, $c(f_0|L \cup K^q) = 0$. Hence $c(f_1) = 0$, and f_1 is extendable over $L \cup K^{q+1} = K$.

37.3. THEOREM. *Let* f_0,f *be cross-sections of* $\mathcal{B}|L$, *and let* f_0 *be extendable over* $L \cup K^{q+1}$. *Then*

$$\bar{c}(f) = \delta\bar{d}(f,f_0).$$

Therefore, f *is extendable over* $L \cup K^{q+1}$ *if and only if* $\delta\bar{d}(f,f_0) = 0$.

Since f_0 is extendable over $L \cup K^{q+1}$, by §35.4, we have $\bar{c}(f_0) = 0$. Then $\delta\bar{d}(f,f_0) = \bar{c}(f)$ follows from §36.7. The last statement follows now from §35.4.

37.4. Second extension theorem. *Let* $\dim (K - L) \leqq q + 1$, *and suppose the bundle* $\mathcal{B}$ *over* (K,L) *has a cross-section* f_0. *Let* $i\colon\ \mathcal{B}|L \to \mathcal{B}$ *be the inclusion map so that*

$$i^*\colon\ H^q(K;\mathcal{B}(\pi_q)) \to H^q(L;\mathcal{B}(\pi_q)).$$

Then a cross-section f *of* $\mathcal{B}|L$ *is extendable to a cross-section of* $\mathcal{B}$ *if and only if there exists an element* d' *in* $H^q(K;\mathcal{B}(\pi_q))$ *such that*

$$i^*d' = \bar{d}(f,f_0|L).$$

Furthermore, for each such d', *there exists an extension* f' *of* f *such that* $\bar{d}(f',f_0) = d'$.

Suppose f is extendable to a cross-section f' of $\mathcal{B}$. If we observe that f and $f_0|L$ are the cross-sections induced by f',f_0, and the inclusion map i, then $i^*\bar{d}(f',f_0) = \bar{d}(f,f_0|L)$ follows from §36.9; and $d' = \bar{d}(f',f_0)$ is the required element.

Conversely, suppose d' given and $i^*d' = \bar{d}(f,f_0|L)$. By §36.2, there exists a homotopy

$$k\colon\ f|L^{q-1} \simeq f_0|L^{q-1}.$$

By §34.9, k extends to a homotopy $k'\colon\ f \simeq f_1$ and $f_1 = f_0$ on L^{q-1}. According to §36.4, $\bar{d}(f,f_0|L) = \bar{d}(f_1,f_0|L)$. Therefore $d(f_1,f_0|L)$ is a cocycle in the class $\bar{d}(f,f_0|L)$. Let d_1 be a cocycle in the class d'. Then $i^\# d_1$ represents $i^*d' = \bar{d}(f,f_0|L)$. It follows that there is a $(q-1)$-cochain c of L such that

$$\delta c = i^\# d_1 - d(f_1,f_0|L).$$

Extend c to a cochain c' of K by defining it arbitrarily on the $(q-1)$-cells of $K - L$. Let $d_2 = d_1 - \delta c'$. Then

$$\begin{aligned} i^\# d_2 &= i^\# d_1 - i^\# \delta c' = i^\# d_1 - \delta i^\# c' \\ &= i^\# d_1 - \delta c = d(f_1,f_0|L). \end{aligned}$$

Thus d_2 *is an extension of* $d(f_1,f_0|L)$ *to a cocycle of* K *belonging to the class* d'.

Define d_3 in $C^q(K,L;\mathcal{B}(\pi_q))$ by

$$d_3 = d_2 \text{ on } K - L, \qquad d_3 = 0 \text{ on } L.$$

Then §33.9 provides an extension f_0' of $f_0|L \cup K^{q-1}$ to $L \cup K^q$ such that

$$d(f_0',f_0) = d_3.$$

Define f_1' on $L \cup K^q$ by

$$f_1'|L = f_1, \qquad f_1'|K^q - L^q = f_0'|K^q - L^q.$$

Since L intersects the closure of $K^q - L^q$ in a subset of L^{q-1}, and $f_1 = f_0 = f_0'$ on L^{q-1}, it follows that f_1' is a continuous extension of f_1. By its definition,

$$d(f_1',f_0) = \begin{cases} d(f_0',f_0) = d_3 & \text{in } K - L, \\ d(f_1,f_0|L) & \text{in } L. \end{cases}$$

Therefore $d(f_1',f_0) = d_2$. Applying §33.5, we have

$$\delta d(f_1',f_0) = c(f_1') - c(f_0).$$

Since d_2 is a cocycle, $\delta d(f_1',f_0) = 0$. Since f_0 is defined over all $(q+1)$-cells, $c(f_0) = 0$. Hence $c(f_1') = 0$. Thus f_1' extends to a cross-section f_1'' over $L \cup K^{q+1} = K$. Then $d(f_1'',f_0) = d(f_1',f_0) = d_2$, and $\bar{d}(f_1'',f_0) = d'$. By §34.9, the reverse of the homotopy $k'\colon\ f \simeq f_1$ extends to a homotopy $f_1'' \simeq f'$. Then f' is an extension of f to all of K; and, by §36.4, $\bar{d}(f',f_0) = \bar{d}(f_1'',f_0) = d'$. This completes the proof.

37.5. Classification theorem. *Let* dim $(K - L) = q$, *and suppose the bundle* $\mathcal{B}$ *over* (K,L) *has a cross-section* f_0. *If we restrict attention to cross-sections* f *of* $\mathcal{B}$ *which coincide with* f_0 *on* L, *and to homotopies relative to* L, *then the assignment of* $\bar{d}(f,f_0)$ *to each such* f *sets up a 1-1 correspondence between homotopy classes of cross-sections and elements of* $H^q(K,L;\mathcal{B}(\pi_q))$.

If $f \simeq f'$ rel. L, §36.4 asserts that $\bar{d}(f,f_0) = \bar{d}(f',f_0)$. Thus each homotopy class corresponds to a single cohomology class.

Suppose $\bar{d}(f,f_0) = \bar{d}(f',f_0)$. By the addition formula §36.6, we have $\bar{d}(f,f') = 0$. Since $K = L \cup K^q$, §36.5 provides a homotopy $f \simeq f'$ rel. L. Thus, distinct homotopy classes correspond to distinct cohomology classes.

Now let $d \varepsilon H^q(K,L;\mathcal{B}(\pi_q))$ be given. By §37.2, there exists an extension f of $f_0|L$ to all of K such that $\bar{d}(f_0,f) = -d$. By §36.6, we have $\bar{d}(f,f_0) = d$. This completes the proof.

37.6. Specialization of results to $K \times Y$. If we specialize $\mathcal{B}$ to be the product bundle $K \times Y$ the preceding results may be given a slightly revised and simpler form. Note first that the coefficient bundle $\mathcal{B}(\pi_q)$ is likewise a product, hence we may deal with ordinary cohomology groups with coefficients in $\pi_q = \pi_q(Y)$.

Any cross-section of $K \times Y$ is the graph of a map $K \to Y$, and any map provides a cross-section. A homotopy k of a map $K \to Y$ provides a homotopy $k'(x,t) = (x,k(x,t))$ of the graph, and conversely. The relation "graph" is a 1-1 correspondence which preserves relations such as equality on L, one function is an extension of another, and homotopic relative to L.

If f: $L \to Y$, we define the primary obstruction $\bar{c}(f)$ (to the extension of f to K) to be the primary obstruction to extending its graph. If f_0,f_1: $K \to Y$ and $f_0|L = f_1|L$, we define the primary difference $\bar{d}(f_0,f_1)$ to be the primary difference of their graphs.

With these conventions, all of the preceding work, beginning with §32, may be divested of the bundle language, and restated in terms of maps of complexes into Y, *and their homotopies. We shall assume any such restatement without further comment.*

37.7. Interpretation for the generalized $\bar{c}$ and $\bar{d}$. In keeping with the remarks of §35.9 and §36.11, the hypothesis on Y and q in the preceding theorems can be weakened. In §37.2, the conditions of §36.11 are adequate. For §37.3 we require that the conditions of §36.11 hold with L in place of (K,L), and the conditions of §35.9 hold for (K,L). In §37.4, it suffices for the conditions of §36.11 to hold with L in place of (K,L) and with K in place of (K,L). For §37.5, the conditions of §36.11 suffice.

37.8. The primary obstruction to contracting Y. We continue with the assumption that q is the least integer such that $\pi_q(Y) \neq 0$. We assume, moreover, that Y *is a complex.*

Let y_0 be a point of Y and let g_0: $Y \to Y$ be the constant map $g_0(y) = y_0$. Let g_1: $Y \to Y$ be the identity map. Define

$$\bar{d}(Y) \; \varepsilon \; H^q(Y;\pi_q), \qquad \pi_q = \pi_q(Y),$$

by

$$\bar{d}(Y) = \bar{d}(g_0,g_1) \qquad\qquad \text{see §37.6.}$$

We call it *the primary obstruction to contracting Y into y_0.*

If $q > 0$, then Y is arcwise connected, and any two constant maps are homotopic. Then §36.4 asserts that $\bar{d}(Y)$ is independent of the choice of y_0. In any case we can assume that y_0 is a vertex of Y.

37.9. Lemma. *Let f_0: $K \to Y$ be the constant map $f_0(x) = y_0$, and let f: $K \to Y$ be any map. Then $\bar{d}(f_0,f) = f^*\bar{d}(Y)$.*

We prove first an elementary fact about the graph relation between maps and cross-sections. Let g: $Y \to Y$ and let g' be its graph: $g'(y) = (y,g(y))$. Let $\tilde{f}$: $K \times Y \to Y \times Y$ be defined by $\tilde{f}(x,y) = (f(x),y)$. Then $\tilde{f}$ is a bundle map inducing the map f of the base space K into the base space Y. Let ϕ' be the cross-section induced by g' and $\tilde{f}$, and let ϕ' be the graph of ϕ. By definition of the induced cross-section, $\tilde{f}\phi' = g'f$. Using this, we find that $\phi = gf$. This may be restated: *Under the graph relationship between cross-sections and maps, induced cross-sections under $\tilde{f}$ correspond to compositions with f.*

Since $g_0(y) = y_0$, we have $g_0f(x) = y_0$. Therefore f_0 is induced by f and g_0. Since $g_1(y) = y$, we have $g_1f(x) = f(x)$. Therefore f is induced by f and g_1. Then §36.4 states

$$\bar{d}(f_0,f) = f^*\bar{d}(g_0,g_1) = f^*\bar{d}(Y).$$

In §§37.3–37.5, we take f_0 to be the constant map; and, using the lemma, we obtain the following three results.

37.10. Extension theorems.

Theorem. *Let L be a subcomplex of K, and f: $L \to Y$, then $\bar{c}(f) = \delta f^*\bar{d}(Y)$. Thus f is extendable to a map $L \cup K^{q+1} \to Y$ if and only if $\delta f^*\bar{d}(Y) = 0$.*

37.11. Theorem. *If $\dim (K - L) \leq q + 1$, i: $L \to K$ is the inclusion map, and f is a map $L \to Y$, then f extends to a map $K \to Y$ if and only if there exists a $d' \; \varepsilon \; H^q(K;\pi_q)$ such that $i^*d' = f^*\bar{d}(Y)$. For each such d', there exists an extension f' of f such that $f'^*\bar{d}(Y) = d'$.*

37.12. Homotopy classification theorem. *Let $\dim K = q$. Then the assignment of $f^*\bar{d}(Y)$ to each map f: $K \to Y$ sets up a 1-1 correspondence between homotopy classes of maps $K \to Y$ and elements of $H^q(K;\pi_q)$.*

In the last theorem, we have applied §37.5 with $L = 0$. This restriction on L can be lifted by using the notion of the difference homomorphism $(f - f_0)^*$: $H^q(Y) \to H^q(K,L)$ where $f = f_0$ on L. This is described in [89].

37.13. The case $Y = S^q$. Assume now that Y is a q-sphere S^q. Then $\pi_q(S^q)$ is infinite cyclic. Thus the cohomology groups appearing in the preceding three theorems may be treated as the ordinary groups with integer coefficients. Furthermore, $H^q(S^q;\pi_q)$ is infinite cyclic and

$$\bar{d}(S^q) \text{ generates } H^q(S^q;\pi_q). \tag{1}$$

To prove this, we represent S^q as a complex consisting of two q-cells E_+,E_- with S^{q-1} as a common boundary. We take an arbitrary cellular decomposition on S^{q-1}. We can suppose $y_0 \varepsilon S^{q-1}$. To compute a representative cocycle of $\bar{d}(S^q)$, we must select a homotopy k of the identity map $g_1|S^{q-1}$ into the constant map g_0. Let k shrink S^{q-1} over E_- into y_0. Constructing F as in §33.4, we find that F maps the boundary of $E_+ \times I$ on S^q with degree 1, and the boundary of $E_- \times I$ on S^q with degree 0. Hence $d(g_1,k,g_0)$ is zero on E_- and is a generator of π_q on E_+; and (1) follows.

It is worth noting in this case that the primary obstructions and differences appearing in §§37.10–37.12 are effectively computable, at least for cellular maps. With a little more effort, using the Hurewicz isomorphism $H_q(Y) \approx \pi_q(Y)$, one can prove the same without restricting Y to be S^q. However we have no general rules for effectively computing the obstructions and differences appearing in the theorems 37.3, 37.4, and 37.5. This is one of the chief problems of the theory.

37.14. The relation $p^*\bar{c}(\mathcal{B}) = 0$.

THEOREM. *If $\mathcal{B} = \{B,p,X,Y\}$ and both B and X are triangulable (i.e. admit cellular decompositions), then*

$$p^*\bar{c}(\mathcal{B}) = 0.$$

Let $\mathcal{B}^2 = \{B^2,p',B,Y\}$ be the bundle induced over B by p: $B \to X$ and the bundle $\mathcal{B}$ over X, and let $\tilde{p}$: $\mathcal{B}^2 \to \mathcal{B}$ be the natural map. By §35.7, we have

$$\tilde{p}^*\bar{c}(\mathcal{B}) = \bar{c}(\mathcal{B}^2).$$

In §10.4 we showed that $\mathcal{B}^2$ has a cross-section. Therefore $\bar{c}(\mathcal{B}^2) = 0$. This gives

$$\tilde{p}^*\bar{c}(\mathcal{B}) = 0$$

which is the form the conclusion of the theorem should have.

37.15. We give now an unpublished result due to G. W. Whitehead which adds a measure of precision to the preceding result. Let $\mathcal{B} =$

$\{B,p,X,Y\}$. Let $y_0 \in B$, $x_0 = p(y_0)$, and $Y_0 = p^{-1}(x_0)$. We assume that X is triangulable with x_0 as a vertex, and that B is triangulable with Y_0 as a subcomplex. Define f by $f(x_0) = y_0$ so that f is a cross-section of the part of $\mathfrak{B}$ over x_0. We obtain the diagram

$$\begin{array}{ccccc}
\bar{d}(Y_0) & \delta & p_1^*\bar{c}(f) & k^* & \\
H^q(Y_0;\pi_q) & \rightarrow & H^{q+1}(B,Y_0;\mathfrak{B}^2(\pi_q)) & \rightarrow & H^{q+1}(B;\mathfrak{B}^2(\pi_q)) \\
\uparrow p_0^* & & \uparrow p_1^* & & \uparrow p^* \\
 & \delta & & j^* & \\
H^q(x_0;\pi_q) & \rightarrow & H^{q+1}(X,x_0;\mathfrak{B}(\pi_q)) & \rightarrow & H^{q+1}(X;\mathfrak{B}(\pi_q)) \\
 & & \bar{c}(f) & & \bar{c}(\mathfrak{B})
\end{array}$$

where k,j are inclusion maps, and p_0,p_1 are maps induced by p. Since the part of $\mathfrak{B}$ over x_0 is a product, the coefficient groups $\mathfrak{B}^2(\pi_q),\mathfrak{B}(\pi_q)$, on the left, reduce to π_q. Then, we have

37.16. THEOREM. *Under the above hypotheses,*

$$\delta\bar{d}(Y_0) = p_1^*\bar{c}(f).$$

Treating X as a pair $(X,0)$, then f and j induce the vacuous cross-section. Hence

$$j^*\bar{c}(f) = \bar{c}(\mathfrak{B})$$

follows from §35.7. Applying §37.14, we obtain $p^*j^*\bar{c}(f) = 0$. Commutativity in the right square of the diagram gives

$$k^*p_1^*\bar{c}(f) = 0.$$

Exactness of the cohomology sequence assures us that there is a $d \in H^q(Y_0;\pi_q)$ such that $\delta d = p_1^*\bar{c}(f)$.

Let g_0 be the cross-section of the part of $\mathfrak{B}^2$ over Y_0 induced by f. Recall (§10.2) that

$$B^2 \subset B \times B$$

consists of pairs (b,b') in B such that $p(b) = p(b')$. Thus the part of B^2 over Y_0 is just $Y_0 \times Y_0$. In this representation g_0 is the graph of the constant map $Y_0 \rightarrow y_0$. The cross-section g_1 of $\mathfrak{B}^2$ is given by $g_1(b) = (b,b)$. Then $g_1|Y_0$ is the graph of the identity map. Hence, by §37.8,

$$\bar{d}(g_0,g_1|Y_0) = \bar{d}(Y_0).$$

By §36.7, we have

$$\delta\bar{d}(Y_0) = \bar{c}(g_0) - \bar{c}(g_1|Y_0).$$

The last term is zero since $g_1|Y_0$ is extendable to the cross-section g_1. Since g_0 is induced by f_0, we have $p_1^*\bar{c}(f) = \bar{c}(g_0)$, and the theorem follows.

37.17. Historical remarks. The theorems 37.11 and 37.12 with $Y = S^q$ are the extension and classification theorems due to Hopf. He stated them in the language of homology. The simpler formulation in terms of cohomology is due to Whitney. An excellent treatment is given in the book of Hurewicz and Wallman [56].

The notion of the characteristic class of a bundle is due to Whitney [103]. Independently, Stiefel treated the characteristic classes of the tangent sphere bundle of a manifold [91].

The primary obstructions and differences for maps $K \to Y$ are due to Eilenberg [31]. The development which we have given of these ideas (in §§29–37) follows closely the treatment given by Eilenberg. The formulation for bundles requires only the complication of a bundle of coefficients.

§38.* THE WHITNEY CHARACTERISTIC CLASSES OF A SPHERE BUNDLE

38.1. Conventions. Throughout this article $\mathfrak{B}$ will denote an $(n-1)$-sphere bundle over a complex K. For notational convenience, we denote by Y^q the Stiefel manifold

$$Y^q = V_{n,n-q} = O_n/O_q, \quad q = 0, 1, \cdots, n-1.$$

And we define $\mathfrak{B}^q$ to be the associated bundle of $\mathfrak{B}$ with the fibre Y^q. Then $\mathfrak{B}^0$ is the principal bundle of $\mathfrak{B}$, and $\mathfrak{B}^{n-1} = \mathfrak{B}$. Since $O_{q-1} \subset O_q$, we have natural projections

$$(1) \qquad O_n = Y^0 \to Y^1 \to \cdots \to Y^{n-1} = S^{n-1}$$

and Y^{q-1} is a $(q-1)$-sphere bundle over Y^q (see §7.8). By §9.6, these projections induce projections

$$(2) \qquad B^0 \to B^1 \to \cdots \to B^{n-1} \to K,$$

and any composition of them is the projection of a bundle structure with a suitable Stiefel manifold as fibre. In particular, B^{q-1} is a $(q-1)$-sphere bundle over B^q. The composition of $B^q \to B^{q+1} \to \cdots \to K$ is the projection of $\mathfrak{B}^q$.

38.2. Definition. The qth *characteristic class of* $\mathfrak{B}$ $(q = 1, \cdots, n)$, in the sense of Whitney [103], is defined to be the characteristic class of $\mathfrak{B}^{q-1}$. We denote it by $c^q(\mathfrak{B})$; thus

$$(3) \qquad c^{q+1}(\mathfrak{B}) = \bar{c}(\mathfrak{B}^q).$$

According to §25.6, π_q is the first non-zero homotopy group of Y^q. Therefore

$$(4) \qquad c^{q+1}(\mathfrak{B}) \in H^{q+1}(K; \mathfrak{B}^q(\pi_q)).$$

* See App. sect. 9 and 10.

Also, by §25.6,

$$(5) \qquad \pi_q(Y^q) = \begin{cases} \infty & \text{if } q \text{ is even, or } q = n-1, \\ 2 & \text{if } q \text{ is odd and } < n-1. \end{cases}$$

In the second case, $\mathfrak{B}^q(\pi_q)$ is a product bundle; for, a cyclic group of order 2 has no non-trivial automorphisms. Thus,

$$(4') \qquad c^{2q}(\mathfrak{B}) \in H^{2q}(K; \bmod 2), \qquad 1 < 2q < n.$$

In the first case, the group of $\mathfrak{B}^q(\pi_q)$ is O_n/R_n which is cyclic of order 2. It operates effectively on $\pi_q(Y^q)$. To see this, choose a generator $f\colon\ S^q \to Y^q$ of the group as described in §25.6. Let $r \in O_n$ have determinant -1 and act as the identity in the space orthogonal to S^q. Then left translation of Y^q by r maps $f(S^q)$ on itself with degree -1; so r reverses sign in $\pi_q(Y^q)$. Thus, by §13.7,

If q is even or $q = n - 1$, $\mathfrak{B}^q(\pi_q)$ is a product bundle if and only if $\chi(\mathfrak{B})\colon\ \pi_1(K) \to O_n/R_n$ is trivial.

For example, let K be a manifold and $\mathfrak{B}$ the tangent sphere bundle. If K is orientable, then $\mathfrak{B}$ is equivalent to a bundle in R_n, so $\chi(\mathfrak{B}) = 0$, and each $\mathfrak{B}^q(\pi_q)$ is a product bundle. If K is non-orientable, $\mathfrak{B}$ is not reducible to R_n; therefore the weakly associated bundle with fibre O_n/R_n is not a product bundle (see §9.5), hence $\chi(\mathfrak{B})$ is non-trivial and $\mathfrak{B}^q(\pi_q)$ is not a product bundle (q even or $q = n - 1$).

38.3. Interpretation. By its definition, $c^{q+1}(\mathfrak{B})$ is the primary obstruction to forming a cross-section of $\mathfrak{B}^q$. Since π_q is the first non-zero homotopy group of Y^q, $\mathfrak{B}^q|K^q$ has a cross-section, and $\mathfrak{B}^q|K^{q+1}$ has a cross-section if and only if $c^{q+1}(\mathfrak{B}) = 0$. Now, for any integer p, §9.5 states that $\mathfrak{B}^q|K^p$ has a cross-section if and only if $\mathfrak{B}|K^p$ is equivalent in O_n to a bundle with group O_q. But O_q operating in S^{n-1} leaves a great $(n - q - 1)$-sphere S' pointwise fixed. Then the subbundle of $\mathfrak{B}|K^p$ corresponding to S' is a product bundle. Thus, we have proved

38.4. Theorem. *For each $q = 0, 1, \cdots, n - 1$, there exists a map*

$$\psi\colon\ K^q \times S^{n-q-1} \to B$$

which, for each x in K^q, maps the fibre $x \times S^{n-q-1}$ orthogonally into the fibre over x in B. And there exists a similar map of $K^{q+1} \times S^{n-q-1}$ if and only if $c^{q+1}(\mathfrak{B}) = 0$.

38.5. The cohomology sequence of a coefficient sequence. The characteristic classes of $\mathfrak{B}$ are not independent. To state the relations requires the use of a little-known operation on cohomology groups which we describe first. Let

$$(6) \qquad 0 \to L \xrightarrow{\lambda} M \xrightarrow{\mu} N \to 0$$

be an *exact* sequence of abelian groups and homomorphisms, i.e. kernel $\lambda = 0$, image λ = kernel μ, and image $\mu = N$. If we pass to cochains in K with coefficients in L, M and N, then λ,μ induce homomorphisms of the cochain groups, and we have the diagram

$$\begin{array}{ccccccccc} 0 & \rightarrow & C^p(K;L) & \overset{\lambda}{\rightarrow} & C^p(K;M) & \overset{\mu}{\rightarrow} & C^p(K;N) & \rightarrow & 0 \\ & & \downarrow\delta & & \downarrow\delta & & \downarrow\delta & & \\ 0 & \rightarrow & C^{p+1}(K;L) & \overset{\lambda}{\rightarrow} & C^{p+1}(K;M) & \overset{\mu}{\rightarrow} & C^{p+1}(K;N) & \rightarrow & 0 \end{array} \tag{7}$$

where each line is an exact sequence. Clearly we have commutativity in each square: $\lambda\delta = \delta\lambda$ and $\mu\delta = \delta\mu$.

Let z in $C^p(K;N)$ be a cocycle. Choose u in $C^p(K;M)$ such that $\mu u = z$. It follows that $\mu\delta u = 0$. Exactness of the lower line provides a w in $C^{p+1}(K;L)$ such that $\lambda w = \delta u$. Furthermore w is a cocycle; for λ: $C^{p+2}(K;L) \rightarrow C^{p+2}(K;M)$ has kernel $= 0$, and $\lambda\delta w = \delta\lambda w = \delta\delta u = 0$. Thus, we have proved

For each cocycle z in $C^p(K;N)$, there exists a cocycle w in $C^{p+1}(K;L)$ and a cochain u in $C^p(K;M)$ such that

$$\lambda w = \delta u, \qquad \mu u = z. \tag{8}$$

Suppose now that z,u',w' is a second triple satisfying analogous relations. Since $\mu u = \mu u' = z$, we have $\mu(u - u') = 0$. Exactness provides a v in $C^p(K;L)$ such that $\lambda v = u - u'$. Then

$$\lambda\delta v = \delta\lambda v = \delta u - \delta u' = \lambda w - \lambda w'.$$

Since λ has kernel $= 0$, we have

$$\delta v = w - w'.$$

Thus the assignment of w to z defines a unique map

$$Z^p(K;N) \rightarrow H^{p+1}(K;L). \tag{9}$$

It is easily proved that (9) is a homomorphism. Suppose now that $z = \delta c$ is a coboundary. Choose d in $C^{p-1}(K;M)$ such that $\mu d = c$. Then $\mu\delta d = z$. Taking $u = \delta d$, we have $\delta u = 0$. Hence we may choose $w = 0$. Thus, under (9), coboundaries are mapped into zero. It follows that (9) induces a homomorphism

$$\delta^*: \ H^p(K;N) \rightarrow H^{p+1}(K;L). \tag{10}$$

It is to be noted that the diagram of (7) is algebraically identical to the diagram (26) of §31.17 obtained from a complex K, a subcomplex L and just one coefficient group. Here $C^p(K;L)$ corresponds to $C^p(K,L)$, $C^p(K;M)$ to $C^p(K)$, $C^p(K;N)$ to $C^p(L)$, etc. A comparison of the definitions shows that δ^* in (10) corresponds to δ: $H^p(L) \to H^{p+1}(K,L)$ defined in §31.17. In §31.18, the exactness of the cohomology sequence of (K,L) was stated and proved in part. This proof was based entirely on the algebraic properties of the diagram (26) of §31.18. Since the diagram (7) above enjoys these properties, it follows that the same argument proves that *the sequence*

$$(11) \qquad \cdots \to H^{p-1}(K;N) \xrightarrow{\delta^*} H^p(K;L) \xrightarrow{\lambda^*} H^p(K;M) \xrightarrow{\mu^*} H^p(K;N) \to \cdots$$

is also exact. We refer to the sequence (11) as *the cohomology sequence of K and the coefficient sequence (6).*

For the application we have in mind the preceding must be generalized to cochains with coefficients in bundles. We replace (6) by

$$(12) \qquad 0 \to \mathfrak{B}(L) \xrightarrow{\lambda} \mathfrak{B}(M) \xrightarrow{\mu} \mathfrak{B}(N) \to 0$$

where $\mathfrak{B}(L)$, etc., are coefficient bundles over K, and λ,μ are bundle homomorphisms in the sense that, for each x in K, the sequence

$$(13) \qquad 0 \to L_x \xrightarrow{\lambda_x} M_x \xrightarrow{\mu_x} N_x \to 0$$

is exact, and the homomorphisms λ_x,μ_x commute with the translations of L_x, M_x, N_x along curves in K. We obtain then the diagram (7) with L, M, and N replaced by $\mathfrak{B}(L)$, $\mathfrak{B}(M)$, and $\mathfrak{B}(N)$ respectively. The generalization of the definition of δ^* and the proof of exactness of the generalized sequence (11) is entirely mechanical.

38.6. The coefficient sequence for the Whitney classes. We intend to show, for a sphere bundle $\mathfrak{B}$, that $\delta^* c^{2q}(\mathfrak{B}) = c^{2q+1}(\mathfrak{B})$. This requires $\mathfrak{B}(N)$ to be the coefficient bundle of $c^{2q}(\mathfrak{B})$, and $\mathfrak{B}(L)$ to be the same of $c^{2q+1}(\mathfrak{B})$. Before the relation makes sense, we must define $\mathfrak{B}(M)$, λ and μ, and prove exactness.

LEMMA. *Let the integer q satisfy $2 \leqq 2q < n$, and let $\mathfrak{B}'$ be the $(2q-1)$-sphere bundle*

$$p': \quad Y^{2q-1} \to Y^{2q} \qquad \text{(see §38.1).}$$

Then, in the section

$$(14) \qquad \pi_{2q}(Y^{2q}) \xrightarrow{\Delta} \pi_{2q-1}(S_0^{2q-1}) \xrightarrow{i_*} \pi_{2q-1}(Y^{2q-1})$$

of the homotopy sequence of $\mathfrak{B}'$, the kernel of Δ is zero and i_* is a homomorphism onto.

By §25.6, a generator of $\pi_{2q-1}(Y^{2q-1})$ is represented by the fibre over the point v_0 of Y^{2q}. Therefore i_* is a homomorphism onto. Since the image of i_* is cyclic of order 2, the kernel of i_* consists of the "even" elements of $\pi_{2q-1}(S_0^{2q-1})$. By the exactness of the homotopy sequence, the latter is the image of Δ. But $\pi_{2q}(Y^{2q}) = \infty$. Hence the kernel of Δ is zero, and the lemma is proved.

38.7. It follows that we may adjoin zero groups on the left and right ends of (14) and have an exact sequence as in (6). But we need a sequence of *bundles* as in (12), i.e. for each $x \varepsilon K$, an exact sequence of groups as in (13). Now, for each $x \varepsilon K$, the bundle projection p': $B^{2q-1} \to B^{2q}$ reduces to a bundle projection

$$p'_x:\ Y_x^{2q-1} \to Y_x^{2q} \tag{15}$$

of the fibres over x. Denote this bundle by $\mathcal{B}'_x$. Now $\mathcal{B}'_x$ is a replica of the bundle $\mathcal{B}'$ of the lemma. Hence the terms of its homotopy sequence corresponding to (14) form the desired sequence (13).

Before this choice is properly defined we must clarify the situation with respect to the base point of the homotopy groups. Since $Y^p = R_n/R_p$, we have, by §16.11, that Y^p is simple in all dimensions. Hence its homotopy groups can be defined, without reference to a base point, as homotopy classes of maps of a sphere. Thus, no base point is needed for the end terms of (14). For the middle term no base point is needed, however, a particular fibre is assumed. We eliminate this choice of a fibre as follows.

Since Y^{2q} is simply-connected, the bundle of groups $\mathcal{B}'(\pi_{2q-1}(S^{2q-1}))$ over Y^{2q} is a product bundle. Thus we have unique isomorphisms connecting the homotopy groups of the various fibres of $\mathcal{B}'$. Using these isomorphisms an element of a homotopy group of one fibre determines a class of equivalent elements—one on each fibre. These equivalence classes form a group isomorphic to $\pi_{2q-1}(S^{2q-1})$. It is this group which we take as the middle term of (14). Then the sequence (14) is assigned to $\mathcal{B}'$ without any choices being necessary.

We now define the sequence

$$L_x \xrightarrow{\lambda_x} M_x \xrightarrow{\mu_x} N_x$$

to be the section of the homotopy sequence of $\mathcal{B}'_x$ corresponding to the section (14) of $\mathcal{B}'$. Using the local product representations for the bundles $\mathcal{B}^{2q-1},\mathcal{B}^{2q}$ over K, it is easy to see that the λ_x,μ_x commute with translations of the groups along a curve lying in a coordinate neighborhood, and then, by composition, along any curve. It follows that this choice of the sequence (13) defines a sequence (12) of coefficient bundles.

It is to be noted that $\mathfrak{B}(M)$ is just the weakly associated bundle of $\mathfrak{B}$ with fibre $\pi_{2q-1}(S^{2q-1})$ and group O_n/R_n operating in such a way that the non-trivial element reverses sign.

38.8. The coboundary relation.

THEOREM. *Let $\mathfrak{B}$ be an $(n-1)$-sphere bundle over K, let q be an integer with $2 \leqq 2q < n$. Then, with respect to the preceding choice of the sequence (12) of coefficient bundles, we have*

$$\delta^* c^{2q}(\mathfrak{B}) = c^{2q+1}(\mathfrak{B}).$$

As shown in §38.3, we may choose a cross-section f of $\mathfrak{B}^{2q-1}|K^{2q-1}$. Then $c(f) \in C^{2q}(K;\mathfrak{B}(N))$ is defined, and is a cocycle representing $c^{2q}(\mathfrak{B})$. Composing f with

$$(16) \qquad p': \quad B^{2q-1} \to B^{2q},$$

we obtain a cross-section $p'f$ of $\mathfrak{B}^{2q}|K^{2q-1}$. Since the fibre of $\mathfrak{B}^{2q}$ is $(2q-1)$-connected, we may extend $p'f$ to a cross-section g of $B^{2q}|K^{2q}$. Then $c(g) \in C^{2q+1}(K;\mathfrak{B}(L))$ is a cocycle representing $c^{2q+1}(\mathfrak{B})$, and

$$(17) \qquad p'f = g|K^{2q-1}.$$

Now g imbeds K^{2q} topologically in B^{2q}. Let $\mathfrak{B}'_g$ denote the bundle of (16) restricted to $g(K^{2q})$. Then we may regard $\mathfrak{B}'_g$ as a $(2q-1)$-sphere bundle over K^{2q}. Then (17) states that f is a cross-section of $\mathfrak{B}'_g|K^{2q-1}$. Let $c(f,g)$ denote the obstruction to extending f to a cross-section of $\mathfrak{B}'_g$. Then the coefficients of $c(f,g)$ are elements of the $(2q-1)$st homotopy groups of fibres of the bundle (16), i.e.

$$c(f,g) \in C^{2q}(K;\mathfrak{B}(M)).$$

It should be kept in mind that $c(f,g)$ need not be a cocycle of K since $\mathfrak{B}'_g$ is only defined on K^{2q}.

Recalling the definition of δ^*, the theorem will follow once we have proved

$$(18) \qquad \mu(-c(f,g)) = c(f),$$

$$(19) \qquad \lambda c(g) = \delta(-c(f,g)).$$

To prove (18), let σ be a $2q$-cell and x its reference point. Shrinking σ to x, deforms $g|\sigma$ into a constant map g_0 with $g_0(\sigma) = y \in Y_x^{2q}$. A covering homotopy deforms $f|\dot{\sigma}$ into a map f_0 of $\dot{\sigma}$ into the fibre S_0^{2q-1} of (16) over y. Then f_0 represents simultaneously

$$c(f,\sigma) \in \pi_{2q-1}(Y_x^{2q-1}), \quad \text{and} \quad c(f,g,\sigma) \in \pi_{2q-1}(S_0^{2q-1}).$$

Since

$$\mu_x\colon\ \pi_{2q-1}(S_0^{2q-1}) \to \pi_{2q-1}(Y_x^{2q-1})$$

is induced by the inclusion map, we have proved that $\mu_x c(f,g,\sigma) = c(f,\sigma)$. Since $\pi_{2q-1}(Y_x^{2q-1})$ is cyclic of order 2, $c(f,\sigma) = -c(f,\sigma)$. This proves (18).

For (19), let σ denote a $(2q+1)$-cell of K. If we apply §35.12 to the bundle $\mathfrak{B}'_g|g(\dot{\sigma})$, we obtain

$$(\delta c(f,g))\cdot\sigma = c(f,g)\cdot\partial\sigma = -\Delta\alpha \tag{20}$$

where $\alpha \in \pi_{2q}(g(\dot{\sigma}))$ is represented by $g|\partial\sigma$. Shrinking $\dot{\sigma}$ over σ into the reference point x of σ, a covering homotopy deforms $g|\partial\sigma$ into a map representing $c(g,\sigma)$. A second covering homotopy deforms $\mathfrak{B}'|g(\dot{\sigma})$ into a bundle map into the bundle Y_x^{2q-1} over Y_x^{2q}. Since Δ commutes with bundle maps, it follows that $\Delta\alpha = \Delta_x c(g,\sigma)$. But Δ_x is λ_x. Combining with (20), we have

$$\delta c(f,g)\cdot\sigma = -\lambda_x c(g,\sigma).$$

This implies (19), and the proof is complete.

38.9. The 0 and 1-dimensional classes. The last theorem states nothing about $c^1(\mathfrak{B})$. We shall remedy this by defining $c^0(\mathfrak{B})$ suitably and proving a similar result.

Let $H = H_0(Y^0)$ with integer coefficients, and let $\tilde{H}$ be the reduced group, i.e. the subgroup of H generated by 0-cycles having a coefficient sum of zero. The operation of forming the coefficient sum of a 0-cycle defines a homomorphism μ of H into the group J of integers. Then

$$0 \to \tilde{H} \xrightarrow{\lambda} H \xrightarrow{\mu} J \to 0$$

is an exact sequence (λ = the inclusion).

For each x in K, let

$$L_x = \tilde{H}_0(Y_x^0), \quad M_x = H_0(Y_x^0), \quad N_x = J.$$

Let λ_x be the inclusion, and μ_x the coefficient sum. We obtain thus an exact sequence of coefficient bundles, as in (12), and $\mathfrak{B}(N)$ is a product bundle. Thus $H^0(K;\mathfrak{B}(N)) = H^0(K;J)$. Define the "unit" 0-cocycle $c \in C^0(K;J)$ by $c(v) = 1$ for each vertex v. Then $\delta c = 0$, and its cohomology class is denoted by c^0. *We set* $c^0(\mathfrak{B}) = c^0$ *for any sphere bundle* $\mathfrak{B}$.

38.10. THEOREM. *With respect to the above sequence of coefficient bundles, we have* $\delta^* c^0 = c^1(\mathfrak{B})$.

Let f be a cross-section of $\mathfrak{B}^0|K^0$. For each vertex v of K, let $c^0(f,v)$ be the element of M_v represented by the point $f(v)$ with coefficient 1. Then $c^0(f)$ is in $C^0(K;\mathfrak{B}(M))$. Then $\mu_v c^0(f,v) = 1$ which proves that $\mu c^0(f)$ is the unit 0-cocycle.

Referring to the definition of δ^*, the theorem will follow once we have proved

$$\mu c(f) = \delta c^0(f). \tag{21}$$

Let σ be an oriented edge, and $\partial\sigma = v - v'$. Let v be the reference point of σ, and w the isomorphism $M_{v'}$ onto M_v obtained by translation along σ. Then

$$\delta c^0(f)\cdot\sigma = c^0(f)\cdot\partial\sigma = c^0(f,v) - wc^0(f,v').$$

The translation w can be achieved by a homotopy of $Y^0_{v'}$ along σ into Y^0_v. This carries $f(v')$ along σ into $f'(v')$. Then $wc^0(f,v')$ is represented by the cycle $1f'(v')$. Hence $\delta c^0(f)\cdot\sigma$ is represented by the cycle $1f(v) - 1f'(v')$. By definition of the obstruction, it also represents $c(f,\sigma)$. Since μ is an inclusion map, (21) follows, and the proof is complete.

38.11. THEOREM. *Every odd dimensional characteristic class of a sphere bundle has order 2.*

For $q > 0$ and $2q < n$, $c^{2q}(\mathfrak{B})$ has coefficients mod 2, hence it is of order 2. Since δ^* is a homomorphism, $c^{2q+1}(\mathfrak{B}) = \delta^* c^{2q}(\mathfrak{B})$ is also of order 2.

For the case $q = 0$, let f_0 be a cross-section of $\mathfrak{B}^0|K^0$. For each vertex v, let $f_1(v)$ be a point in the component of Y^0_v not containing $f_0(v)$. ($Y^0 = O_n$ has two components.) Then f_1 is also a cross-section of $\mathfrak{B}^0|K^0$. It is clear that $c(f_1,\sigma) = -c(f_0,\sigma)$ for each edge σ. Then §33.5 gives

$$\delta d(f_0,f_1) = c(f_0) - c(f_1) = 2c(f_0)$$

which implies $2c^1(\mathfrak{B}) = 0$.

REMARK. The above result shows that every characteristic class of an n-sphere bundle $\mathfrak{B}$ has order 2 except $c^0(\mathfrak{B})$ and, possibly, $c^n(\mathfrak{B})$ when n is even (note the exception in (4) and (4′)).

38.12. THEOREM. *The following conditions on an n-sphere bundle $\mathfrak{B}$ are equivalent:*

(i) $c^1(\mathfrak{B}) = 0$,

(ii) $\chi(\mathfrak{B})\colon\ \pi_1(K) \to O_n/R_n$ *is trivial,*

(iii) $\mathfrak{B}$ *is equivalent to a bundle with group* R_n,

(iv) *for each q, the coefficient bundle of $c^q(\mathfrak{B})$ is a product.*

The equivalence of (ii) and (iv) was noted in §38.2. By definition, $\chi(\mathfrak{B}) = \chi(\mathfrak{B}')$ where $\mathfrak{B}'$ is the weakly associated bundle with fibre

O_n/R_n. Since the group of $\mathcal{B}'$ is discrete, $\chi(\mathcal{B}') = 0$ if and only if $\mathcal{B}'$ is a product bundle. Since $\mathcal{B}'$ is a principle bundle, it is a product bundle if and only if it has a cross-section. By §9.5 this last condition is equivalent to (iii). Thus (ii) and (iii) are equivalent.

Since $c^1(\mathcal{B})$ is the primary obstruction of $\mathcal{B}^0$, condition (i) implies that $\mathcal{B}^0|K^1$ has a cross-section. Since $\mathcal{B}^0$ is a principal bundle, this implies that $\mathcal{B}^0|K^1$ is a product bundle. Then the associated bundle $\mathcal{B}'|K^1$ is also a product; hence $\chi(\mathcal{B}'|K^1) = 0$. But $\chi(\mathcal{B}'|K^1)$ is the composition of

$$\pi_1(K^1) \xrightarrow{f_*} \pi_1(K) \xrightarrow{\chi(\mathcal{B}')} O_n/R_n$$

where f is the inclusion map. Since f_* is onto, it follows that $\chi(\mathcal{B}') = 0$. Thus (i) implies (ii).

Suppose (iii) holds and $\mathcal{B}$ is represented as a bundle with group R_n. Then the same holds for $\mathcal{B}^0$. Since the left translations of O_n by R_n map R_n on itself, R_n determines a subbundle of $\mathcal{B}^0$ with fibre R_n. Since R_n is arcwise connected, the portion of this subbundle over K^1 has a cross-section. This provides a cross-section of $\mathcal{B}^0|K^1$; so (i) holds.

38.13. Remarks. The preceding results are due to Whitney [106]. We have recast the results somewhat and taken full account of the fact that the coefficient bundles may not be product bundles. For example, Whitney states the relation $c^{2q+1} = \delta^* c^{2q}$ in the form $c^{2q+1} = \frac{1}{2}\,\delta\omega c^{2q}$. He regards c^{2q+1} as having integer coefficients, c^{2q} as having coefficients mod 2, and ω as the inverse of reduction mod 2. Thus ω corresponds to i_*^{-1} in §38.6 and 1/2 to Δ^{-1}.

Whitney has announced [106] a "duality theorem" for sphere bundles. Let $\mathcal{B},\mathcal{B}'$ be sphere bundles over K with groups O_m, O'_n respectively. We may regard $O_m \times O'_n$ as a subgroup of O''_{m+n}. If we set $g''_{ji}(x) = g_{ji}(x)g'_{ji}(x)$, we obtain coordinate transformations for a sphere bundle $\mathcal{B}''$ over K with group O''_{m+n}. Then the duality theorem reads

$$c^r(\mathcal{B}'') = \sum_{p+q=r} c^p(\mathcal{B}) \smile c^q(\mathcal{B}'). \tag{22}$$

As Whitney has shown, it has numerous important applications.

No proof of (22) in full generality has been published.* The proposition is somewhat ambiguous. The use of the cup product presupposes that the coefficient groups $\pi_{p-1}(Y^{p-1})$ and $\pi_{q-1}(Y'^{q-1})$ are paired to $\pi_{r-1}(Y''^{r-1})$. A clarification of (22) would present such a pairing in a natural geometric fashion. W. T. Wu [108] has proved the special case obtained by reducing everything mod 2. Reduction mod 2

* See App. sect. 9.

eliminates the ambiguity. It also eliminates difficulty with the coefficient bundles: all such become product bundles when reduced mod 2. A clarification and proof of (22), in full generality, is needed.

§39.* The Stiefel Characteristic Classes of Differentiable Manifolds

39.1. Definitions and interpretation. Let M denote a compact, connected, n-dimensional, differentiable manifold. Using the result that M is triangulable, the concepts of §38 can be applied to the tangent sphere bundle $\mathfrak{B}$ of M (defined in §12.10). The *Stiefel characteristic classes of M* are defined to be the Whitney classes of $\mathfrak{B}$.

$$c^q(M) = c^q(\mathfrak{B}), \qquad q = 0, 1, \cdots, n.$$

The result of §38.4 translates as follows:

For each q, there exists a continuous field of tangent, orthogonal $(n-q)$-frames (see §7.7) defined over the q-dimensional skeleton M^q of M. There exists such a field over M^{q+1} if and only if $c^{q+1}(M) = 0$.

The translation is effected by selecting a fixed orthogonal $(n-q)$-frame spanning S^{n-q-1}.

39.2. Theorem. *$c^1(M) = 0$ if and only if M is orientable.*

This follows from §38.12; for the condition (iii) of §38.12 is equivalent to the orientability of M.

39.3. If M is orientable, then §38.12 asserts that the Stiefel classes belong to cohomology groups of M with ordinary coefficients (either infinite cyclic or cyclic of order 2). In particular, the coefficients of $c^n(M)$ lie in the infinite cyclic group $\pi_{n-1}(V_{n,1})$ $(V_{n,1} = S^{n-1})$. As is well known, the group $H^n(M;\pi_{n-1})$ is also infinite cyclic.

If M is non-orientable, the ordinary group $H^n(M)$, with integer coefficients, is cyclic of order 2. However §38.12 asserts that $\mathfrak{B}(\pi_{n-1})$ is not a product bundle. We will show that $H^n(M;\mathfrak{B}(\pi_{n-1}))$ is infinite cyclic. This requires a mild digression.

39.4. Homology with coefficients in a bundle. If π is an infinite cyclic group, we define a *pairing* of π with itself to the group J of integers by choosing an isomorphism ψ: $\pi \approx J$ and defining the *product* $\alpha\beta$, for α,β in π, by

$$(1) \qquad \alpha\beta = \psi(\alpha)\psi(\beta).$$

Then $\alpha\beta$ is an integer, the product is bilinear, and $\alpha^2 = 1$ if α generates π. There are just two possible choices for ψ and they differ in sign; hence the form of (1) shows that the product is independent of ψ. From this it follows that, if π' is also infinite cyclic and w: $\pi \approx \pi'$, then

$$(2) \qquad \alpha\beta = w(\alpha)w(\beta).$$

* See App. sect. 9 and 10.

Let $\mathcal{B}(\pi)$ be a bundle of coefficients over a complex K. We shall deal with both chains and cochains in $\mathcal{B}(\pi)$. The former we have not defined. The group $C_q(K;\mathcal{B}(\pi))$ of q-chains is defined to be the group $C^q(K;\mathcal{B}(\pi))$ of q-cochains. If $u \in C_q(K;\mathcal{B}(\pi))$, we define $\partial u \in C_{q-1}(K;\mathcal{B}(\pi))$ by

(3) $$(\partial u)(\sigma) = \sum_\tau [\sigma:\tau]w_{\sigma\tau}^{-1}(u(\tau)) \qquad \text{(see §31.2).}$$

The sum is taken over q-cells τ having σ as a face. One proves $\partial\partial = 0$, and defines the homology group $H_q(K;\mathcal{B}(\pi))$ in the usual way.

Assuming π to be infinite cyclic, we define the *Kronecker index* of a q-cochain c and a q-chain u to be the integer

(4) $$c\cdot u = \sum_\sigma c(\sigma)u(\sigma).$$

The sum is taken over all q-cells—one term for each cell. It is clear that $c(\sigma)u(\sigma)$ is independent of the orientation of σ. Using (2) with $w = w_{\sigma\tau}$, one proves easily that

(5) $$c\cdot\partial v = \delta c\cdot v$$

for any q-cochain c and $(q + 1)$-chain v. Then the Kronecker index of a cocycle and a cycle depends only on their respective homology and cohomology classes. This yields a pairing of $H^q(K;\mathcal{B}(\pi))$ with $H_q(K;\mathcal{B}(\pi))$ to J which is also called the Kronecker index. It is clearly bilinear.

39.5. The fundamental n-cycle of a manifold. Now let $K = M$ be a differentiable n-manifold (orientable or non-orientable), let $\mathcal{B}$ be the tangent sphere bundle, $\pi = \pi_{n-1}(S^{n-1})$, and $\mathcal{B}(\pi)$ the associated coefficient bundle. We suppose the subdivision K of M is so fine that any two adjacent cells are contained in a coordinate neighborhood of M. Define $z \in C_n(K;\mathcal{B}(\pi))$ as follows. An orientation of an n-cell σ determines a concordant orientation of the tangent plane at the reference point x_σ. The latter determines a concordant orientation of the unit $(n - 1)$-sphere Y_σ (the fibre of $\mathcal{B}$ over x_σ). This in turn determines a generator $z(\sigma)$ of $\pi_\sigma = \pi_{n-1}(Y_\sigma)$. Clearly $z(-\sigma) = -z(\sigma)$. An $(n - 1)$-cell τ is a face of just two n-cells, σ, σ' say. An orientation of a neighborhood V of $\sigma \cup \sigma'$ determines concordant orientations of σ and σ', i.e. $[\tau:\sigma] = -[\tau:\sigma']$, and translation of $z(\sigma)$ along a path in V carries $z(\sigma)$ into $z(\sigma')$. It follows that $\partial z = 0$. We call z the *fundamental n-cycle* of M.

One proves now in the standard manner that any n-cycle of $C_n(M;\mathcal{B}(\pi))$ is a multiple of z. It follows that *$H_n(M;\mathcal{B}(\pi))$ is cyclic infinite.*

A cochain on the cell τ is one which is zero on all cells except possibly τ; it is called an *elementary* cochain. An n-cochain is always a cocycle due to the absence of $(n + 1)$-cells. If c is an n-cocycle on σ, σ' is an adjacent n-cell, and τ is the common $(n - 1)$-face, then there is an $(n - 1)$-cochain d on τ such that $c - \delta d$ is an n-cocycle on σ' (one uses here the fact that σ, σ' are the only n-cells having τ as a face). Any n-cell σ can be connected to a reference n-cell σ_0 by a "path" of successively adjacent n-cells. Using a succession of steps as above, one shows that any n-cocycle on σ is cohomologous to one on σ_0. Since any n-cocycle is a sum of elementary cocycles, it follows that *any n-cocycle is cohomologous to a cocycle on* σ_0.

As π_{σ_0} is infinite cyclic, it follows that $H^n(M;\mathcal{B}(\pi))$ is a factor group of the infinite cyclic group of n-cocycles on σ_0. Hence H^n is a cyclic group. Define $c(\sigma) = 0$ for $\sigma \neq \sigma_0$, and $c(\sigma_0) = z(\sigma_0)$. Then c generates the group of n-cocycles on σ_0; and the cohomology class of c generates $H^n(M;\mathcal{B}(\pi))$. Since

$$c \cdot z = c(\sigma_0)z(\sigma_0) = 1$$

we have $(mc)\cdot z = m$ for any integer m. But $mc = \delta d$ implies, by (5),

$$mc \cdot z = \delta d \cdot z = d \cdot \partial z = 0,$$

so $m = 0$. Therefore $H^n(M;\mathcal{B}(\pi))$ *is infinite cyclic.*

39.6. The n-dimensional class of an n-manifold.

THEOREM. *If $n = \dim M$ is odd, then $c^n(M) = 0$. Therefore the tangent sphere bundle of M has a cross-section, i.e. M has a continuous field of non-zero tangent vectors.*

Since n is odd, $c^n(M)$ has order 2 (see §38.11). But $H^n(M;\mathcal{B}(\pi))$ is infinite cyclic. Hence $c^n(M) = 0$.

There is a more general result which holds for manifolds of arbitrary dimension:

39.7 THEOREM. *If z is the fundamental n-cycle of M (see §39.5), then*

$$c^n(M)\cdot z = \text{the Euler number of } M.$$

It is known that the Euler number e of M is zero for any manifold of odd dimension. By the results of §39.5, $c^n(M)\cdot z = 0$ implies $c^n(M) = 0$. Thus, the above theorem generalizes the preceding one.

We shall omit the proof. It can be found in the book of Alexandroff and Hopf [1, p. 549]. It is a long proof and we have nothing to add. We shall give, however, a brief intuitive discussion which suggests the truth of the result.

Let K be a simplicial triangulation of M. Let K' and K'' denote.

respectively, the first and second barycentric subdivisions of K. A vertex v of K'' lies in the interior of just one simplex of K, let $\phi(v)$ denote the barycenter of that simplex. This vertex assignment determines a unique simplicial map ϕ: $K'' \to K'$. It is easily shown that the fixed points of ϕ are the barycenters of the simplexes of K.

Now x and $\phi(x)$ lie on a single simplex of K' and are joined by a unique line segment of the simplex. We assume that the triangulation is differentiable. Then the segment has a tangent direction at each point, and we can define $f(x)$ to be the unit tangent vector at x. It follows that f is defined and continuous except at the barycenters of the simplexes of K.

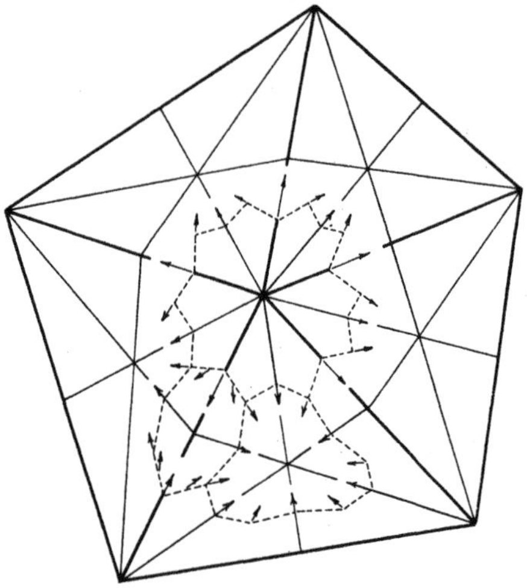

FIG. 8.

Let K'^* be the cellular decomposition of M dual to K' so that, for each q-simplex τ of K', there is a dual (transverse) $(n - q)$-cell of K'^* which is the union of those simplexes of K'' having the barycenter of τ as vertex of least order. Then the singularities of f occur at the centers of the n-cells of K'^*; so f provides a cross-section of $\mathfrak{B}|K'^{*^{n-1}}$, and $c(f)$ in $Z^n(K'^*;\mathfrak{B}(\pi_{n-1}))$ is defined.

Figure 8 illustrates the case of a 2-manifold. The heavy lines are the edges of K, these and the light lines are the edges of K'. Three of the 2-cells of K'^* are outlined by dotted lines. One is dual to a vertex of K' which is also a vertex of K; a second is dual to the barycenter of an edge of K; and the third is dual to the barycenter of a 2-simplex of K.

In general, K'^* has one n-cell for each simplex σ of K, namely, the dual, σ^*, of the barycenter of σ. Let z be the fundamental n-cycle of

K'^* (§39.5). Then

$$(6) \qquad c(f,\sigma^*)\cdot z(\sigma^*) = (-1)^{\dim \sigma}.$$

To prove this we take the barycenter x_σ of σ as the reference point of σ^*. Let S_σ be the fibre of $\mathfrak{B}$ over x_σ. For $v \in S_\sigma$, let $g(v)$ be the point in which the line segment from x_σ in the direction v meets $\partial\sigma^*$. If the triangulation is sufficiently fine and *smooth*, g will provide a topological map of S_σ on $\partial\sigma^*$. Furthermore, g^{-1} will represent the generator $z(\sigma^*)$ of $\pi_{n-1}(S_\sigma)$. If f': $\partial\sigma^* \to S_\sigma$ is homotopic to $f|\partial\sigma^*$, it follows that (6) is just the degree of the map gf' of $\partial\sigma^*$ on itself (= the index of the singularity). Let $q = \dim \sigma$. Now σ^* is the product of the q-cell $\sigma \cap \sigma^*$ and the $(n - q)$-cell $\tilde{\sigma} \cap \sigma^*$ where $\tilde{\sigma}$ is the dual cell of σ. All vectors on $\partial(\tilde{\sigma} \cap \sigma^*)$ head "outward" so gf' is the identity on it. All vectors on $\partial(\sigma \cap \sigma^*)$ head "inward," so gf' is the antipodal transformation on it. The antipodal transformation has degree $(-1)^q$. Then the composite map gf' on the boundary of the product cell has degree $(-1)^q$.

If α_q denotes the number of q-simplexes of K, (6) implies that

$$c(f)\cdot z = \sum_{q=0}^{n} (-1)^q \alpha_q.$$

The right side is sometimes taken as the definition of the Euler number e of M. The alternative definition is $e = \Sigma_{q=0}^{n}(-1)^q R_q$ where R_q is the qth Betti number of M (i.e. the rank of $H_q(M)$). In any case, the equality of the two sums is a standard theorem. This completes our intuitive proof.

39.8. Corollary. *A differentiable manifold admits a continuous field of non-zero tangent vectors if and only if its Euler number is zero.*

39.9. Remarks. The preceding results provide satisfactory "computations" of $c^1(M)$ and $c^n(M)$. Corresponding results for the other Stiefel classes have not been obtained. Whitney [105] has announced that a representative cocycle for $c^q(M)$ mod 2 is obtained by assigning the value 1 to each q-cell of the subdivision K'^* (§39.7) of M. No proof of this has appeared.

Stiefel [91] has given a proof that $c^2(M) = 0$ if M is an orientable 3-manifold. We will not attempt to reproduce it here. His argument is sketchy in a major detail. He asserts that any mod 2 homology class of M is representable by a 2-manifold M' differentiably imbedded in M without singularities, and the structure of the "normal" bundle of M' in M is independent of the imbedding. This seems to be highly likely; but a full proof would be quite awkward. (See App. sect. 9.)

Granting that $c^2(M) = 0$, Stiefel goes on to prove that any orientable 3-manifold is parallelizable (i.e. its tangent bundle is a product bundle). The vanishing of $c^2(M)$ means that a 2-field f can be constructed on the 2-skeleton of M. Then the 3-cocycle $c(f)$ is defined with coefficients in $\pi_2(V_{3,2}) = \pi_2(R_3) = 0$; i.e. the secondary obstruction vanishes identically. Hence f extends to a 2-field f' over all of M. The 2-field f' is extended to a 3-field over M by adjoining, at each point x, a unit vector perpendicular to the pair of vectors $f'(x)$ so that the three vectors give a prescribed orientation of M.

Recently Thom [92, 93] and Wu [112, 113] have announced interesting results relating the Stiefel classes to products of other basic cohomology classes of the manifold. These offer methods for computing the Stiefel classes. They also show that the Stiefel classes are topological invariants of the manifold, i.e. they are independent of the differential structure. It is not known whether or not there exists a topological manifold having two differential structures with inequivalent tangent bundles. (See App. sect. 7.)

§40. QUADRATIC FORMS ON MANIFOLDS

40.1. Formulation of the problem. In §12.12 we proved that any differentiable manifold M admits a Riemannian metric. We consider now the problem of constructing *indefinite* quadratic forms, precisely, a covariant, second order, symmetric, tensor function which, at each point, has a non-zero determinant. The *signature* k of the quadratic form at a point is the number of negative characteristic values of the matrix. A simple continuity argument shows that k is independent of the point.

The problem is to construct a cross-section of a suitable tensor bundle over M. Just as the vector field problem led to a study of the Stiefel manifolds $V_{n,k}$, the present problem requires a preliminary study of the fibres involved.

As before, L_n denotes the group of non-singular, real matrices of order n, O_n the orthogonal group, and R_n the rotation group. Let S_n be the subset of L_n of symmetric matrices, and let $S_{n,k}$ be the subset of S_n of matrices of signature k. It is easily seen that $S_{n,k}$ is an open set in S_n, and $S_n = \bigcup_{k=0}^{n} S_{n,k}$.

If the problem is to construct a quadratic form of signature k over a differentiable n-manifold, then we are seeking a cross-section of the bundle, associated with the tangent bundle, with fibre $S_{n,k}$ and group O_n where σ in O_n operates on τ in $S_{n,k}$ by the similarity $\tau \to \sigma\tau\sigma'$ (σ' = transpose of σ). (We are assuming that the reduction of the tangent bundle to the group O_n has already been carried out, then the covariant, contravariant and mixed variance problems become equivalent (see

§12.11).) We proceed to study the structure of $S_{n,k}$ under these operations of O_n.

40.2. The imbedding of $M_{n,k}$ in O_n. Choose as a reference point of $S_{n,k}$ the matrix

$$\sigma_k = \begin{vmatrix} -I_k & 0 \\ 0 & I_{n-k} \end{vmatrix} \tag{1}$$

where I_k denotes the unit diagonal matrix of order k. Define ϕ: $O_n \to S_{n,k}$ by

$$\phi(\sigma) = \sigma\sigma_k\sigma' \qquad (\sigma' = \text{transpose of } \sigma). \tag{2}$$

The subgroup of O_n carried into σ_k by ϕ is $O_k \times O'_{n-k}$. We may therefore identify $\phi(O_n)$ with the Grassmann manifold $M_{n,k}$ (see §7.9). Since any orthogonal element of $S_{n,k}$ is orthogonally equivalent to σ_k, we have

$$M_{n,k} = \phi(O_n) = O_n \cap S_{n,k}. \tag{3}$$

It is to be observed that, under this identification of the coset space $M_{n,k}$ with a set of matrices, the left translation of $M_{n,k}$ by σ in O_n corresponds to the similarity $\tau \to \sigma\tau\sigma'$ of the matrices.

40.3. The deformation retraction of $S_{n,k}$ into $M_{n,k}$. Any complex non-singular matrix τ can be factored in one and only one way into a product $\tau = \sigma\alpha$ where σ is unitary and α is positive definite Hermitian (see Chevalley [12; p. 14]). This decomposition is continuous and sets up a homeomorphism of the complex linear group with the product space of U_n and the space of positive definite Hermitian matrices.

An examination of the factorization shows that, if τ is real, so also are σ and α. Then $\sigma \in O_n$ and $\alpha \in S_{n,0}$. Therefore the function

$$\psi(\sigma,\alpha) = \sigma\alpha \tag{4}$$

defines a homeomorphism ψ: $O_n \times S_{n,0} \to L_n$.

40.4. Lemma. *If $\sigma \in O_n$, and $\alpha \in S_{n,0}$, then $\sigma\alpha \in S_{n,k}$ if and only if*

$$\sigma \in M_{n,k} \quad \textit{and} \quad \sigma\alpha = \alpha\sigma. \tag{5}$$

If $\sigma\alpha \in S_{n,k}$, then $(\sigma\alpha)' = \sigma\alpha$ implies $\alpha = \sigma^2(\sigma'\alpha\sigma)$. By the uniqueness of the factorization ψ^{-1}, we have $\sigma^2 = I_n$ and $\alpha = \sigma'\alpha\sigma$. Therefore $\sigma = \sigma'$ and $\sigma\alpha = \alpha\sigma$. Regarding σ as a linear transformation, let C (C') be the space of negative (positive) characteristic vectors of σ. Since $\sigma\alpha = \alpha\sigma$, α transforms C into C and C' into C'. Thus we may choose characteristic vectors for α, each lying in C or in C'. Then each will be a characteristic vector for $\sigma\alpha$. Since the characteristic values of α are positive it follows that σ and $\sigma\alpha$ have the same signature. Hence $\sigma \in M_{n,k}$.

Conversely, if (5) holds, then $(\sigma\alpha)' = \alpha'\sigma' = \alpha\sigma = \sigma\alpha$. So $\sigma\alpha$ is symmetric. The argument above shows that σ and $\sigma\alpha$ have the same signature; hence $\sigma\alpha \varepsilon S_{n,k}$.

40.5. Define $S'_{n,k}$ to be the subset of pairs (σ,α) in $M_{n,k} \times S_{n,0}$ such that $\sigma\alpha = \alpha\sigma$. Let μ in O_n operate on $M_{n,k} \times S_{n,0}$ by

$$\mu\cdot(\sigma,\alpha) = (\mu\sigma\mu',\mu\alpha\mu').$$

It is clear that μ transforms $S'_{n,k}$ into itself. Then the preceding lemma can be restated:

40.6. THEOREM. *The map ψ of (4) is a homeomorphism of $S'_{n,k}$ with $S_{n,k}$, and it commutes with the operations of O_n.*

Since ψ is an equivalence of the pair $(S'_{n,k},O_n)$ with the pair $(S_{n,k},O_n)$, we shall identify them under ψ and omit the ψ: $(\sigma,\tau) = \sigma\tau$.

40.7. Define p: $S_{n,k} \rightarrow M_{n,k}$ by

$$p(\sigma\alpha) = \sigma. \tag{6}$$

Then p is a continuous retraction of $S_{n,k}$ into $M_{n,k}$. Define a homotopy k by

$$k(\sigma\alpha,t) = \sigma[tI_n + (1-t)\alpha]. \tag{7}$$

Since α commutes with σ, so also does $tI_n + (1-t)\alpha$. Hence k deforms $S_{n,k}$ over itself. It is easy to verify the relations

$$k(\sigma\alpha,0) = \sigma\alpha, \qquad k(\sigma\alpha,1) = \sigma, \tag{8}$$

$$pk(\sigma\alpha,t) = \sigma, \tag{9}$$

$$k(\sigma,t) = \sigma, \tag{10}$$

$$\mu k(\sigma\alpha,t)\mu' = k(\mu\sigma\alpha\mu',t), \qquad \mu \varepsilon O_n. \tag{11}$$

In words:

40.8. THEOREM. *The homotopy k is a deformation retraction of $S_{n,k}$ into $M_{n,k}$ which commutes with the operations of O_n. For each σ in $M_{n,k}$, k contracts $p^{-1}(\sigma)$ over itself into σ.*

40.9. Reduction of the problem. Let M be a differentiable n-manifold, and let $\mathfrak{B}$ be the bundle over M with fibre $Y = S_{n,k}$ and group O_n associated with its tangent sphere bundle (i.e. $\mathfrak{B}$ is the bundle of quadratic forms of signature k at the various points of M). Let $\mathfrak{B}'$ be the subbundle of $\mathfrak{B}$ corresponding to the subspace $Y' = M_{n,k}$ of Y. (It must be noted that each σ in O_n maps $M_{n,k}$ on itself.)

For each x in M, choose an admissible ξ: $Y \rightarrow Y_x$, and define k' by

$$k'(b,t) = \xi k(\xi^{-1}b,t), \qquad b \varepsilon Y_x. \tag{12}$$

From (11) it follows that k' is independent of the choice of ξ. Setting $\xi = \phi_{i,x}$, where ϕ_i is a coordinate function of $\mathfrak{B}$, we find that k' is continuous. Thus:

40.10. THEOREM. *The homotopy k' is a deformation retraction of B into B', and for each x it contracts Y_x over itself into Y'_x. Then, for any cross-section f of $\mathcal{B}$, k' provides a homotopy of f into a cross-section of $\mathcal{B}'$ Thus $\mathcal{B}$ has a cross-section if and only if $\mathcal{B}'$ has a cross-section.*

If we return to the interpretation of $M_{n,k}$ as the manifold of k-planes in n-space, we may restate the result in the form

40.11. THEOREM. *A compact differentiable manifold admits an everywhere defined, continuous, non-singular, quadratic form of signature k if and only if it admits a continuous field of tangent k-planes.*

40.12. Applications. (See App. sect. 7.)

COROLLARY. *The results of Theorems 27.14 through 27.18 concerning fields of tangent k-planes over spheres are equally valid if "tangent k-plane" is replaced by "quadratic form of signature k."*

40.13. THEOREM. *If M is compact and dim M is odd, then M admits a quadratic form of signature 1. If dim M is even, then this holds if the Euler number of M is zero.*

In either case M has a tangent vector field (§39.6 and §39.7) which provides a field of tangent line elements (1-planes).

For 2-manifolds, the theorem may be completed by saying "if and only if the Euler number is zero." In this case, the vector field problem and the line element problem have 1-spheres as fibres and the first fibre is a 2-fold covering of the second. Choose a 1-field f over the 1-skeleton of M, and let f' be the induced field of line elements. For any 2-cell σ, the double covering maps $c(f,\sigma)$ onto $c(f',\sigma)$. It follows that the sum of the indices of $c(f')$ is twice that of $c(f)$, i.e. twice the Euler number. Thus, we have

The only compact 2-manifolds which admit a quadratic form of signature 1 are the torus and Klein bottle.

The above argument can be extended to arbitrary manifolds of even dimensions. One obtains an obstruction $c(f')$ having a sum of indices equal to twice the Euler number. However it is not the *primary* obstruction. The primary obstruction has dimension 2 and is zero. Instead $c(f')$ is a *secondary* obstruction. It may be possible to alter its cohomology class by an alteration of f' by a 1-cocycle on the 1-skeleton. So far as the author knows, this problem is unsolved. It may be worth noting that, if $H^1(M; \text{mod } 2) = 0$, then no alteration of f' on the 1-skeleton can affect the class of $c(f')$. (See App. sect. 11 and 12.)

If an n-manifold admits a quadratic form of signature k, a change of sign provides one of signature $n - k$. Therefore, a compact 3-manifold admits a quadratic form of any possible signature.

Further results along these lines should be obtainable without too great an effort, at least for 4-manifolds. The lower dimensional homotopy groups of the fibre $M_{n,k}$ are known (§25.8). An interesting special case is the problem of constructing a form of signature 2 on the complex projective plane.

40.14. The bundle $S_{n,k} \to M_{n,k}$. It may be of interest to show that the map p: $S_{n,k} \to M_{n,k}$ given in §40.7 admits a bundle structure.

The fibre Y is defined to be $p^{-1}(\sigma_k)$. These are the symmetric matrices which commute with σ_k. They have the form $\begin{vmatrix} A & 0 \\ 0 & B \end{vmatrix}$ where A is a negative definite $k \times k$ matrix and B is positive definite. Since the space of definite matrices of a fixed order is a cell, it follows that Y is a product space of two cells; hence Y is a cell.

The elements of O_n which map Y on itself are those which commute with σ_k. They form the subgroup denoted by $O_k \times O'_{n-k}$. The subgroup H of the latter which commutes with all elements of Y is the 4-group of matrices of the form $\begin{vmatrix} \pm I_k & 0 \\ 0 & \pm I_{n-k} \end{vmatrix}$. The group of Y is the factor group $G = O_k \times O'_{n-k}/H$.

If $b \in O_n$ and $\tau \in S_{n,k}$, we will adopt the notation $b\cdot\tau$ for $b\tau b'$. Then we may use the symbolism of §7.4. Let f be a local cross-section of $O_k \times O'_{n-k}$ in O_n. Then f is defined in a neighborhood V of σ_k in $M_{n,k}$, and $f(\sigma)\cdot\sigma_k = \sigma$. For b in O_n set $V_b = b\cdot V$, and

$$f_b(\sigma) = bf(b^{-1}\cdot\sigma), \qquad (\sigma \in V_b).$$

Define

$$\phi_b\colon\ V_b \times Y \to p^{-1}(V_b)$$

by

$$\phi_b(\sigma,\alpha) = f_b(\sigma)\cdot\alpha$$

and define p_b: $p^{-1}(V_b) \to Y$ by

$$p_b(\tau) = [f_b p(\tau)]^{-1}\cdot\tau.$$

Now $\alpha \in Y$ implies $\alpha = \sigma_k\beta$ where β is positive definite. If $\mu = f_b(\sigma)$, then

$$\begin{aligned} p\phi_b(\sigma,\alpha) &= p(\mu\sigma_k\beta\mu') = p(\mu\sigma_k\mu'\mu\beta\mu') \\ &= \mu\sigma_k\mu' \end{aligned}$$

since $\mu\beta\mu'$ is positive definite. Then

$$\begin{aligned} p\phi_b(\sigma,\alpha) &= f_b(\sigma)\cdot\sigma_k = [bf(b^{-1}\cdot\sigma)]\cdot\sigma_k \\ &= b\cdot[f(b^{-1}\cdot\sigma)\cdot\sigma_k] = b\cdot(b^{-1}\cdot\sigma) = \sigma. \end{aligned}$$

Likewise

$$p_b\phi_b(\sigma,\alpha) = f_b(\sigma)^{-1}\cdot\phi_b(\sigma,\alpha) = f_b(\sigma)^{-1}\cdot[f_b(\sigma)\cdot\alpha] = \alpha.$$

Therefore ϕ_b is a homeomorphism. Computing coordinate transformations, we obtain

$$g_{cb}(\sigma)\cdot\alpha = p_c\phi_b(\sigma,\alpha) = f_c(\sigma)^{-1}\cdot[f_b(\sigma)\cdot\alpha].$$

Therefore $g_{cb}(\sigma)$ is the image in G of $f_c(\sigma)^{-1}f_b(\sigma)$ in $O_k \times O'_{n-k}$.

If the foregoing is compared with the proof of §7.4, one observes that $\{f_c(\sigma)^{-1}f_b(\sigma)\}$ are the coordinate transformations of the bundle $O_n \to M_{n,k}$. It follows that the bundle $S_{n,k}$ over $M_{n,k}$ is weakly associated with the bundle O_n over $M_{n,k}$ under the natural homomorphism of $O_k \times O'_{n-k}$ into G. It is easily shown that the principal bundle of $S_{n,k} \to M_{n,k}$ is the bundle $O_n/H \to M_{n,k}$. Since the latter is not a product for $0 < k < n$, neither is the former.

§41. Complex Analytic Manifolds and Exterior Forms of Degree 2

41.1. Quasi-complex manifolds. If M is a complex analytic manifold of n complex dimensions, its tangent bundle has, for fibre Y, the n-dimensional complex vector space, and, for group, the complex linear group CL_n. Passage to real and imaginary parts of the coordinates in M represents M as a real analytic manifold of $2n$ dimensions, Y becomes a real $2n$-space of variables $(x_1, \cdots, x_n, y_1, \cdots, y_n)$, and CL_n is imbedded in L_{2n}. If γ is a complex matrix in CL_n, and $\gamma = \alpha + i\beta$ where α,β are real $n \times n$ matrices, then, as an element of L_{2n}, γ is represented by the $2n \times 2n$ real matrix $\begin{vmatrix}\alpha & -\beta \\ \beta & \alpha\end{vmatrix}$. Conversely any matrix of L_{2n} of the latter form belongs to CL_n. Thus the tangent bundle of the real manifold is represented as a bundle in the subgroup CL_n of L_{2n}.

Let M be a real, differentiable $2n$-manifold. The foregoing shows that a necessary condition for M to be differentiably equivalent to the real form of a complex analytic manifold is that the tangent bundle of M be equivalent, in its group L_{2n}, to a bundle in the subgroup CL_n. A manifold satisfying this necessary condition will be called *a quasi-complex manifold.*

We shall restrict our attention to real manifolds, and derive conditions for such to be quasi-complex. Our results, so far as complex analytic manifolds are concerned, will be of the negative form: A particular real manifold is not quasi-complex, so it does not admit a complex analytic structure. It seems highly unlikely that every quasi-complex manifold has a complex analytic structure.

41.2. Unitary sphere bundles. Let M be a real differentiable $2n$-manifold, and $\mathfrak{B}$ its tangent bundle. We have seen in §12.9 that $\mathfrak{B}$ is equivalent in L_{2n} to a bundle $\mathfrak{B}'$ with group O_{2n}. If M is quasi-complex, $\mathcal{B}$ is equivalent in L_{2n} to a bundle $\mathcal{B}_1$ with group CL_n. By the analog of §12.9 for the complex case, $\mathcal{B}_1$ is equivalent in CL_n to a bundle $\mathcal{B}_1'$ in the unitary group $U_n = (CL_n) \cap O_{2n}$. It follows that $\mathcal{B}'$ and $\mathcal{B}_1'$ are equivalent in L_{2n}. According to §12.9, this equivalence holds also in O_{2n}. This proves: *M is quasi-complex if and only if its tangent sphere bundle is equivalent in O_{2n} to a bundle in the unitary group U_n.*

In general a sphere bundle will be called a *unitary sphere bundle* if its fibre is an odd dimensional sphere S^{2n-1} and its group is U_n.

41.3. The Chern characteristic classes. Let $\mathcal{B}$ denote a unitary $(2n-1)$-sphere bundle over a cell complex K. For notational convenience let Y'^q denote the "complex" Stiefel manifold (§25.7)

$$Y'^q = W_{n,n-q} = U_n/U_q, \quad q = 0, 1, \cdots, n-1.$$

Define $\mathcal{B}'^q$ to be the associated bundle of $\mathcal{B}$ with fibre Y'^q. Then $\mathcal{B}'^0$ is the principal bundle of $\mathcal{B}$, and $\mathcal{B}'^{n-1} = \mathcal{B}$. Since $U_{q-1} \subset U_q$, we have natural projections

$$U_n = Y'^0 \to Y'^1 \to \cdots \to Y'^{n-1} = S^{2n-1}, \tag{1}$$

and Y'^{q-1} is a unitary $(2q-1)$-sphere bundle over Y'^q. By §9.6, these projections induce projections

$$B'^0 \to B'^1 \to \cdots \to B'^{n-1} \to K, \tag{2}$$

and any composition of them is the projection of a bundle structure. In particular B'^{q-1} is a unitary $(2q-1)$-sphere bundle over B'^q. The composition of $B'^q \to B'^{q+1} \to \cdots \to K$ is the projection of $\mathcal{B}'^q$.

41.4. With $\mathcal{B}$ as above, we define the $2q$th *characteristic class of* $\mathcal{B}$ $(q = 1, \cdots, n)$, in the sense of Chern [7], to be the characteristic class of $\mathcal{B}'^{q-1}$. We denote it by $c'^{2q}(\mathcal{B})$. Thus

$$c'^{2q}(\mathcal{B}) = \bar{c}(\mathcal{B}'^{q-1}). \tag{3}$$

Since U_n is connected, $\pi_0(U_n) = 0$; hence all coefficient bundles are products, and we may use ordinary coefficients. According to §25.7, the first non-zero homotopy group of Y'^{q-1} is π_{2q-1} and it is infinite cyclic, thus

$$c'^{2q}(\mathcal{B}) \in H^{2q}(K; \pi_{2q-1}). \tag{4}$$

We state the analog of §38.4; the proof is similar.

41.5. THEOREM. *For each $q = 1, \cdots, n$, there exists a map*

$$\psi\colon\ K^{2q-1} \times S^{2n-2q+1} \to B$$

which, for each x, maps the fibre $x \times S^{2n-2q+1}$ by a unitary transfor-

mation into the fibre over x in B. And there exists a similar map of $K^{2q} \times S^{2n-2q+1}$ if and only if $c'^{2q}(\mathfrak{B}) = 0$.

The form this theorem takes in the complex analytic case is of interest:

41.6. THEOREM. *If K is a complex analytic manifold of n complex dimensions, then, for each $q = 1, \cdots, n$, there exist $n - q + 1$ fields of tangent complex vectors which are independent at each point of K^{2q-1}; and there exist $n - q + 1$ such fields independent at each point of K^{2q} if and only if c'^{2q}(tangent bundle) $= 0$.*

41.7. Relations between the Chern and Whitney classes. The Chern classes are related to the Whitney classes. To exhibit the relationship, we adopt several conventions. The fibre of $\mathfrak{B}$ is the unit sphere in the space of n complex variables $(z_1, \cdots, z_n)$. The subgroup U_q of U_n operates trivially in the subspace $z_1 = \cdots = z_q = 0$. We pass to real coordinates by setting

$$x_{2i-1} = \mathfrak{R}z_i, \qquad x_{2i} = \mathfrak{J}z_i$$

Let O_q be the subgroup of O_{2n} operating trivially in the subspace $x_1 = \cdots = x_q = 0$. Then we have the obvious relations

$$U_n \cap O_{2q} = U_n \cap O_{2q+1} = U_q. \tag{5}$$

Thus, distinct cosets of U_q in U_n are contained in distinct cosets of O_{2q} and O_{2q+1} in O_{2n}. These coset inclusions induce natural imbeddings of the coset spaces:

$$U_n/U_q \subset O_{2n}/O_{2q}, \qquad U_n/U_q \subset O_{2n}/O_{2q+1}.$$

In the notations of §41.3 and §38.1, these become

$$Y'^q \subset Y^{2q}, \qquad Y'^q \subset Y^{2q+1}. \tag{6}$$

It is important to note that these imbeddings conform with the left translations of $U_n \subset O_{2n}$.

Let $\mathfrak{B}'$ be a unitary $(2n - 1)$-sphere bundle over K, and let $\mathfrak{B}$ denote the corresponding orthogonal bundle under the imbedding $U_n \subset O_{2n}$. Let $\mathfrak{B}^q$ be defined as in §38.1, and $\mathfrak{B}'^q$ as in §41.3. Then the inclusions (6) induce the relations in the following diagram

$$\begin{array}{ccccccccccccccc} B'^0 & = & B'^0 & \to & B'^1 & = & B'^1 & \to & B'^2 & \cdots & B'^{n-1} & = & B'^{n-1} & \to & K \\ \downarrow & & \downarrow & & \downarrow & & \downarrow & & \downarrow & & \downarrow & & \downarrow & & \downarrow \\ B^0 & \to & B^1 & \to & B^2 & \to & B^3 & \to & B^4 & \cdots & B^{2n-2} & \to & B^{2n-1} & \to & K \end{array} \tag{7}$$

Each vertical arrow is an inclusion. The last two (on the right) are equalities.

The fibre of $B'^{q-1} \to B'^q$ is

$$U_q/U_{q-1} = S^{2q-1} = O_{2q}/O_{2q-1}$$

and it therefore coincides with the fibre of $B^{2q-1} \to B^{2q}$. It follows that $B'^{q-1} \to B'^q$ is the bundle obtained by restricting $B^{2q-1} \to B^{2q}$ to the subspace B'^q. The fibre S^{2q-1} generates both $\pi_{2q-1}(Y^{2q-1}) = 2$ and $\pi_{2q-1}(Y'^{q-1}) = \infty$. Thus the homomorphism

$$\lambda\colon\ \pi_{2q-1}(Y'^{q-1}) \to \pi_{2q-1}(Y^{2q-1}) \tag{8}$$

induced by the inclusion is onto. When $q = n$, λ is an equality.

Let f' be a cross-section of $\mathfrak{B}'^{q-1}|K^{2q-1}$. Then f' is also a cross-section f of $\mathfrak{B}^{2q-1}|K^{2q-1}$. It follows that $\lambda c(f') = c(f)$. We note also that f' is a cross-section of $\mathfrak{B}^{2q-2}|K^{2q-1}$. Since the first non-zero homotopy group of Y^{2q-2} is π_{2q-2}, it follows that the primary obstruction of $\mathfrak{B}^{2q-2}$ is zero. Thus we have proved

41.8. THEOREM. *If $\mathfrak{B}'$ is a unitary $(2n-1)$-sphere bundle over K and $\mathfrak{B}$ the corresponding orthogonal sphere bundle, then, for $q = 0, 1, \cdots, n-1$,*

$$c^{2q+1}(\mathfrak{B}) = 0, \qquad c^{2q}(\mathfrak{B}) = \lambda c^{2q}(\mathfrak{B}').$$

and

$$c^{2n}(\mathfrak{B}) = c^{2n}(\mathfrak{B}').$$

Since λ is reduction mod 2, the relation $\delta^* c^{2q}(\mathfrak{B}) = c^{2q+1}(\mathfrak{B})$, and the exactness of the sequence (11) of §38.5 show that the two sets of relations in the above theorem are not independent. Either set implies the other.

41.9. COROLLARY. *In order that a real, differentiable, compact $2n$-manifold be quasi-complex, it is necessary that each odd dimensional Stiefel class be zero and each even ($<2n$) dimensional class be the mod 2 image of a cohomology class with integer coefficients. In particular the manifold must be orientable (see §39.2).*

These appear to be rather strong conditions. Just how strong is not known. Does there exist an orientable differentiable manifold with a non-zero odd dimensional Stiefel class? Any guess is useless. We need effective methods for computing Stiefel classes, and applications to many examples.

41.10. Skew matrices. We approach the present problem from a new angle. Let $\mathfrak{B}$ denote a $(2n-1)$-sphere bundle with group O_{2n}. According to §9.5, $\mathfrak{B}$ *is equivalent to a bundle in U_n if and only if the weakly associated bundle with fibre O_{2n}/U_n has a cross-section.* We proceed to a study of this fibre.

Let W'_n be the set of real, non-singular, $2n \times 2n$, skew symmetric matrices, and let $W_n = O_{2n} \cap W'_n$. Let

$$(9) \qquad \sigma_0 = \begin{vmatrix} 0 & -I_n \\ I_n & 0 \end{vmatrix}$$

be a reference point of W_n. Define ϕ: $O_{2n} \to W_n$ by

$$(10) \qquad \phi(\sigma) = \sigma\sigma_0\sigma' \qquad (\sigma' = \text{transpose of } \sigma).$$

Now $\phi(\sigma) = \sigma_0$ if and only if σ has the form

$$(11) \qquad \sigma = \begin{vmatrix} \alpha & -\beta \\ \beta & \alpha \end{vmatrix}.$$

If we pass from complex coordinates $(z_1, \cdots, z_n)$ to real coordinates $(x_1, \cdots, x_n, y_1, \cdots, y_n)$ where $z_i = x_i + iy_i$, then orthogonal matrices of the form (11) correspond exactly to unitary matrices. This imbedding of U_n in O_{2n} is equivalent to the one in §41.7 under conjugation by the orthogonal transformation carrying $(x_1, \cdots, x_n, y_1, \cdots, y_n)$ into $(x_1, y_1, x_2, y_2, \cdots, x_n, y_n)$.

It follows that ϕ induces an identification

$$(12) \qquad O_{2n}/U_n = W_n.$$

Under this identification a left translation of the coset space by σ in O_{2n} corresponds to conjugation of W_n by σ (i.e. $\tau \to \sigma\tau\sigma'$).

Consider now the effect of the factorization $\tau = \sigma\alpha$ of §40.3 when τ is in W'_n. The analog of §40.4 is

41.11. Lemma. *If $\sigma \in O_{2n}$ and $\alpha \in S_{2n,0}$, then $\sigma\alpha \in W'_n$ if and only if*

$$(13) \qquad \sigma \in W_n \quad \text{and} \quad \sigma\alpha = \alpha\sigma.$$

Suppose $\sigma\alpha \in W'_n$. Then $(\sigma\alpha)' = -\sigma\alpha$ implies $\alpha = -\sigma^2(\sigma'\alpha\sigma)$. Since σ^2 is orthogonal and $\sigma'\alpha\sigma$ is positive definite, the uniqueness of the factorization ψ of §40.3 yields

$$-\sigma^2 = I_{2n}, \qquad \alpha = \sigma'\alpha\sigma.$$

But these conditions are equivalent to (13). The converse argument is trivial.

In analogy with §40.5, we shall identify W'_n with the subspace of $W_n \times S_{2n,0}$ consisting of pairs (σ,α) satisfying (13). Define the projection p: $W'_n \to W_n$ by

$$(14) \qquad p(\sigma\alpha) = \sigma.$$

Then p is a continuous retraction of W'_n into W_n. Define the homotopy k by the formula (7) of §40.7. Then the formulas (8), (9), (10) and (11) of §40.7 continue to hold, and we have

41.12. THEOREM. *There is a deformation retraction k of the manifold W'_n of skew matrices into the submanifold W_n of orthogonal skew matrices. The homotopy k commutes with the operations of O_{2n} on W'_n* (i.e. $\tau \to \sigma\tau\sigma'$). *For each σ, k contracts $p^{-1}(\sigma)$ over itself into σ.*

One may prove more here, namely: p: $W'_n \to W_n$ admits a bundle structure with fibre $S_{2n,0}$ and group U_n/H where H is the group of two elements $\pm I_{2n}$. The proof is similar to that of §40.14 for the analogous case $S_{n,k} \to M_{n,k}$.

41.13. Reduction of the problem. Let M be a real, differentiable $2n$-manifold. Let $\mathcal{B}$ denote the tangent sphere bundle of M, and let $\mathcal{B}'_s$ be the weakly associated bundle with fibre W'_n. Then $\mathcal{B}'_s$ is the bundle of 2nd order skew symmetric tensors over M having non-zero determinants. (Since the group of $\mathcal{B}$ is O_{2n}, the variance of the tensors is irrelevant; see §12.11.) Let $\mathcal{B}_s$ denote the subbundle corresponding to W_n.

For each x in M, choose an admissible ξ mapping W'_n onto the fibre over x, and define the homotopy k' by (12) of §40.9. In analogy with §40.10 we have

41.14. THEOREM. *The homotopy k' is a deformation retraction of B'_s into B_s and contracts each fibre over itself. Then k' deforms any cross-section of $\mathcal{B}'_s$ into one of $\mathcal{B}_s$. Thus $\mathcal{B}'_s$ has a cross-section if and only if $\mathcal{B}_s$ has a cross-section.*

Referring to the first paragraph of §41.10, we have

41.15. COROLLARY. *The real, differentiable $2n$-manifold M is quasi complex if and only if it admits a 2nd order, skew symmetric, tensor field which is non-singular at each point.*

A tensor field of the type prescribed in the corollary is otherwise known as a non-singular exterior form of degree 2.

41.16. Applications. Various facts about the topology of the compact manifolds W_n are readily available. Since O_{2n} has two components and U_n is connected, W_n has two components. They are homeomorphic since O_{2n} operates transitively. This means that the primary obstruction to finding a cross-section of a bundle with fibre W_n is 1-dimensional. In the tensor problem of §41.15, the vanishing of this obstruction is equivalent to orientability. In general, the vanishing is equivalent to the reducibility of the bundle to the group R_{2n}. When this happens, the bundle is the union of two disjoint isomorphic subbundles; and the problem reduces to finding a cross-section of one of them, say, the one with fibre R_{2n}/U_n which we will denote by Z_n.

Since $U_1 = R_2$, Z_1 is a point. This gives:

A 2-manifold is quasi-complex if and only if it is orientable.

It is well known, of course, that any orientable 2-manifold admits a complex analytic structure.

In the decomposition of R_4 into the product space of the symplectic group Sp_1 and R_3, we have $Sp_1 \subset U_2$ and $U_2 \cap R_3 = R_2 = U_1$. Then U_2 is the product space of Sp_1 and U_1. This implies that

$$R_4/U_2 = R_3/R_2 = S^2.$$

Therefore Z_2 is a 2-sphere. Hence $\pi_2(Z_2)$ and $\pi_3(Z_2)$ are infinite cyclic and $\pi_4(Z_2)$ is cyclic of order 2. If M is an orientable 4-manifold, the skew-tensor problem of §41.15 leads to a 3-dimensional primary obstruction with coefficients in an infinite cyclic group.

For a general n, the lower homotopy groups of Z_n can be deduced from the homotopy sequence of the bundle $R_{2n} \to Z_n$:

$$\pi_4(R_{2n}) \to \pi_4(Z_n) \to \overset{\lambda}{\pi_3(U_n)} \to \pi_3(R_{2n}) \to \pi_3(Z_n) \to \pi_2(U_n)$$
$$\to \pi_2(R_{2n}) \to \pi_2(Z_n) \to \pi_1(U_n) \xrightarrow{\mu} \pi_1(R_{2n}) \to \pi_1(Z_n).$$

When $n > 2$, $\pi_4(R_{2n}) = 0$ and λ is an isomorphism onto (see §§24.6, 25.1, 25.4). Exactness of the sequence implies $\pi_4(Z_n) = 0$. Since λ is onto, and $\pi_2(U_2) = 0$, we have $\pi_3(Z_n) = 0$. Since μ is onto, $\pi_1(U_n)$ is infinite cyclic, $\pi_1(R_{2n})$ is cyclic of order 2, and $\pi_2(R_{2n}) = 0$, it follows that $\pi_2(Z_2)$ is infinite cyclic. Since U_n is connected and μ is onto, we have $\pi_1(Z_n) = 0$. Thus, for $n > 2$, we have

$$\pi_i(Z_n) = \begin{cases} 0 & i = 1, \\ \infty & i = 2, \\ 0 & i = 3, \\ 0 & i = 4. \end{cases}$$

41.17. Quasi-complex spheres. We turn to the problem of determining the dimensions of spheres which are quasi-complex. In the euclidean space C^{2n} of coordinates $(x_1, \cdots, x_{2n})$, let S be the $(2n - 2)$-sphere defined by $x_{2n} = 0$ and $\Sigma_1^{2n-1} x_i^2 = 1$. Let $\mathfrak{B}_s$ be the bundle of orthogonal skew-tensors of order 2 over S. If $b \in B_s$ lies over $x \in S$ and T_x is the tangent $(2n - 2)$-plane to S at x, then b is an orthogonal transformation of T_x on itself which carries each vector v of T_x into a vector perpendicular to v (this follows from the skew-symmetry of the matrix representation). We assign to b a linear transformation $\psi(b)$ of C^{2n} as follows: it carries the vector $x_0 = (0, \cdots, 0, 1)$ into x, it carries x into $-x_0$, and, in the $(2n - 2)$-space L_x parallel to T_x, the operation $\psi(b)$ is obtained by parallel translation of the operation b in T_x. Then $\psi(b)$ carries each vector into an orthogonal vector;

hence $\psi(b)$ is in $W_n \subset R_{2n}$. If $\sigma \in W_n$, then $x = \sigma x_0$ is in S and $\sigma|L_x$ is skew-symmetric. It follows that ψ maps B_s topologically onto W_n. If p: $R_{2n} \to S^{2n-1}$ is defined by $p(\sigma) = \sigma x_0$, then p maps W_n onto S, and, for each x in S, ψ maps the fibre of $\mathcal{B}_s$ over x onto $p^{-1}(x) \cap W_n$. We have proved:

41.18. THEOREM. *With respect to the projection* p: $W_n \to S^{2n-2}$ *given by* $p(\sigma) = \sigma(x_0)$, W_n *is a bundle over* S^{2n-2} *with fibre* W_{n-1}. *It is equivalent to the bundle of orthogonal skew-tensors of order 2 over* S^{2n-2}.

If the bundle $R_{2n} \to S^{2n-1}$ is restricted to the hemisphere $x_{2n} \geqq 0$, we obtain a product bundle. Hence the same is true of the bundle restricted to S^{2n-2}. Thus ψ imbeds B_s in $S^{2n-2} \times R_{2n-1}$. We cannot conclude that $\mathcal{B}_s$ is a product bundle since ψ is, in no sense, a bundle mapping.

It is to be noted that the relation "W_n is a bundle over S^{2n-2} with fibre W_{n-1}" is analogous to the relation "R_n is a bundle over S^{n-1} with fibre R_{n-1}." The latter was used to compute homotopy groups of R_n. One may do likewise for the homotopy groups of W_n and obtain the results already given in §41.16.

As to the problem of which spheres are quasi-complex; the main results are embodied in the following theorem of Kirchhoff [61].

41.19. THEOREM. *If* S^{2n-2} *is quasi-complex, then the bundle* $R_{2n} \to S^{2n-1}$ *admits a cross-section; and is therefore equivalent to a product bundle.*

By assumption, there is a cross-section f of the bundle $W_n \to S^{2n-2}$ of §41.18. Any vector x in S^{2n-1} is uniquely expressible in the form

$$x = \lambda x_0 + \mu y, \quad y \in S^{2n-2}, \quad \mu \geqq 0, \quad \lambda^2 + \mu^2 = 1.$$

Set

$$\sigma(x) = \lambda I_{2n} + \mu f(y).$$

Since $f(y)$ is skew-orthogonal, $f(y)^2 = -I_{2n}$.
Hence

$$\begin{aligned}\sigma(x)\sigma(x)' = [\lambda I_{2n} + \mu f(y)][\lambda I_{2n} - \mu f(y)] &= \lambda^2 I_{2n} - \mu^2 f(y)^2 \\ &= (\lambda^2 + \mu^2) I_{2n} = I_{2n}.\end{aligned}$$

Therefore $\sigma(x) \in R_{2n}$. Also

$$\sigma(x)\cdot x_0 = \lambda I_{2n}\cdot x_0 + \mu f(y)\cdot x_0 = \lambda x_0 + \mu y = x.$$

It follows that $\sigma(x)$ is a cross-section of $R_{2n} \to S^{2n-1}$, and the theorem is proved.

In §24.8, we have shown that the characteristic map T_{4m+2} of the bundle $R_{4m+2} \to S^{4m+1}$ is not homotopic to a constant; hence the bundle is not a product. Thus, we have

41.20. Corollary. *For $m = 1, 2, \cdots$, the sphere of dimension $4m$ is not a quasi-complex manifold.* (See App. sect. 7.)

It should be kept in mind that this asserts only that S^{4m}, *with its usual differential structure,* is not differentiably equivalent to the real form of a complex analytic manifold. The question here is a special case of a general one: If M and M' are two differentiable manifolds on the same space, are their tangent bundles equivalent?

41.21. Since S^2 admits a complex analytic structure it is quasi-complex. We have also that: *S^6 is quasi-complex.* To prove this, let S^7 be the set of Cayley numbers of norm 1 (see §20.5). Let S^6 be the equator of points c of S^7 which are orthogonal to the Cayley unit 1. Since right multiplication by $b \varepsilon S^7$ is orthogonal, c orthogonal to 1 implies cb orthogonal to b. Hence left multiplication of S^7 by c in S^6 carries each point into an orthogonal point. Then the matrix $f(c)$ of this left multiplication lies in W_4. Since $c1 = c$, f is a cross-section of the bundle $W_4 \rightarrow S^6$. Then §41.18 asserts that S^6 is quasi-complex.

41.22. It is interesting to note that *W_n is contractible to a point in R_{2n}.* This is obtained by modifying the proof of §41.19. Set

$$k(\sigma,t) = tI_{2n} + (1 - t^2)^{1/2}\sigma, \quad \sigma \varepsilon W_n, \ 0 \leqq t \leqq 1.$$

As before it follows that $k(\sigma,t)$ is orthogonal; then k is the required homotopy.

Appendix

This appendix has been added (November 1956) to call attention to some of the important advances in the theory of fibre bundles since 1951, and to show how they answer, wholly or in part, questions raised in the text. The order of the following material approximates that of the related subjects in the text.

1. Local cross-sections of a subgroup. A generalization of the conjecture made at the end of §7.5, p. 33 has been proved by P. S. Mostert: Local cross sections in locally compact groups, *Proc. Amer. Math. Soc.* *4* (1953), 645–649. He shows that, if B is a locally-compact and finite-dimensional group, and G is a closed subgroup, then G has a local cross-section in B.

2. The covering homotopy theorem. The hypotheses of the covering homotopy theorems, §11.3, p. 50, and §11.7, p. 54, can be weakened without affecting the conclusions by replacing the condition "X is a normal, locally-compact C_σ-space" by "X is normal and paracompact." See W. Huebsch, On the covering homotopy theorem, *Annals of Math.* *61* (1955), 555–563.

3. The existence of cross-sections. The hypotheses of the existence theorem §12.2, p. 55, may be relaxed. The conclusion still holds if the base space X is normal and paracompact, and the fibre Y is solid. This improvement is of the same nature as that made in the covering homotopy theorem; and the modifications in the proof are similar.

4. Homotopy groups. At the ends of §15.10, p. 80, and §21.7, p. 114, it is stated that very few homotopy groups have been successfully computed. The situation in 1956 is entirely different. Major advances in the theory have been made. It has been shown that the homotopy groups of finite, simply-connected complexes are finitely generated, and are effectively computable. The computations have been made in numerous special cases. It is notable that the concept of *fibre space* (a somewhat broader notion than *fibre bundle*) played a vital role in this development. A complete review of these results and their implications for fibre bundles would be too long. The following references give a substantial indication of the progress.

J.-P. Serre, Homologie singulière des espaces fibrés, *Annals of Math.* *54* (1951), 425–505.

——, Groupes d'homotopie et classes de groupes abéliens, *ibid.* *58* (1953), 258–294.

H. Cartan, *Algèbres d'Eilenberg-MacLane et homotopie*, Seminar notes 1954/1955, Paris.

H. Toda, Calcul des groupes d'homotopie des spheres, *C. R. Acad. Sci. Paris 240* (1955), 147–149.

——, Le produit de Whitehead et l'invariant de Hopf, *ibid. 241* (1955), 849–850.

In this last note Toda announces that there is no mapping $S^{31} \to S^{16}$ of Hopf invariant 1. Hence there is no real division algebra of dimension 16 (see §20.7, p. 110).

5. Homotopy groups of Lie groups. Articles 22 to 25, pp. 114–134, of the text are devoted to the computation of a few of the homotopy groups of the classical Lie groups and their coset spaces. Far more extensive results have been obtained by better methods. For a survey of these see A. Borel, Topology of Lie groups and characteristic classes, *Bull. Amer. Math. Soc. 61* (1955), 397–432.

6. Sphere bundles over spheres. In §26.6 to 26.10 various sphere bundles over spheres are exhibited which are not equivalent to product bundles but have the homotopy groups and homology structure of products. I. M. James and J. H. C. Whitehead have devised a homotopy invariant of such bundles which enables them to distinguish many of these spaces from products and one another. See: The homotopy theory of sphere bundles over spheres I and II, *Proc. London Math. Soc. 4* (1954), 196–218, and *5* (1955), 148–166.

7. The tangent bundle of S^n. In a paper by J. H. C. Whitehead and the author (Vector fields on the n-sphere, *Proc. Nat. Acad. Sci. 37* (1951), 58–63), many of the results of §§27, 40, and 41 are generalized and proved by easier methods. For example, Theorems 27.8 and 27.9 are special cases of the following: If n and k are related by $n + 1 = 2^k(2r + 1)$, then any set of 2^k continuous vector fields tangent to S^n are somewhere dependent. Theorems 27.18 and 40.12 are included in: *If n and k are as above, and $2^k \leqq q \leqq n - 2^k$, then S^n does not admit a continuous field of tangent q-planes nor a continuous quadratic form which is nonsingular of signature q.*

Theorem 41.20 becomes: *If S^n admits an almost (= quasi) complex structure, then n must be of the form $2^k - 2$.* This last result has been greatly improved by A. Borel and J.-P. Serre (Groupes de Lie et puissances réduites de Steenrod, *Amer. Jour. Math. 75* (1953), 409–448) as follows: *The only spheres which admit an almost complex structure are S^2 and S^6.*

The result 27.16 has been improved by I. M. James: Note on factor spaces, *Jour. London Math. Soc. 28* (1953), 278–285.

8. The fibering of spheres by spheres. The results of §28 have been considerably improved. In the paper by Whitehead and the author, referred to above, it is shown that, if S^{n+r} is an r-sphere bundle over S^n, then $n = 2^k$ and $r = 2^k - 1$ for some k. The impossibility of fibering a sphere by spheres has been shown in many other cases by J. Adem: Relations on iterated reduced powers, *Proc. Nat. Acad. Sci.* *39* (1953), 636–638.

9. Characteristic classes of sphere bundles. Knowledge of the Stiefel and Whitney characteristic classes (§38, 39) has been greatly extended by the work of R. Thom: Espaces fibrés en sphères et carrés de Steenrod, *Ann. Sci. École Norm. Sup.* *69* (1952), 109–182. His first main result is a formula which characterizes the Whitney classes W^i $(i = 0, 1, \cdots, n)$ of an $(n-1)$-sphere bundle $B \to X$, namely:

$$W^i = \phi^{-1}\mathrm{Sq}^i\phi W^0$$

In this formula ϕ is an isomorphism $H^q(X) \approx H^{q+n}(A,B)$ where $A \to X$ is the associated n-cell bundle; and Sq^i: $H^n \to H^{n+i}$ is the squaring operation on cohomology mod 2 which this author defined for other purposes (Cyclic reduced powers of cohomology classes, *Proc. Nat. Acad. Sci.* *39* (1953), 213–223).

Using this formula, Thom obtains simple proofs of the various properties of the W^i. In particular the Whitney duality theorem, stated but not proved in §38.13, is an easy consequence of the Cartan identity

$$\mathrm{Sq}^i(u \smile v) = \Sigma_{j=0}^{i}\mathrm{Sq}^j u \smile \mathrm{Sq}^{i-j} v.$$

Next Thom considers a differentiable imbedding of the differentiable r-manifold X in a differentiable $(r+n)$-manifold M. Letting B be the normal bundle of X in M, he realizes the associated bundle A as a tubular neighborhood of X in M, and obtains a new formulation of the isomorphism ϕ which is purely topological in character. This provides him with a definition of normal classes for *any* topological imbedding of X in M, and it gives the usual normal classes when the imbedding is differentiable. He considers next the diagonal imbedding of X in $X \times X$. When X is differentiable, he shows that the tangent bundle of X is isomorphic to its normal bundle in $X \times X$. When X is an arbitrary manifold, he defines its *tangent* classes to be its normal classes in $X \times X$. This extends the definition of the Stiefel characteristic classes to arbitrary manifolds; and at the same time it proves that the Stiefel classes, defined in terms of a differential structure, are independent of that structure.

Wu Wen-Tsun has improved on this result (Classes caracteristiques et i-carrés d'une variété, *C. R. Acad. Sci. Paris* *230* (1950), 508). Basing his work on that of Thom, he derives formulas for the Stiefel classes of a manifold X which involve only the cohomology ring of X, and the squaring operations. These formulas enable one to compute easily the Stiefel classes in special cases. In particular they give a quick proof that $c^2(X) = 0$ when X is an orientable 3-manifold (see §39.9).

10. The theory of characteristic classes. Recent developments have produced a change in the point of view on characteristic classes. They are no longer regarded primarily as obstructions to cross-sectioning suitable bundles. The new attitude is based on the theorem of §19 that a bundle over a complex K with group G is uniquely determined by the universal bundle $B_G \rightarrow X_G$ and a mapping $f\colon\ K \rightarrow X_G$. Letting H^* denote the cohomology ring, the image of $H^*(X_G)$ in $H^*(K)$ under f^* is called the characteristic ring of the bundle. A set of generators of $H^*(X_G)$ are called *universal* characteristic classes, and their images in $H^*(K)$ are called the characteristic classes of the bundle.

This procedure presupposes the ability to compute successfully $H^*(X_G)$. The work of A. Borel in this direction has been very important (Sur la cohomologie des espaces fibrés principaux et des espaces homogenes de groupes de Lie compacts, *Annals of Math. 57* (1953), 115–207). He has extended greatly the results of Hopf concerning the structure of $H^*(G)$, and has applied the spectral sequence technique of Leray to obtain theorems on the structure of $H^*(X_G)$. In case G is an orthogonal group, then $H^*(X_G)$ with coefficients mod 2 is a polynomial ring whose generators are the universal Whitney classes. If G is a unitary group, then $H^*(X_G)$ with integer coefficients is a polynomial ring generated by the universal Chern classes. Other special cases have led to new characteristic classes such as the Pontrjagin classes associated with the special orthogonal groups.

This new approach has been fruitful in the applications of fibre bundle theory to differential geometry, complex manifolds and algebraic varieties. For a survey of these see the paper of A. Borel referred to in §5 above, and the monograph of F. Hirzebruch: Neue topologische Methoden in der algebraischen Geometrie, *Ergeb. der Math.*, Springer (1956), Berlin.

11. Secondary obstructions. In case the primary obstruction to finding a cross-section of a bundle is zero, the secondary obstruction is defined and is a set of cohomology classes. For sphere bundles, the secondary obstruction has been analysed successfully by S. D. Liao: On the theory of obstructions of fiber bundles, *Annals of Math. 60* (1954), 146–191. Another special case has been treated by E. G. Kundert: Über Schnittflächen in speziellen Faserungen und Felder reeller und komplexer Linienelemente, *Annals of Math. 54* (1951),

215–246. A general treatment of primary and secondary obstructions is given in the monograph of V. A. Boltyanskii: Homology theory of mappings and vector fields, (in Russian), *Trudy Mat. Inst. Steklov*, no. 47 (1955).

12. Fields of line elements. The subject of tangent fields of line elements has been thoroughly analysed by L. Marcus: Line element fields and Lorentz structures on differentiable manifolds, *Annals of Math. 62* (1955), 411–417. In particular, the statement on p. 207, lines 12b and 13b, must be corrected by deleting the word "twice." See also H. Samelson, A theorem on differentiable manifolds, *Portugaliae Math. 10* (1951), 129–133.

Bibliography

Papers specifically concerned with fibre bundles are indicated by *

ALEXANDROFF, P. and HOPF, H.

1. Topologie, Berlin, J. Springer, 1935.

ARENS, R.

2. *Topologies for homeomorphism groups*, Amer. Jour. Math., 68 (1946), 593–610.

BORSUK, K.

3. *Sur les rétracts.* Fund. Math., 17 (1931), 152–170.

CHERN, S. S.

4. **Integral formulas for the characteristic classes of sphere bundles.* Proc. Nat. Acad., 30 (1944), 269–273.
5. **The Gauss-Bonnet formula.* Annals of Math., 45 (1944), 747–752.
6. **On the curvatura integra.* Ibid., 46 (1945), 674–684.
7. **Characteristic classes of Hermitian manifolds.* Ibid., 47 (1946), 85–121.
8. **On the multiplication in the characteristic ring of a sphere bundle.* Ibid., 49 (1948), 362–372.
9. **Some new viewpoints in differential geometry in the large*, Bull. Amer. Math. Soc., 52 (1946), 1–30.
10. *(with Y. Sun) *The imbedding theorem for filre bundles.* Trans. Amer. Math. Soc., 67 (1949), 286–303.
11. *(with S. T. Hu) *Parallelisability of principal fibre bundles*, Ibid., 67 (1949), 304–309.

CHEVALLEY, C.

12. Theory of Lie groups. Princeton Univ. Press, 1946.

DICKSON, L.

13. Linear algebras. Cambridge Tract, Cambridge Univ. Press, 1914.

ECKMANN, B.

14. **Zur Homotopietheorie gefaserter Räume.* Comm. Math. Helv., 14 (1942), 141–192.
15. **Über die Homotopiegruppen von Gruppenräumen.* Ibid., 14 (1942), 234–256.
16. **Systeme von Richtungsfelder in Spharen.* Ibid., 15 (1942), 1–26.
17. **Stetige Lösungen linearer gleichungssyteme.* Ibid., 15 (1942), 318–339.
18. **Beweis des Satzes von Hurwitz-Radon.* Ibid., 15 (1942), 358–366.
19. *(with H. Samelson and G. Whitehead) *On fibering spheres by toruses.* Bull. Amer. Math. Soc., 55 (1949) 433–438.

EHRESMANN, C.

20. **Sur la topologie de certains espaces homogenes.* Annals of Math., 35 (1934), 396–443.
21. **Espaces fibrés de structures comparables.* C. R. Acad. Sci. Paris, 214 (1942), 144–147; also 213 (1941), 762; and 216 (1943), 628–630.
22. *(with J. Feldbau) *Sur les propriétés d'homotopie des espaces fibrés.* Ibid., 212 (1941), 945–948.
23. **Sur les espaces fibrés associés à une variété différentiable.* Ibid., 216 (1943), 628–630.

24. *(with G. Reeb) *Sur les champs d'éléments de contact. . . .* Ibid., 218 (1944), 955–957.
25. **Sur les applications d'un espace dans in espace fibré.* Bull. Soc. Math. France, 72 (1944), 27–54.
26. **Sur les sections d'un champ d'éléments de contact.* C. R. Acad. Sci. Paris, 224 (1947), 444–445.
27. **Sur les espaces fibrés differentiables.* Ibid., 224 (1947), 1611.
28. **Sur les variétés plongées dans une variété differentiable.* Ibid., 226 (1948), 1879–1880.
29. **Sur la théorie des espaces fibrés.* Colloque International de Topologie Algébrique. Paris (1949), 3–15.

EILENBERG, S.

30. *On the relation between the fundamental group of a space and the higher homotopy groups.* Fund. Math., 32 (1939), 167–175.
31. *Cohomology and continuous mappings.* Ann. of Math., 41 (1940), 231–251. See also, *Lectures in Topology*, Univ. of Mich. Press, 1941, 57–100.

FELDBAU, J.

32. **Sur la classification des espaces fibrés.* C. R. Acad. Sci. Paris, 208 (1939), 1621–1623.
See also [22].

FOX, R. H.

33. *On topologies for function spaces.* Bull. Amer. Math. Soc., 51 (1945), 429–432.
34. *On fibre spaces I.* Ibid., 49 (1943), 555–557.
35. *On fibre spaces II.* Ibid., 49 (1943), 733–735.

FREUDENTHAL, H.

36. *Über die Klassen der Sphärenabbildungen.* Comp. Math., 5 (1937), 299–314.

GLEASON, A. M.

37. *Spaces with a compact Lie group of transformations.* Proc. Amer. Math. Soc., 1 (1950), 35–43.

GYSIN, W.

38. **Zur Homologietheorie der Abbildungen und Faserungen der Mannigfaltigkeiten.* Comm. Math. Helv., 14 (1941), 61–122.

HIRSCH, G.

39. **Sur les groupes d'homologie des espaces fibrés.* Bull. Soc. Royale Sci. de Liége, 4 (1941), 246–260.
40. **Sur une théorème de Hopf-Rueff.* Bull. Acad. royale de Belgique, 29 (1943), 516–524.
41. **Sur la signification topologique des Axiomes de la geométrie projective.* C. R. Acad. Sci. Paris, 223 (1946), 528–530.
42. **Un isomorphisme attaché aux structures fibrées.* Ibid. 227 (1948), 1328–1330.
43. **Sur les groupes d'homologie des espaces fibrés.* Bull. Soc. Math. de Belgique, (1947–1948), 24–33.
44. **La géomètrie projective et la topologie des espaces fibrés.* Colloque International de Topologie Algébrique. Paris (1949), 35–42.

HOTELLING, H.

45. *Three dimensional manifolds of states of motion.* Trans. Amer. Math Soc., 27 (1925), 329–344.

HOPF, H.

46. *Über die Curvatura Integra geschlossener Hyperflächen.* Math. Annalen, 95 (1925), 340–367.
47. **Vektorfelder in Mannigfaltigkeiten.* Ibid., 96 (1927), 225–250.
48. *Über die Abbildungen der 3-Sphäre auf die Kugelfläche.* Ibid., 104 (1931), 637–665.
49. **Über die Abbildungen von Sphären auf Sphären neidrigerer Dimension.* Fund. Math., 25 (1935), 427–440.
50. *(with M. Rueff) *Über faserungstreue Abbildungen der Sphären.* Comm. Math. Helv., 11 (1938), 49–61.
51. **Ein topologischer Beitrag zur reelen Algebra.* Ibid., 13 (1940), 219–239.
52. *(with H. Samelson) *Ein Satz über die Wirkungsräume geschlossener Liescher Gruppen.* Ibid., 13 (1940), 240–251.
53. **Bericht über einige neue Ergebnisse in der Topologie.* Revista Mat. Hisp.-Amer. (4) 6 (1946), 147–159.
54. **Zur Topologie der komplexen Manigfaltigkeiten.* Studies and Essays Presented to R. Courant on His 60th Birthday. New York, Interscience, 1948.

See also [1].

HU, S. T.

55. *An exposition of the relative homotopy theory.* Duke J. Math., 14 (1947), 991–1033.

See also [11].

HUREWICZ, W.

56. (with H. Wallman) Dimension Theory. Princeton Univ. Press, 1941.
57. *Beitrage zur Topologie der Deformationen I–IV.* Proc. Akad. Amsterdam, 38 (1935), 112–119, 521–528; also 39 (1936), 117–126, 215–224.
58. (with N. Steenrod) *Homotopy relations in fibre spaces.* Proc. Nat. Acad. Sci., 27 (1941), 60–64.

IWASAWA, K.

59. *On some types of topological groups.* Ann. of Math., 50 (1949), 507–558.

KIANG, T.

60. **The manifolds of linear elements of an n-sphere.* Bull. Amer. Math. Soc., 51 (1945), 417–428.

KIRCHHOFF, A.

61. **Sur l'existence de certains champs tensoriels sur les spheres à n dimensions.* C. R. Acad. Paris, 223 (1948), 1258–1260.

KOSZUL, J.

62. *Sur les opérateurs de derivation dans un anneau.* C. R. Acad. Sci. Paris, 225 (1947), 217–219.
63. **Sur l'homologie des espaces homogènes.* Ibid., 225 (1947), 477–479.

LEFSCHETZ, S.

64. Algebraic topology. Colloq. Publ. Amer. Math. Soc., 1942.

LERAY, J.

65. *L'anneau d'homologie d'une representation.* C. R. Acad. Paris, 222 (1946), 1366–1368, 1419–1422.
66. **Propriétés de l'anneau d'homologie de la projection d'un espace fibre sur sa base.* Ibid., 223 (1946), 395–397, 412–415.

LICHNEROWICZ, A.

67. **Sur les nombres de Betti de certaines . . . espaces fibré.* C. R. Acad. Paris 226 (1948), 212–214.

68. **Une théorème sur l'homologie dans les espaces fibrés.* Ibid., 227 (1948), 711–712.

MONTGOMERY, D.

69. *(with H. Samelson) *Fiberings with singularities.* Duke Math. Jour., 13 (1946), 51–56.

MOSTOW, G. D.

70. *A new proof of E. Cartan's theorem on the topology of semi-simple groups.* Bull. Amer. Math. Soc., 55 (1949), 969–980.

PONTRJAGIN, L.

71. Topological groups. Princeton Univ. Press, 1939.
72. *A classification of continuous transformations of a complex into a sphere 2.* C. R. (Doklady) Acad. Sci. USSR, 19 (1938), 361–363.
73. **Characteristic cycles on manifolds.* Ibid., 35 (1942), 34–37.
74. **On some topologic invariants of Riemannian manifolds.* Ibid., 43 (1944), 91–94.
75. **Classification of some skew products.* Ibid., 47 (1945), 322–325.
76. **Characteristic cycles.* Ibid., 47 (1945), 242–245.
77. **Characteristic cycles on differentiable manifolds.* Mat. Sbornik N. S. 21 (63) (1947), 233–284.
78. **Vector fields on manifolds.* Ibid., 24 (66) (1949), 129–162.
79. **Some topological invariants of closed Riemannian manifolds.* Izvestiya Akad. Nauk SSSR. Ser. Mat., 13 (1949), 125–162.

REEB, G.

80. **Sur les variétés integrales des champs d'elements de contact completement intégrables.* C. R. Acad. Sci. Paris, 220 (1945), 236–237; see also 218 (1944), 955–957; and 224 (1947), 1613–1614.

REIDEMEISTER, K.

81. Topologie der Polyeder. Leipzig, Akad. Verlagsges, 1938.

SAMELSON, H.

82. **Über die Sphäre die als gruppenräume auftreten.* Comm. Math. Helv., 13 (1940), 144–155.
See also [52] and [69].

SEIFERT, H.

83. **Topologie 3-dimensionaler gefaserter Räume.* Acta Math., 60 (1932), 147–238.
84. *Algebraische Approximation von Mannigfaltigkeiten.* Math. Zeit., 41 (1936), 1–17.
85. (with W. Thelfall) *Lehrbuch der Topologie.* Teubner, Leipzig, 1934.

STEENROD, N. E.

86. **Topological methods for construction of tensor functions.* Annals of Math., 43 (1942), 116–131.
87. **Homology with local coefficients.* Ibid., 44 (1943), 610–627.
88. **Classification of sphere bundles.* Ibid., 45 (1944), 294–311.
89. *Products of cocycles and extensions of mappings.* Ibid., 48 (1947), 290–320.
90. **Cohomology invariants of mappings.* Ibid., 50 (1949), 954–988.
See also [58].

STIEFEL, E.

91. **Richtungsfelder und Fernparallelismus in Mannigfaltigkeiten.* Comm. Math. Helv., 8 (1936), 3–51.

SUN, Y.

See [10].

THOM, R.

92. *Classes caracteristiques et i-carrés.* C. R. Acad. Sci. Paris, 230 (1950), 427–429.
93. *Variétés plongées et i-carrés.* Ibid., 230 (1950), 508–511.

THRELFALL, W.

94. *Räume aus Linienelemente.* Jahr. Deutschen Math. Verein., 42 (1932), 88–110.
See also [85].

VEBLEN, O.

95. Invariants of quadratic differential forms. Cambridge Tracts No. 24, Cambridge Univ. Press, 1933.
96. (with J. H. C. Whitehead) The foundations of differential geometry. Cambridge Tracts No. 29, Cambridge Univ. Press, 1932.

WALLMAN, H.

See [56].

WANG, H. C.

97. *Homogeneous spaces with non-vanishing Euler characteristics.* Ann. of Math., 50 (1949), 925–953.
98. *The homology groups of the fibre bundles over a sphere.* Duke Math. J., 16 (1949), 33–38.

WHITEHEAD, G. W.

99. *On the homotopy properties of the real orthogonal groups.* Annals of Math., 43 (1942), 132–146.
100. *On families of continuous vector fields over spheres.* Ibid., 47 (1946), 779–785; also 48 (1947), 782.
101. *On spaces with vanishing low dimensional homotopy groups.* Proc. Nat. Acad. Sci., 34 (1948), 207–211.

WHITEHEAD, J. H. C.

102. **On the groups* $\pi_r(V_{nm})$ *and sphere bundles.* Proc. London Math. Soc., 48 (1944), 243–291.
See also [96].

WHITNEY, H.

103. *Sphere spaces.* Proc. Nat. Acad. Sci., 21 (1935), 462–468.
104. *Topological properties of differentiable manifolds.* Bull. Amer. Math. Soc., 43 (1937), 785–805.
105. *On the theory of sphere bundles.* Proc. Nat. Acad. Sci., 26 (1940), 148–153.
106. *On the topology of differentiable manifolds.* Lectures in Topology, Univ. of Mich. Press, 1941.

WU, W. T.

107. *Sur l'existence d'un champ d'elements de contact on d'une structure complex sur une sphere.* C. R. Acad. Sci. Paris, 226 (1948), 2117–2119.
108. *On the product of sphere bundles and the duality theorem modulo two.* Ann. of Math., 49 (1948), 641–653.
109. *Sur le structure presque complexe d'une variété differentiable réele de dimension* 4. C. R. Acad. Paris, 227 (1948), 1076–1078.
110. *Sur la second obstacle d'un champ d'elements de contact dans une structure fibrée spherique.* Ibid., 227 (1948), 815–817.
111. *Sur les classes caracteristiques d'un espace fibrées en spheres.* Ibid., 227 (1948), 582–584.
112. *Class caracteristiques et i-carrés d'une variété.* Ibid., 230 (1950), 508–511.
113. *Les i-carrés dans une variété grassmannienne.* Ibid., 230 (1950), 918–920.

Index

Milton Keynes UK
Ingram Content Group UK Ltd.
UKHW020048310824
447656UK00005B/229

9 780691 005485